The Political Economy of Oil and Gas in Africa

The evolution of the Nigerian oil and gas industry spanned nearly one hundred years, during which time several challenges were encountered and surmounted by major International Oil Companies (IOCs). This book provides a thoroughly researched guide to the Nigerian oil and gas industry.

The author examines the increasing role of Africa in the contribution of oil and gas resources to the global energy market and provides an overview of oil and gas exploration and production activities in Algeria, Libya, Egypt and Angola. The book presents an in-depth review of the growth and challenges of the Nigerian oil and gas industry and also highlights the geological features of the oil and gas bearing regions of the country. In particular, the emerging prominence of the Gulf of Guinea as a prolific hydrocarbon bearing zone is extensively evaluated. There are chapters devoted to environmental issues both in Nigeria and globally, while relevant petroleum laws are brought into focus with a view to guiding potential investors. It culminates with a detailed account of investment opportunities in the dynamic Nigerian oil and gas industry.

This book offers students, potential investors, academics and policy makers the opportunity to get acquainted with various dimensions of the oil and gas industry. It is relevant to subject areas such as environmental pollution, gas monetisation, and oil and gas exploration and production.

Soala Ariweriokuma is an economist and former lecturer at the University of Port Harcourt. He joined the Nigerian National Petroleum Corporation (NNPC) in 1992 and has worked in key divisions of the corporation. He is currently General Manager of NIDAS International (an NNPC/DSME Joint Venture Company).

Routledge Studies in International Business and the World Economy

1 States and Firms
Multinational enterprises in institutional competition
Razeen Sally

2 Multinational Restructuring, Internationalization and Small Economies
The Swedish case
Thomas Andersson, Torbjörn Fredriksson and Roger Svensson

3 Foreign Direct Investment and governments
Catalysts for economic restructuring
Edited by John H. Dunning and Rajneesh Narula

4 Multinational Investment and Economic Structure
Globalization and competitiveness
Rajneesh Narula

5 Entrepreneurship in a Global Context
Edited by Sue Birley and Ian MacMillan

6 The Global Structure of Financial Markets
An overview
Edited by Dilip K. Ghosh and Edgar Ortiz

7 Alliance Capitalism and Global Business
John H. Dunning

8 Multinational Enterprises from the Netherlands
Edited by Roger van Hoesel and Rajneesh Narula

9 Competition, Growth Strategies and the Globalization of Services
Real estate advisory services in Japan, Europe and the United States
Terrence LaPier

10 European Integration and Foreign Direct Investment in the EU
The case of the Korean consumer electronics industry
Sang Hyup Shin

11 New Multinational Enterprises from Korea and Taiwan
Beyond export-led growth
Roger van Hoesel

12 Competitive Industrial Development in the Age of Information
The role of co-operation in the technology sector
Edited by Richard J. Braudo and Jeffrey G. MacIntosh

13 The Global Restructuring of the Steel Industry
Innovations, institutions and industrial change
Anthony P. D'Costa

14 Privatisation and Liberalisation in European Telecommunications
Comparing Britain, the Netherlands and France
Willem Hulsink

15 Multinational Corporations
Emergence and evolution
Paz Estrella Tolentino

16 Foreign Direct Investment in Emerging Economies
Corporate strategy and investment behaviour in the Caribbean
Lou Anne A. Barclay

17 European Integration and Global Corporate Strategies
Edited by François Chesnais, Grazia Ietto-Gillies and Roberto Simonetti

18 The Globalisation of Corporate R & D
Implications for innovation systems in host countries
Prasada Reddy

19 Globalization of Services
Some implications for theory and practice
Edited by Yair Aharoni and Lilach Nachum

20 A Century of Foreign Investment in the Third World
Michael J. Twomey

21 Global Capitalism at Bay
John H. Dunning

22 Foreign Direct Investment
Research Issues
Edited by Bijit Bora

23 Ford and the Global Strategies of Multinationals
The North American auto industry
Isabel Studer Noguez

24 The World Trade Organization Millennium Round
Freer trade in the next century
Klaus Deutsch / Bernhard Speyer

25 Consultancy and Innovation
The business service revolution in Europe
Edited by Peter Wood

26 Knowledge Economies
Clusters, learning and co-operative advantage
Philip Cooke

27 The Role of Resources in Global Competition
John Fahy

28 Globalization, Employment and the Workplace
Diverse impacts
Edited by Yaw A. Debrah and Ian G. Smith

29 Transnational Corporations
Fragmentation amidst integration
Grazia Ietto-Gillies

30 Growth Theory and Growth Policy
Edited by Harald Hagemann and Stephan Seiter

31 International Business and the Eclectic Paradigm
Developing the OLI framework
Edited by John Cantwell and Rajneesh Narula

32 Regulating Global Trade and the Environment
Paul Street

33 Cultural Industries and the Production of Culture
Edited by Dominic Power and Allen J. Scott

34 Governing Interests
Business associations facing internationalization
Edited by Wolfgang Streeck, Jürgen Grote, Volker Schneider and Jelle Visser

35 Infrastructure Development in the Pacific Region
Edited by Akira Kohsaka

36 Big Business and Economic Development
Conglomerates and economic groups in developing countries and transition economies under globalisation
Edited by Alex E. Fernández Jilberto and Barbara Hogenboom

37 International Business Geography
Case studies of corporate firms
Edited by Piet Pellenbarg and Egbert Wever

38 The World Bank and Global Managerialism
Jonathan Murphy

39 Contemporary Corporate Strategy
Global perspectives
Edited by John Saee

40 Trade, Globalization and Poverty
Edited by Elias Dinopoulos, Pravin Krishna, Arvind Panagariya and Kar-yiu Wong

41 International Management and Language
Susanne Tietze

42 Space, Oil and Capital
Mazen Labban

43 Petroleum Taxation
Sharing the oil wealth: a study of petroleum taxation yesterday, today and tomorrow
Carole Nakhle

44 International Trade Theory
A critical review
Murray C. Kemp

45 The Political Economy of Oil and Gas in Africa
The case of Nigeria
Soala Ariweriokuma

The Political Economy of Oil and Gas in Africa

The case of Nigeria

Soala Ariweriokuma

Routledge
Taylor & Francis Group
LONDON AND NEW YORK

First published 2009 by Routledge
2 Park Square, Milton Park, Abingdon, Oxon OX14 4RN

Simultaneously published in the USA and Canada
by Routledge
605 Third Avenue, New York, NY 10017

Routledge is an imprint of the Taylor & Francis Group, an informa business

Typeset in Times New Roman by
RefineCatch Limited, Bungay, Suffolk

British Library Cataloguing in Publication Data
A catalogue record for this book is available from the British Library

Library of Congress Cataloging-in-Publication Data
Ariweriokuma, Soala.
The political economy of oil and gas in Africa : the case of Nigeria / Soala Ariweriokuma.
p. cm.
Includes bibliographical references and index.
1. Petroleum industry and trade—Nigeria. 2. Gas industry—Nigeria. 3. Petroleum industry and trade—Africa. 4. Gas industry—Africa. I. Title.
HD9577.N52A73 2008
338.2′72809669—dc22 2008008364

ISBN 13: 978–0–415–46484–0 (hbk)
ISBN 13: 978–0–203–89199–5 (ebk)

To my wife Sabinah, and children Somina, Sopriye, Nemi and Soala Jr

Contents

Figures

Tables

Abbreviations and acronyms

AENR	Agip Energy and Natural Resources Limited
AFE	Authorisation for expenditure
AGO	Automotive gas oil (diesel)
ALSCON	Aluminum Smelting Company of Nigeria
API	American Petroleum Institute
APPA	African Petroleum Producers Association
APRM	African peer review mechanism
b/d	Barrels per day
bcf	Billion cubic feet
bcm/yr	Billion cubic metres per year
BG	British Gas
BGT	Bonny Gas Transport
BOT	Built, operate and transfer
BP	British Petroleum
BPE	Bureau for Public Enterprises
CABGOC	Cabinda Gulf Oil Company
CAC	Corporate Affairs Commission
CCG	Combined cycle generation
CDU	Crude distillation unit
CIF	Cost insurance freight
CNG	Compressed natural gas
CTP	Corporate transformation programme
CRU	Catalytic reforming unit
CS	Corporate services
DPK	Dual purpose kerosene
DPR	Department of Petroleum Resources
DRSTP	Democratic Republic of São Tomé and Príncipe
DWT	Dead weight
E&P	Exploration and production
EEPNL	Esso Exploration and Production Nigeria Limited
EEZ	Exclusive economic zone
EGAS	Egyptian Natural Gas Company
EGP	Escravos gas pipeline

EGPC	Egyptian General Petroleum Corporation
EITI	Extractive Industry Transparency Initiative
EL	Exploration licence
ENI	Ente Nazionale Idrocarburi
EOR	Enhanced oil recovery
EPCL	Eleme Petrochemical Company Limited
ERHC	Environmental Remediation Holding Corporation
F&A	Finance and accounts
FCC	Fluid catalytic cracker
FDI	Foreign direct investment
FEED	Front end engineering design
FID	Final investment decision
FOB	Free on board
FPSO	Floating production storage offload
FSO	Floating storage and offload
GATT	General arrangement on trade and tariffs
GDP	Gross domestic product
GHG	Greenhouse gases
GOPA	Geregu, Omotosho, Papalanto and Alaoji
GTL	Gas-to-Liquid
GUPCO	Gulf of Suez Petroleum Company
HDPE	High Density Polyethylene
HPFO	High Pour Fuel Oil
IDSL	Integrated Data Services Limited
IEOC	International Egyptian Oil Company
IOCs	International oil companies
IPP	Independent power plant
IRR	Internal rate of return
JDA	Joint Development Authority
JDMs	Joint development models
JDZ	Joint development zone
JOA	Joint operating agreement
JV	Joint venture
KHU	Kero hydro treating unit
KRPC	Kaduna Refining and Petrochemical Company Limited
LCD	Local content development
LDCs	Less developed countries
LDPE	Low Density Polyethylene
Lipetco	Libyan General Petroleum Corporation
LNG	Liquefied natural gas
LPFO	Low Pour Fuel Oil
LPG	Liquefied petroleum gas
MEG	Maghreb–Europe pipeline
MF	Marginal Field
mm scf/d	Million standard cubic feet per day

mmbd	Million barrels per day
MNC	Multi-national company
MON	Mobil Oil Nigeria plc
MOU	Memorandum of understanding
MPN	Mobil production unlimited
MT/year	Metric tonnes per year
MW	Megawatts
MWD	Measurements while drilling
₦	Nigerian Naira
NAE	Nigerian Agip Exploration Limited
NAFCON	National Fertilizer Company of Nigeria
NAOC	Nigeria Agip Oil Company
NAPIMS	National Petroleum Investment Management Services
NBC	National Boundary Commission
NCD	Nigerian Content Development
NDDC	Niger Delta Development Commission
NDT	Non-destructive testing
NEITI	Nigeria extractive industry transparency initiative
NEPA	National Electric Power Authority
NEPAD	New partnership for African development
NETCO	National Engineering and Technical Company Limited
NGC	Nigeria Gas Company
NGL	Natural gas liquid
NHU	Naphtha hydrotreating unit
NICON	National Insurance Company of Nigeria
NIPPs	National integrated power plants
NITEL	Nigerian Telecommunication Company
NLNG	Nigeria Liquefied Natural Gas Company
NNOC	Nigerian National Oil Corporation
NNPC	Nigerian National Petroleum Corporation
NOC	National oil company
NPDC	Nigerian Petroleum Development Company
NPV	Net present value
OECD	Organisation for Economic and Cultural Development
OK-LNG	Olokola LNG
OML	Oil mining lease
OMPADEC	Oil Mineral Producing Area Development Commission
ONGC	Oil and Natural Gas Corporation
OPEC	Organisation of Petroleum Exporting Countries
OPL	Oil prospecting licence
OSCs	Oil service companies
OSP	Official selling price
PA	Participation agreement
PACE	Positioning, aligning, creating and empowering
PANAM	Pan American (airline)

PETAN	Petroleum Technology Association of Nigeria
PHCN	Power Holding Company of Nigeria
PP	Polypropylene
PPMC	Pipelines and Products Marketing Company
PPPRA	Petroleum Products Prices Regulatory Agency
PPT	Petroleum profit tax
PPTA	Petroleum Profit Tax Act
PSA	Production sharing agreement
PSC	Production sharing contract
PSEs	Public sector enterprises
PTDF	Petroleum Technology Development Fund
R & P	Refineries and petrochemicals
ROI	Return on investment
RON	Research octane number
ROR	Rate of return
ROT	Refurbish, operate, transfer
$	US Dollars
SBU	Strategic Business Unit
SC	Service contract
scf	Standard cubic feet
scf/d	Standard cubic feet per day
SNEPCO	Shell Nigeria Exploration and Production Company
SNG	Shell Nigeria Gas Limited
SNOP	Shell Nigeria Oil Products Limited
SPC	Sale and Purchase Contract
SPDC	Shell Petroleum Development Company
TAM	Turnaround Maintenance
TCF	Trillion cubic feet
TOPCON	Texaco Overseas (Nigeria) Petroleum Company
UAE	United Arab Emirates
ULCC	Ultra large crude carrier
UMC	United Meridian Corporation
UN	United Nations
UNCLOS	United Nations Convention on the Laws of the Sea
UNDP	United Nations development programme
VDU	Vacuum distillation unit
VLCC	Very large crude carrier
WAGP	West African gas pipeline
WRPC	Warri Refining and Petrochemical Company Limited

About the author

Soala Ariweriokuma is on the staff of the Nigerian National Petroleum Corporation (NNPC) and works in the Commercial and Investment Directorate. His background is in business administration and economics, and he obtained his doctorate degree from the University of Nebraska, Lincoln. He was at one time a senior lecturer at the University of Port Harcourt and joined NNPC in 1992. In 1995 he was appointed technical assistant to the Honourable Minister of Petroleum Resources and in that capacity interacted extensively with key players of the oil and gas industry. His areas of interest include energy economics, petroleum investment analysis and hydrocarbon transportation economics. For relaxation he enjoys fishing, golf and gospel music.

Foreword

Africa has great energy potentials which remained unexplored for many years and the oil and gas industries in the various countries present unique opportunities and challenges. Oil exploration in the continent derived its roots from the early 1900s and in the mid-1950s oil was discovered in commercial quantities in Nigeria, Algeria and Libya. In recent years Angola, Gabon, Equatorial Guinea, Chad and Sudan have joined the ranks of oil producing countries, but the low technological development of the continent hindered the progress of the oil and gas industry. Consequently, the pace and scope of development of the industry has depended on external entrepreneurial participation and the interest of the international oil companies (IOCs) has steadily increased, thereby expanding crude oil production from a million barrels per day (mmbd) in the 1950s to over 10 mmbd in 2006. Africa currently accounts for about 10% and 8% of global oil and gas reserves respectively. The oil and gas producing countries in the continent have coalesced to form the African Petroleum Producers Association (APPA) guided by the fundamental objective of sharing valuable information and experiences that could maximise the benefits derived from oil and gas resources. Revenues from these resources have substantially enhanced the economies of producing countries and some, such as Nigeria, Algeria, Libya and Angola, belong to the OPEC family and contribute significant volumes to the aggregate production of the organisation.

The Nigerian oil and gas industry started in 1908 and has made huge progress. With the discovery of oil at Oloibiri in 1956 production commenced at a modest level of 5,100 b/d and subsequently escalated to about 2.4 mmbd in 2006. The industry covers a broad terrain spanning land, swamp, shallow continental shelf and the Deep Offshore. The ranks of the IOCs have increased and in recent years the pioneer oil companies – Shell, Mobil, ChevronTexaco, Total, Agip and Panocean – have been joined by the National Oil Companies of China, Brazil, Norway and Korea. The industry is characteristically vibrant and adjudged the biggest in Africa. Nigeria's oil and gas reserves are estimated to be 36 billion barrels and 187 trillion cubic feet (TCF) respectively. There is a strong prospect for further expansion in reserves in view of the aggressive exploration and production activities in the

Deep Offshore. The collaborative joint development of the overlapping maritime boundaries of Nigeria and The Democratic Republic of São Tomé and Príncipe (DRSTP) in the Gulf of Guinea offers additional opportunities of reserves expansion.

Although the industry has been widely explored, the evolution, challenges and numerous investment opportunities have not been adequately captured in books and scholarly literature. Against this background, the publication of this book is most timely. It is based on themes, namely: Regional oil and gas activities; Evolution of the Nigerian hydrocarbon industry; The upstream sector; The downstream sector; Gas monetisation; Privatisation and liberalisation of the industry; Environmental pollution; Elements of petroleum law; and Investment opportunities. Regional upstream activities focus on the exploration and production of oil and gas in Algeria, Libya, Egypt and Angola. The evolution of the industry in Nigeria presents a succinct account of the pioneer activities and the discovery of the first oil, while the section on the upstream sector presents an elaborate account of the achievements of the sector. Similarly the theme on the downstream sector features an interesting account of the dynamics and growth of the activities in the sector. Gas monetisation has become a major activity in the industry leading to the establishment of the world-class Nigeria Liquefied Natural Gas Company (NLNG) plant at Bonny. The success of the pioneer LNG programme has paved the way for Brass LNG and OK-LNG projects which are at advanced stages of execution. The book also presents an insightful discussion of pollution at the national and global levels.

In order to guide potential investors, two chapters are devoted to Nigerian Petroleum Law and Memorandum of Understanding (MOU). These chapters provide useful information on applicable taxes, royalties and incentives for investors. The book culminates with a chapter which clearly outlines the vast investment opportunities in the Nigerian oil and gas industry. The content of the book is rich in well-researched information and statistical data on the Nigerian oil and gas industry and some selected producing countries in Africa. It naturally presents itself to a wide spectrum of readers, both locally and internationally. As an interested participant in the global oil and gas industry, it gives me great pleasure to introduce this book and I feel certain that every reader will benefit from the vast amount of information captured in it.

RILWANU LUKMAN
Former Minister of Petroleum Resources of Nigeria, former President of OPEC and former Secretary General of OPEC

Preface

The starting point in the search for oil and gas in Africa is often traced to the late nineteenth century and over the years oil and gas have been discovered in commercial quantities in countries including Nigeria, Algeria, Libya, Egypt and Angola. The growing significance of these natural resources in Africa warrants the evaluation of the evolutionary pattern of the regional hydrocarbon industries and the challenges confronting the producing countries. Africa has made significant progress and contributes substantial volumes to the aggregate production of OPEC member countries. This notwithstanding, information on the oil and gas activities in the above countries, especially Nigeria, is limited. In view of this, an effort will be made to provide a fairly detailed account of the evolutionary pattern of the oil and gas industry in Nigeria. In addition the text will examine and analyse key issues which include gas monetisation, marginal field development, oil assets privatisation, petroleum products subsidy, environment, revenue generation etc. Nigeria is currently rated as the leading oil and gas producer in Africa; the industry is dynamic and has experienced rapid development in the past two decades, while its current level of reserves and contributions in OPEC distinguishes it as a force to be reckoned with in the African region. The oil and gas industries in Algeria, Libya, Egypt and Angola will be briefly examined and the aim will be to delineate areas of similarities in development and other challenges experienced by the producing countries referred to.

The text will also provide a detailed analysis of the impact of the oil and gas industry on the Nigerian economy in terms of its contributions to the GDP, employment and federal government earnings. The oil and gas business is a high revenue earner; therefore numerous interests impinge on the activities in the African region. Political power in the continent tends to be absolute in nature and therefore unilateral actions which are often not in the interest of the public are taken in order to satisfy personal and political interests. The impacts of such actions on the oil and gas industry are evaluated. The industry in Africa is confronted by major challenges, including low technological development, corruption and low value addition. The above challenges are profound in nature and cause major distortions in the growth and earnings from the oil and gas industries.

For purposes of fluid reading the text is divided into eight themes:

- oil and gas in Africa;
- the Nigerian oil and gas industry;
- the upstream sector;
- the downstream sector;
- petroleum law;
- the Niger Delta;
- shipping activities;
- industry re-engineering.

Oil and gas in Africa

This theme provides an overview of oil and gas exploration and production activities in key producing countries and members of the OPEC family such as Algeria, Libya and Angola. The Egyptian oil and gas industry is also examined. The analysis of the upstream activities in the aforementioned countries provides an opportunity to evaluate the trend of development of oil and gas activities and the contributions of the region to the global energy market.

Nigerian oil and gas industry

This provides an in-depth review of the evolution of the Nigerian, oil and gas industry. It provides an insight into the geological characteristics of the various regions of the country. Different rock types – granite series, volcanic rocks, Albian age formation, Maestrichtian equivalent, marine formation etc. are discussed in order to isolate distinct geological epochs and oil- bearing rocks in the country. It also examines the formation and role of the Nigerian National Petroleum Corporation (NNPC) in the industry. Chapter 4 provides a brief account of the formation of the Joint Ventures (JVs) between NNPC and International Oil Companies (IOCs), Production Sharing Contracts (PSCs) and Services Contracts (SCs). The Corporate Transformation Programme (CTP) is also discussed.

Upstream sector

Under the domain of exploration and production, the Nigerian upstream sector which depends on a gamut of cutting edge technology is analysed. Chapter 6 reviews the Marginal oil Field (MF) development programme of the Nigerian government. Chapter 8 examines the Nigerian Content Development (NCD) which derives from serious agitations for the domiciliation of a significant proportion of the expenditures of the IOCs in Nigeria.

Downstream sector

This theme evaluates the performance of the refineries and petrochemical companies. The low performance of these companies and the associated petroleum products scarcity are discussed in Chapters 10 and 11. Also examined is gas monetisation, which has in recent times become an important issue in the context of environmental pollution mitigation and revenue generation.

Petroleum law

Chapters 13 and 14 focus on the laws governing the Nigerian oil and gas industry. The discussions cover the origins of Nigerian petroleum law, Petroleum Profits Tax (PPT) ordinance, National Petroleum Investment Management Services (NAPIMS) and Joint Operating Agreements (JOAs). The Memorandum of Understanding (MOU) is also examined.

Niger Delta

The Niger Delta is the pivot of oil and gas activities in Nigeria. In recent years it has become volatile due to various agitations emanating from the region. In Chapters 15 and 16 some major issues concerning the plight of the communities in which oil is produced and the impact of oil and gas activities on the environment are critically examined.

Shipping activities

This theme evaluates the origin of global tanker transportation. It also focuses on the shipping business in the National Oil Companies (NOCs) and more particularly on the shipping opportunities in Nigeria.

Industry re-engineering

This examines the restructuring activities in the industry which include privatisation of oil and gas assets as well as liberalisation of the industry. Chapter 19 explicates various investment opportunities in the oil and gas industry which are available to potential investors.

Acknowledgements

I wish to express special thanks to NNPC for its support and the privilege of my exposure to the oil and gas industry. I wish to thank in particular Engineer Abubakar L. Yar' Adua, the Group Managing Director of NNPC, for his able leadership and commitment to professional development of staff. My profound appreciation goes to Dr Rilwanu Lukman for granting me audience amidst tight schedules and for writing the preface to the first edition of this book. I wish to thank M. S. Barkindo, S. I. Lawson, A. Babakusa, G. Meheux and Dr S. M. O. Amachree for their support in various forms which facilitated the completion of the book. I am also grateful to P. O. Makinde, A. O. Adelakun, Paul Ugbong and Friday Ebute for IT and secretarial support.

1 Oil and gas in Africa

Introduction

Africa is abundantly endowed with oil, gas and other energy resources. Exploration of these resources in the continent can be traced to early 1900; however, commercial discoveries were only recorded in the 1950s. In the 1970s consuming countries relied on oil from the Middle East for major industrial and domestic activities, but recent events in the Middle East have necessitated the shift of activities of IOCs to oil bearing areas in Africa. A hot spot in the region is the Gulf of Guinea, which is estimated to have 5–12 billion barrels of crude oil. The continent is technologically backward; therefore oil and gas exploration and exploitation depends on external entrepreneurial initiatives. Experts are of the view that the continent at current levels of production accounts for 10 per cent and 8 per cent of global oil and gas reserves respectively. Global interest in the industry has steadily increased, accounting for the expansion in crude oil production from less than 1 mmbd in the 1950s to well over 10 mmbd in 2006. The degree of contribution of each country varies, which in part determines the different levels of inflow of capital into the upstream sectors of the region. The industry in each of the producing countries presents unique opportunities and challenges: in the case of Nigeria it is observed that the terrain covers land, swamp, shallow continental shelf and Deep Water. Nigeria, Libya and Algeria have long been associated with hydrocarbon production and also belong to the OPEC family. Egypt, on the other hand, is actively involved in the African Petroleum Producers Association (APPA). In recent years Angola, Sudan and Equatorial Guinea joined the ranks of oil producing countries in the region. In view of the diversity of the continent, it can be contended that the political economy of the oil and gas industry in Africa covers a broad spectrum, with each shade of the spectrum exhibiting distinct characteristics which demand thorough analysis. Nigeria serves as the primary focus of the discussion; however, it can be posited that contextually the various oil and gas industries in the continent have political, economic and social/cultural links. The formation of APPA is an eloquent attestation of these links. In view of these relationships it would be necessary to briefly examine the upstream activities of selected countries, namely Algeria, Libya, Egypt and Angola, in order to establish basic characteristics in the evolutionary and operational patterns of the oil and gas industries in Africa. Such an analysis would provide an opportunity to estimate the potentialities of the various industries and the underlying political forces that shape them.

REGIONAL CRUDE OIL PRODUCTION

Algeria

Oil and gas activities

Algeria started oil and gas production around 1956 and currently has proven reserves of 12.3 billion barrels of crude oil (Figure 1.1). It is ranked third largest producer in the continent. An estimated 70 per cent of the proven reserves are located in the Hassi Messaoud Basin, while about 30 per cent are found in Berkine Basin. In 2006, average daily production amounted to 1.4 million barrels. In addition 440,000 b/d of lease condensate and 305,000 b/d of natural gas liquids (NGL) were produced from active fields. Available data (Figure 1.2) indicates that aggregate production of hydrocarbons (i.e. crude oil, condensates and NGLs) in 2006 amounted to 2.13 mmbd. Algerian Sahara Blend is rated high grade hydrocarbon with a sulphur content of 0.1 per cent and has a 45° API rating. Prior to 2005, the industry was dominated by Sonatrach, the NOC. The role of Sonatrach was modified through the enactment of the hydrocarbon reform bill. The bill paved the way for foreign

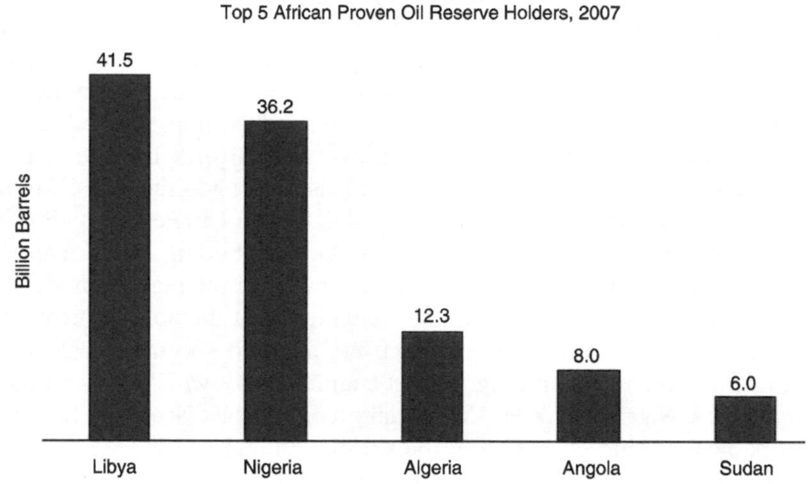

Figure 1.1 Proven oil reserves of reference countries.

Source: http://www.eia.doe.gov. EIA Country Analysis Publication, 2007

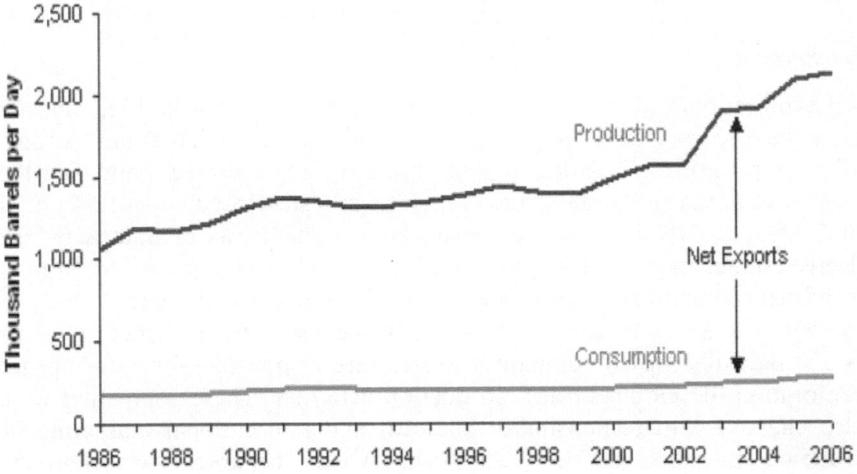

Figure 1.2 Algerian crude oil production and consumption.
Source: http://www.eia.doe.gov. EIA Algeria Country Analysis 2007

participation and empowered the NOC to acquire at least 51 per cent equity interest in new oil and gas concessions or joint venture (JV) companies in the industry.

Sixth licensing round

In 2005 the NOC executed its sixth licensing round which placed on offer ten Blocks for IOC participation. On the whole, 54 companies expressed interest and took part in the bid process. BP won three concessions while Shell, BHP-Billiton and the UAE-US consortium won two concessions each. Sonatrach attained a production level of 440,000 b/d at the Hassi Messaoud field in 2006, by far the highest individual company. It is also associated with the Hassi R'Mel field (north of Hassi Messaoud) which has an estimated crude oil production of 18,000 b/d. Other fields are located at Zarzaitine, Ben Kahla, Ait Kheir, Tin Fouye and Tabankort. The enactment of the hydrocarbon reform bill paved the way for active foreign participation in the industry and first among the foreign producers was Anadarko, with a production capacity of 500,000 b/d. The company also operates in Ourhound (Eastern Algiers) and Hassi Berkine South fields which collectively account for 450,000 b/d. The NOC continues to inject new investment capital into its operations, thereby paving the way for the simultaneous development of seven fields in Block 208 of the Berkine Basin. Production from these fields was due to come on stream in 2008.[1]

Libya

Background

Oil exploration and production activities started in Libya in 1953, shortly after the discovery of oil in Algeria. The Libyan General Petroleum Corporation (Lipetco) was founded in 1968 through a royal decree. Following the overthrow of the monarchy in 1969 Lipetco was restructured under Law no. 24 of 3 March 1970 to form the Libyan NOC, which was mandated by the decree establishing it to engage in exploration and production of oil through its affiliate companies or in collaboration with IOCs. The dominant mode of operation was the Production Sharing Agreement (PSA). Structurally the NOC had fully-owned companies which were responsible for carrying out exploration, development and production activities. These companies were also charged with responsibility for local and international marketing of crude oil and products. The NOC's primary export markets were Germany and Spain. Initially the NOC signed a participation agreement with selected IOCs but these agreements were subsequently converted to PSAs. The oil and gas industry in Libya progressed smoothly until the PANAM Flight 103 bombing incident over Lockerbie, Scotland in 1988. Libya was accused of sponsoring terrorist activities which led to the bombing of the passenger aircraft. Following this incident UN sanctions were imposed on Libya on 31 March 1992. Consequently, oil and gas activities in the country suffered a serious setback and the NOC did not enter into new collaborative activities with foreign companies in the 1990s.

Lipetco in the 1960s

The activities of Lipetco in the 1960s and 1970s were primarily defined by political and economic events. In the early 1950s Libya was essentially a subsistence agrarian economy with modest to low income from the sector. The discovery of oil in 1957 dramatically changed the fortunes of the country and annual growth rate progressed to about 20 per cent in the 1960s. The new revenue stream from oil became the vehicle of growth, thereby necessitating elaborate structural changes in the economy. The outcome of one of these changes was the creation of Lipetco. In 1969 the monarchy was overthrown, paving the way for Colonel Muammar Gadaffi to become head of State. The new government espoused self-reliance and Socialist ideologies. These initial manifestations of the new regime indicated government intention to participate actively in economic planning, policy formulation and other broader issues of national interest. The need for the new regime to actively participate in governance was signalled by the introduction of more aggressive policies targeted at ownership of oil assets and a new pricing policy. The strategy of price control through production cuts was introduced by OPEC and was widely embraced by the members. Libya joined OPEC in 1962 and progressed to be an influential member and the seventh largest producer in the organisation in 1977, but this position could not be sustained in the era of UN sanctions. JV agreements were signed between Lipetco and the IOCs, the first with French companies ERAP (later ElF), and SNPA (Aquitaine). In 1969 additional JV agreements were signed with Royal Dutch/Shell, ENI's Agip and Ashland Refining.

JV activities in the 1970s

As pointed out earlier, Lipetco was transformed into Libyan NOC through Law no. 24 of 1970. The law restricted the formation of new JVs with IOCs. Alternatively, Production Sharing Agreements (PSAs) were introduced as the new mode of engagement of foreign oil companies. Production sharing was at a ratio of 85:15 onshore and 81:19 offshore. In July 1970 a new law was enacted vesting in the NOC the authority to market all oil and gas products in Libya. In order to carry out this mandate Brega Petroleum Marketing Company was established as a subsidiary of the NOC. The foreign owned companies – Shell, Ente Nazionale Idrocarburi (ENI) marketing subsidiaries and Petrolibya were transferred to the NOC. The operations of Brega (the marketing company) would under these circumstances be responsible for importing, distributing and marketing of petroleum products in the country. The NOC aggressively pursued the policies of the government including the new higher oil prices policy and PSAs. These policies were objectionable to IOCs, who put up stout resistance. The government was resolute and companies were initially given the opportunity to surrender voluntarily participatory interest in their concessions in compliance with the new partial nationalisation policy of the government. Some companies voluntarily complied while others continued on the path of resistance. Non-compliant companies were subjected to stiff political pressure to relinquish the concessions.[2]

Crude oil production

Aggressive crude oil exploration in Libya commenced in 1953 and the first oil was discovered at West Fezzan in 1957. However, Esso (later Exxon) made the first commercial discovery in 1959 at Zaltan. The Zaltan field was linked with export facilities at Marsa al Burayqa in 1961. The early discoveries were followed by others which included major strikes in Sirtica Basin field, classified as one of the largest oil fields southeast of the Gulf of Sidra. The Sirtica Basin production remained a major source of crude oil until 1987. In 1969 another major discovery was recorded at Sarir, southeast of Sirtica Basin field. In addition to the major fields some other oil deposits were discovered in fields located in Northwest Tripolitania. The intense exploration and development activities led to the discovery of new oil deposits at the Ghadamis Basin, about 400 km southwest of Tripoli. Similar strikes were recorded in 1974 at fields located about 29 km northwest of Tripoli. It is important to note that in 1977 major oil exploration activities were localised in the offshore fields. In 1987 NOC and Agip collaborated to put on stream the Bouri field. It is significant to note also that the settlement of the maritime boundary disputes between Libya and Tunisia in 1982 and that of Malta in 1983 expanded the scope of offshore exploration activities. The settlement of these disputes was considered strategic in an area believed to hold about 7.5 billion barrels of extractable crude oil. Oil production in 1984 was principally governed by the Petroleum Law of 1955 which was subsequently amended in 1961, 1965, and 1971. In an effort to expedite national development, the concession contracts had enshrined in them progressive nationalisation of foreign operations in the industry within a period of ten years. In this regard the government placed its share of operations at 25 per cent, with a provision for rising to 75 per cent. The PSA with Esso (first exporter of Libyan crude oil in 1961) being among the first, it served as a litmus test for the profitability of the PSA model. The Esso experiment proved successful and this encouraged many companies from Europe and the US to sign similar agreements with the Libyan government. Available records indicate that in 1969 about 32 companies agreed concession agreements with the NOC. The government intensified its nationalisation objectives in the industry and the NOC actively served as the vehicle for the execution of the nationalisation agenda.

The post-revolutionary nationalisation programme commenced in December 1971. The first casualty in the exercise was British Petroleum in the BP-Bunker Hunt Sarir field. Industry experts described the action against BP as a retaliation for Britain's failure to prevent Iran from seizing three small islands in the Persian Gulf believed to belong to the United Arab Emirates. In 1972 the NOC requested a 50 per cent participatory interest in the Bunker Hunt operations. The request was denied, which led to total nationalisation of all Bunker-Hunt assets in 1973. In 1972 ENI and the NOC mutually settled for 50 per cent government participation. Similar discussions took place between the NOC, Occidental Petroleum Corporation and Oasis Group. Occidental conceded to the NOC the purchase of 51 per cent of the assets. In 1973 Oasis Group owned by Continental Oil (33.3 per cent), Marathon (33.3 per cent), Amereda (16.6 per cent) and Shell (16.6 per cent) agreed to a 51 per cent assets acquisition by the government through the NOC. The government pressed ahead with the nationalisation programme and on 1 September 1973 it made a blanket announcement confirming the acquisition of 51 per cent interest in all the remaining companies in the industry. Shell opposed the government acquisition of its interest in Oasis and initiated legal proceedings against the Libyan government. The government took exception to the action of Shell and as a result nationalised all its assets in 1974. The Libyan-American Oil Company, Asiatic Company and Texaco had their assets nationalised and were paid compensation in 1977. The unfavourable posture of the government to IOCs forced Exxon to pull out of Libya in 1981. Mobil took a similar action in 1982 by withdrawing from its operations in the Ras al Unuf system. The withdrawal of these companies from the Libyan upstream sector indirectly expanded the scope of operations and control of the NOC. In 1987 the total equity of the NOC in the industry was estimated to be about 70 per cent. The Libyan oil industry suffered a serious setback during the period of isolation emanating from the UN sanctions against the oil rich country. The sanctions imposed in 1992 lasted until April 1999. Upon the lifting of the sanctions the country initiated revisions of the petroleum regulatory laws. Available data also indicates that about 135 Blocks were earmarked for bid/offer to the IOCs. The situation in the country has improved and a good number of IOCs have returned to Libya to reactivate the upstream sector. In 2004 Libyan crude oil production stabilised at about 1.2 mmbd. In view of the enhanced production activities, it was projected that production would attain 2 mmbd in 2007.[3]

Egypt

Oil and gas production

Egypt is a significant oil and gas producer and long standing member of the APPA. It has aggregate crude oil reserves of 3.7 billion barrels. Average daily production increased over the years and peaked at 576,000 b/d in 2005, but recent trends indicate a decline in production which is taken seriously by the government. In this regard, appropriate steps were taken to introduce cutting edge technology to the exploration, and production programmes and Enhanced Oil Recovery (EOR) techniques have also been adopted as options for slowing down the declining rate of production. Oil is derived from four main territories, namely: the Suez Canal, which accounts for 50 per cent of recoverable oil; the Sinai Peninsula; and the eastern and western deserts. The Gulf of Suez Petroleum Company (GUPCO) is the producer in the Gulf of Suez Basin under a PSA arrangement between BP and the Egyptian General Petroleum Company (EGPC). Production in the GUPCO fields commenced in the 1960s and increased until about the mid-1980s when regression in production was apparent. Petrobel, ranked the second largest producer in the country, is a JV company involving EGPC and ENI of Italy. Its active fields are located at Belayim, proximate to the Gulf of Suez. It is also actively engaged in the implementation of EOR programmes in order to stem production decline in the fields. Exploration and production activities in the industry are also undertaken by the Suez Oil Company (a JV involving EGPC and Deminex), Badr El Din (EGPC and Shell) Petroleum Company, El Zaafarana Oil Company (EGPC) and British Gas-BG JV. As part of efforts to reverse decline in production, BP embarked on a broader programme aimed at discovering rich oil fields. The campaign led to the discovery of a new robust oil field at Saqqara, located offshore in the vicinity of El-Morgan field and by far the largest since 1989. It was programmed to commence commercial production in 2007 and expected to attain a peak production range of 40–51,000 b/d. Exploration and production are currently being targeted at offshore fields in the Mediterranean. Shell was successful in the bidding round in 1999 and was therefore awarded a Deep Water Block off the Mediterranean coast. Similarly, Total, ENI and BP secured offshore Blocks in the 1999 bidding round. All these concessions are geared toward the enhancement of hydrocarbon reserves in the Egyptian oil and gas industry.[4]

Angola

Gas development

Angola was embroiled in a 27 year civil war which destroyed various facets of the society, but the country is now fast emerging as a significant regional producer of gas. Proven reserves were estimated to be 2.0 TCF in 2007. Aggressive campaigns being carried out in the offshore segment of the industry have led to the discovery of gas fields at Takula, Kokongo and Numbi, which could increase proven reserves to 9.5 TCF. Angola is associated with 85 per cent of gas flare, with the balance of 15 per cent being re-injected to boost the performance of the reservoirs or extracted as Liquefied Petroleum Gas (LPG). In an effort to comply with global environmental standards, Angola has drawn up programmes to end gas flaring in the various fields. The government targeted fields north of the Congo River to be zero flare compliant in 2005 and other fields to attain the same standard in subsequent periods. In 2007 zero flare programmes were being pursued in Nemba, Lomba and Kuito. The reduction and indeed subsequent elimination of gas flaring will boost availability of gas resources as feed stock in the industrial sector. More importantly, availability of gas will pave the way for the conversion of gas to LNG, NGLs and LPG. Angola LNG Limited was established as a JV company between Sonangol, Chevron, ExxonMobil, Norsk Hydro, Total and BP with a view to monetising the gas reserves. The proposed plant is earmarked to cost $5.00 billion and will depend on associated gas which will be derived from Blocks 1–3 and 15–18 respectively. The Ministry of Urbanism and Environment has approved EIA studies and necessary legislation has been enacted to pave the way for the construction of the plant. A contract was awarded to Boskalis International BV and Jan de Nul Dredging Limited. The establishment of LNG plants will expand the revenue base, which if properly applied should lead to enhanced development and improvement of the quality of life of the citizens.

The NOC (Sonangol) was established in 1976 and appointed sole concessionaire by the government in 1978. It achieved daily production of about 900,000 b/d in 2002, ranking second in oil and gas production in the Sub-Saharan region behind Nigeria. It joined OPEC as the twelfth member country on 1 January 2007 and currently has total reserves of about 8.2 billion barrels. Exploration and production activities are carried out through JVs and PSAs with Multi-National Companies (MNCs). The Angolan oil and gas industry is growing rapidly and the contributions of the sector account for about 50 per cent and 90 per cent of the GDP and government revenues respectively. The key players include Shell, Chevron, Total and ENI. Crude oil production averaged 280,000 b/d in 1986 but further increased to about 900,000 b/d in 2005. The upward trend continued and in 2006 a 1.4 mmbd level was attained (Figure 1.3). Experts opine that production would further escalate to 2.5 mmbd in 2008. Angola's oil and gas industry is dominantly offshore with Blocks 15 and Zero featuring the most prolific fields. The crude oil stream derived from the industry measures 30°–40° API and a sulphur content spectrum of 0.125 to 0.14 per cent. The low sulphur content in the crude stream makes it a highly demanded commodity in the US and other environment conscious European countries. On average, 476,000 and 470,000 b/d of Angolan crude oil are exported to China and the US respectively.

Field development activities

The Angolan oil and gas industry embodies onshore and offshore fields. The offshore segment is divided into three functional zones, namely Bands A, B, and C. Band A is located in the shallow water and comprises Blocks 0–13 while Band B in the Deep Water features Blocks 14–30. Band C corresponds with the Ultra Deep segment and contains Blocks 31–40. Block Zero, located offshore in the Cabinda axis, is generally classified as prolific. It accounts for 370,000 b/d or one third of the aggregate daily crude oil production in the industry. The highly viable fields in Block Zero are located at Numbi, Kokongo, and Takula. In addition to the preceding fields, Chevron in 2005 brought on stream the Bomboco oil and Sanha gas fields respectively. Both fields produce oil, condensate and LPG. Available data indicate that total production from the fields would peak at 100,000 b/d in 2007.

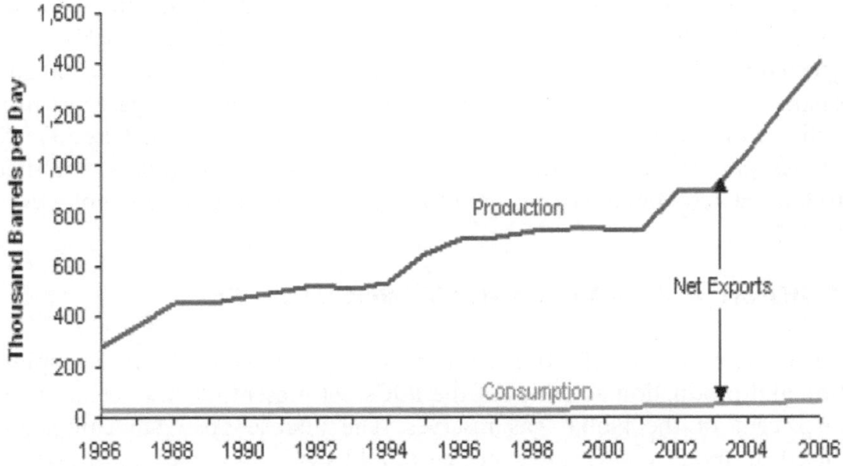

Figure 1.3 Angolan oil production and consumption trend.

Source: EIA Country Analysis March 2007

Cabinda Gulf Oil Company (CABGOC)

Cabinda Gulf Oil Company is the operator of Block 14 which has been associated with nine oil discoveries of commercial quantity. In 1997 the first oil discovery was recorded at Kuito and production of 80,000 b/d was achieved in 2000. Production at the field declined to about 55,000 b/d in subsequent years. Although initial production at the field showed significant promise, reservoir studies isolated fast hydrocarbon depletion trends. In addition to this problem the stream of crude oil from Kuito was found to be of lower grade than other crude streams from fields in Block 14 and is usually traded at a $5.00/bbl discount in the international market. CABGOC is actively involved in the development of fields at Benguela, Belise, Lobito and Tomboco which are collectively referred to as the BBLT project. The fields have commenced production and it is projected that 200,000 b/d will be attained in 2008.

ExxonMobil

ExxonMobil is the operator of Block 15 located in Deep Water off the Angolan coast. Block 15 is classified as a large field with recoverable hydrocarbon estimated to be 4.7 billion barrels. Production from the Block is robust and in 2003, a 750,000 b/d production level was achieved. The company has brought on stream the Xikomba field with about 105 million barrels of reserves. In 2006 the field produced 68,000 b/d. Hugo and Chocalho fields grouped under the $3.37 billion Kizomba 'A' project were produced through the deployment of the FPSO (Floating Production Storage and Offload) vessel, and production from the fields attained 250,000 b/d in 2005. Kizomba project 'B' fields are reasonably prolific and are estimated to hold 1 billion barrels of recoverable oil and operate at a peak of 25,000 b/d. Other production activities from Kizomba 'C' – Mondo, Saxi and Batuque fields – are expected to boost daily production with an estimated 200,000 b/d. Total and Sonangol have made similar discoveries at Blocks 4 and 17. Aggregate production from these Blocks will boost daily production by as much as 500,000 b/d. As part of the government upstream development agenda, multiple production activities are being vigorously pursued at other licensed blocks.[5]

REGIONAL GAS DEVELOPMENT PROGRAMMES

The energy potentials of Africa continue to expand due to enhanced exploration and production activities of the IOCs. At present, Africa accounts for 8 per cent of the global gas reserves. The intense upstream campaigns embarked upon by IOCs are expected to add significant volumes to existing reserves, thereby elevating the continent's share of world gas reserves. Global environmental concerns and the relatively low price of gas both created favourable conditions for the expansion of global demand for gas. The commodity is generally considered the energy of the future, replacing dependence on fuel oil, diesel and coal for the firing of furnaces, power plants and domestic consumption. Gas is widely used in Combined Cycle Generating (CCG) plants and as feedstock in methanol, fertilizer and LNG plants. The profile of gas in revenue generation for the national economies of developing countries continues to rise. Consequently, oil and gas producing countries have taken definite steps to harness the resources. This section will briefly evaluate and analyse the level of gas production activities in reference countries, namely Algeria, Libya, Egypt and Angola. The purpose of the analysis is to examine the approach adopted by these countries in the development of gas. It would be necessary to identify from the analysis areas of commonalities in the practice (gas development) and draw vital lessons from the strategies adopted by the reference countries.

Algeria

Gas development and consumption

Algeria commenced oil and gas production in 1956 and in early 2007 gas reserves were estimated to be about 162.5 TCF, second only to Nigeria's reserves of about 187 TCF. Globally, Algeria's gas reserves are ranked eighth. Several gas fields exist in the country but Hassi R'Mel, discovered in 1956, is by far the biggest, accounting for one quarter of the total dry natural gas production. About 75 per cent of the gas occurs as associated gas in a commingled form with oil and non-associated gas which occurs alone in fields located in the south and southeastern segments of the country. Available data indicates that fields at the Rhourde Nouss territory (the southeastern region of the country) accounts for 12.8 TCF of gas reserves. In Amenas region, proximate to the Libyan border, there are the Tin Fouye, Tabankort, Quan Dimeta, Qued Noumer and Alrar fields which collectively account for over 10 TCF. The In Salah region in Southern Algeria is also designated as having aggregate reserves of 6–9.8 TCF. The gas reserves of Algeria are quite robust and in 2004 it produced 2.7 TCF of natural gas, thereby making it the eighth largest producer and second in OPEC, Iran being the leader at the time. Domestic gas utilisation is on the increase and total consumption reached 0.7 TCF, being 24 per cent of total production. Gas accounts for 62 per cent of total energy consumption in the country. In 1997 gas production was intense and it momentarily overtook oil production although in subsequent years the trend was reversed. The NOC Sonatrach is the major producer of gas as well as the wholesale distributor of gas in Algeria, and Sonel Gas controls retail distribution of the commodity. Following some policy reforms of the government, foreign companies (BP, Repsol, BHP-Billiton, Total and Statoil) were given the opportunity to invest in the gas sector.

Exploration and production

Algeria has for some years targeted natural gas development as a major source of revenue for the government. Consequently, plans were articulated to expand gas production through the formation of JV companies between Sonatrach and reputable international companies. In this regard Sonatrach, BP, and Statoil formed a consortium. The Algerian government granted the resultant company – In Salah Gas Consortium – development rights over seven out of twelve fields in the In Salah region. The mandate of the consortium included the appraisal of existing wells and to explore for new gas reserves in the territory. The fields controlled by the consortium hold proven reserves of about 6.2 TCF and the prospect for an additional 9.5 TCF of recoverable reserves. Natural gas production in the fields commenced in July 2004 and stabilised at approximately 875 million scf/d. The gas produced was targeted at European markets – Enel (Italy), Turkey and some North African countries. Other gas projects linked with three Blocks in Illizi province (South Eastern Algeria), namely Gassi Touil, Ohanet and In Amenas, were developed by the Sonatrach–BHP-Billiton consortium. Production of NGLs and LPG began in 2003. The full scope of the project embodies a natural gas processing plant capable of delivering about 29,000 b/d and 25,000 b/d of condensate and LPG respectively. Gas development projects continue to expand and in 2004 Repsol-YPF and Gaz Natural were awarded a project located at Gassi Touil. The affected field was estimated to have 8.9 TCF of reserves. The project which was earmarked to cost $2.0 billion consisted of 52 development wells, a 630 mm scf/d natural gas pipeline, 500 mm scf/d gas liquefaction terminal at Arzew and a 780 mm scf/d natural gas processing plant. Initial production at Gassi Touil is programmed to commence in 2009 with the bulk of the gas targeted at markets in Spain and Europe. In 2006 the In Amenas field operated by the Sonatrach, BP and Statoil consortium also started operations. The field was evaluated to have peak potential of about 895 million scf/d of natural gas and additional capacity of 52,000 b/d of condensate and LPG respectively.

Pipeline gas transportation

Algeria relies on a wide network of pipelines for domestic transmission and export of gas and Hassi R'Mel, with proven reserves of 84 TCF, serves as the centre of all gas activities in the country. The field is connected to liquefied natural gas terminals by an elaborate network of pipelines. Arzew and Skikda are connected to Hassi R'Mel by pipelines measuring about 312 and 359 miles respectively. Each segment (i.e. Arzew and Skikda) has daily transmission capacity of 4.36 bcf and 1.95 bcf respectively. Furthermore, Isser terminal is connected by a 270 mile, 690 mm scf/d capacity pipeline to Hassi R'Mel which serves as the nerve centre of the gas transportation system. Gas producing fields in the Amenas and In Salah regions are similarly connected to Hassi R'Mel through pipelines measuring 605 and 330 miles long respectively. These pipelines transport about 3.4 bcf/d and 770 mm scf/d of natural gas respectively. Furthermore, Gassi Touil is connected to Hassi R'Mel with a 90 mile, 610 mm scf/d capacity pipeline from the surrounding fields. Other pipeline networks include the Medgas, Galsi and Trans-Sahara pipelines.

Export pipeline network

Algeria is connected to the European market by the Trans-Med and Maghreb–Europe gas pipelines. The Trans-Med gas pipeline measures 670 miles and has a capacity of 2.30 bcf/d. The pipeline, which was completed in 1983, originates from Hassi R'Mel and links central Italy through Tunisia and Sicily. The pipeline was de-bottlenecked in 2004, which enhanced the installed capacity to a higher volume. The 1,000 mile, 820 mm scf/d capacity Maghreb–Europe (MEG) pipeline is operated by a consortium of companies comprising Enagas of Spain, Sonatrach and SNPP of Morocco. It was completed in 1996 and connects Hassi R'Mel with Cordoba in Spain via Morocco, where it linked the Portuguese and Spanish gas transmission facilities. Subsequently, Sonatrach awarded a $92.6 million contract to ABB for the construction of a gas compressor station designed to expand the capacity of MEG to 1.76 bcf/d.

Medgaz pipeline

Sonatrach and Cepsa of Spain in July 2001 reached a mutual agreement to construct a gas pipeline connecting Algeria and Europe. The project often referred to as Medgaz (a 120 mile gas transportation pipeline) is designed to link Beni Saf in Algeria to Almenia, Spain and has a provision to connect France at a future date. A feasibility study for the project was concluded in 2002 with initial construction programmed to commence in June 2007. The project, which is estimated to cost about $1.25 billion, would be completed in 2009. Medgas will have an initial capacity of 385 mm scf/d and attain a peak volume of 1.6 bscf/d. Cepsa and Iberdrola have both signed contracts with Medgaz to purchase part of the gas.

Galsi pipeline

Sonotrach in 2002 struck an accord with Wintershall of Germany and Enel of Italy and formed the Galsi Consortium. The consortium was designed to build a gas pipeline from Algeria to Italy. The pipeline would run onshore from Hassi R'Mel and progress in an underwater path to Cagliari in Sardinia. The pipe would continue onshore to Olbia, Sardinia. The last portion of the pipeline was designed to be onshore terminating at Pescara in Italy. The project, which has an initial capacity of 775–992 mm scf/d, is scheduled for completion in the third quarter of 2009.

Trans-Sahara pipeline

The Trans-Sahara gas pipeline is a collaborative project between NNPC and Sonatrach. In order to effectively execute the project NNPC and Sonatrach formed the NIGEL consortium dedicated for the construction of about 2,500 miles of natural gas pipeline originating from the Warri Delta State of Nigeria through Niger to Hassi R'Mel. Both countries intend to take advantage of the Trans-Sahara pipeline project and construct a road as well as lay optic fibre cables for communication. The pipeline is designed to link up with the existing Trans-Med pipeline and the proposed Medgaz pipeline through which gas would be transmitted to the European markets. The project, designed to operate at an initial capacity of 1 bcf/d and peak capacity of 1.3 bcf/d, is estimated to cost about $3.6 billion.

In recent years oil and gas assets have become targets of rogue elements of Nigerian society. The extensive length of pipeline, especially through the swampy mangrove forests, tropical forests and desert, is a matter of concern to the primary sponsors of the project. However, completion in about 2015 is considered to be a high priority.

Liquefied natural gas

Algeria started LNG production in 1964 at the Arzew GL4Z plant. It is currently the fourth largest exporter of LNG with Indonesia, Malaysia and Qatar leading in volume of output. Total exports from Algeria account for 13 per cent of world demand. Western Europe (Spain, France and Turkey) is the primary market for LNG. Enagas (Spain), Botas (Turkey), Gaz de France, DEPA of Greece and Distrigaz (Belgium) are major takers of the LNG produced.

Libya

Gas production

Libya has significant gas reserves estimated to be about 50 TCF but expansion of reserves remains a high priority of the Libyan NOC. The government plans to increase domestic consumption in order to free more oil for export. Expansion of gas reserves will create a conducive situation for large gas volumes to be exported. Substantial quantities of associated gas were discovered at Sirte Basin (mainly from six offshore fields) as well as Bouri, Ghamades and Mursuq Basins. Limited quantities of LPG produced from the gas fields are channelled to domestic refineries for storage and distribution for domestic consumption. In 1996 NOC signed an agreement with ENI of Italy to execute the $4.0 billion West Libyan development plan designed to develop and export natural gas to Italy. The project envisaged an annual export of 275 bcf to Italy through an underwater pipeline across the

Mediterranean to southern Sicily terminating at the Italian mainland. Similar plans were made for onshore wells at Wafaa, proximate to the Algerian Border, and bilateral talks were concluded for the construction of a pipeline linking Libya and Egypt. A proposal was also considered for the construction of a 900 mile natural gas pipeline from Egypt to southern Europe. The pipeline would traverse Egypt, Tunisia and Algeria. A section of the pipeline is earmarked to link the existing pipeline connecting Morocco and Spain.

Libya exported its first LNG cargo from the Marsa el Brega plant which came on stream in 1971. It was the second country in the world after Algeria (1964) to export LNG. The plant was built by Esso (Exxon) in 1960 but technical limitations constrained production to one-third of installed capacity for export to Spain.

Sanctions imposed on the country drastically affected the gas monetisation programme. With all sanctions lifted, investors have returned to the oil rich country to undertake vigorous exploration and production activities in the upstream sector. Global demand for gas has increased; therefore the Libyan NOC will be in a position to provide appropriate incentives to IOCs to take advantage of investment opportunities.[6]

Egypt

Oil was first discovered in Egypt in 1886 but commercial production only commenced in 1913. Output remained low until 1950 when JV companies were established with IOCs. Its oil reserves are currently 4.7 billion barrels. In the gas sector the reserves have risen to 57 TCF while daily production attained about 5.0 bcf/d in 2007. The EGPC was established in 1962 as a vehicle for the expansion of the upstream sector and the 1974 oil price increase prompted the government to open up the sector through JV companies. Natural gas discoveries were made in the 1970s and 1980s and gas exploration gained greater significance in the 1990s. In 1989 the government signed 39 contracts to explore for gas in the western desert, most of the exploration activities being executed through JV agreements between the EGPC and IOCs. The decision to establish JVs was necessitated by debt pressure, the fall of oil prices in the 1980s and the large investment capital required to execute the projects. Payment in cash and oil was extended to the IOCs as an incentive for companies to invest in the upstream sector of the economy. In the early 1990s IOCs, which included BG, ENI, Shell and Apache, engaged in gas exploration in the western desert. In order to supervise gas related activities effectively the government, in August 2001, created the Egyptian Natural Gas Company (EGAS). Gas development programmes are currently being executed vigorously and the government intends to increase gas reserves by as much as 31 TCF in 2011. The expansion of the reserves will provide an opportunity for the government to double its exports: new gas discoveries will emanate mainly from offshore fields (Port Faud, South Temsah and Wakah) in the Nile Delta, while the Obeiyed field in the western desert is an important field under development. Gas exploration activities in the Gulf of Suez are mainly undertaken by the International Egyptian Oil Company (IEOC), a subsidiary of ENI, and in recent years IEOC has collaborated with BP to execute exploration and production programmes in the Nile Delta region.

Gas production and utilisation

Gas demand at the domestic level has increased substantially, driven mainly by enhanced use of gas for power generation by IPP operators. The franchise for the construction of IPPs was awarded in 1998 to BG, Orascom and Petronas and these companies built gas distribution networks to Asyut, located in the Southern territory. The steady increase in gas reserves has warranted the identification of target markets in Europe as export revenues are crucial for the execution of key national projects as well as to meet mandatory obligations to foreign creditors. In order to guarantee availability of gas reserves at the domestic level the proportion of gas reserves earmarked for export was reduced from 33 per cent to 25 per cent. A considerable amount of gas is exported through pipelines to neighbouring countries and as a result mutual agreements have been reached between Egypt and Israel for the supply of gas. To accomplish this objective EGPC formed a consortium with Merhav of Israel and an indigenous entrepreneur to construct a $2.6 billion pipeline network which would be completed in about 2008. The first export of gas from Egypt was to Jordan through the pipeline. The interconnecting export pipelines were built from Arish to Aquaba and were further extended to Northern Jordan. The network which was completed in 2006 had an initial capacity of 1.15 bcf/d.

LNG

Egypt has expanded its gas monetisation programme with a view to supplying LNG to the US, Spain and France. The first single-Train LNG plant was built at Damietta by Union Fenosa of Spain which, in the same year, delivered the maiden cargo to an off-taker. In the face of increasing demand for LNG the company in collaboration with BP and ENI as partners signed a contract for the expansion of the pioneer plant from 770 mm scf/d to 1.4 bcf/d by the year 2009. The feed gas for the expansion will not come from upstream natural gas; therefore Union Fenosa contracted with EGPC to supply gas for the plant. Union Fenosa in pursuance of its power supply obligations signed to off-take 60 per cent of the LNG from the Damietta plant to fire its IPP plants in Spain. A Sales and Purchase Agreement has already been signed between BP and Union Fenosa for the supply of natural gas to Train 2. The first phase of the LNG project was successful, paving the way for the establishment of the second LNG plant project (Egyptian LNG) at Idku by the BP and Petronas consortium. The feed gas for the project was sourced from Sienna and Simian offshore fields and the two-Train 1.75 bcf/d plant commenced operations in 2005. In 2006 BG de-bottlenecked the plant and expanded the capacity by 10 per cent. Gaz de France and BG LNG services signed off-take agreements with the BG and Petronas consortium for all the LNG liquids derived from the two Train of the second plan. The bulk of the LNG from the phase two plant was targeted at the US market and was effective in the second quarter of 2006. In an effort to expand the scope of the gas monetisation programme discussions were undertaken with Shell to establish a 75,000 b/d gas-to-liquid (GTL) plant in Egypt.

CHALLENGES OF GAS MONETISATION

Low technological development

It may be fair to posit that African oil producing countries are rather overwhelmed by the huge revenues derived from oil and gas. The revenue streams from the up and downstream sectors by far surpass revenues from the agricultural and solid minerals sectors. For this reason development in the agricultural and solid minerals sector is low. It is important to note, however, that oil and gas resources are depletable but agriculture is a renewable sector because a farm can be replanted many times by several generations. Over the years, huge sums of money have been derived from oil in Nigeria but a good proportion of these revenues has been expended on food importation. A similar situation occurs in Angola, Equatorial Guinea, Gabon and other African oil and gas exporting countries. It is imperative for African oil and gas producing countries to reduce the tempo of exports and recoup the shortfalls in revenues through development and exploitation of the agricultural and solid minerals sectors. The growth of oil revenues in the continent is devoid of technological development. This implies that personal incomes are engaged in the procurement of imported goods. More importantly, all technologies used in the upstream sector are imported, so this practice erodes the advantages created by the existence of oil in the continent. Annual expenditure in the upstream sector of the Nigerian oil and gas industry amounts to $12.0 billion. In 2005 about 95 per cent of expenditure in the sector was used by IOCs to procure equipment and other technical services from vendors and consultants offshore. Only about 5 per cent of the expenditure was retained in the Nigerian economy. The practice is injurious to the national economy thereby necessitating the introduction of the Nigerian Content Policy. The policy aimed at retaining about 45 per cent of upstream expenditures in the economy by 2008 and progress to 75 per cent in 2010. Angola, Equatorial Guinea and Libya are in similar situations such that the earnings from crude oil export are significantly depleted by huge offshore procurements. The solution to this problem is the development of the capacity to manufacture technology to produce certain grades of equipment, materials and tools which are widely used in the oil and gas industry.

Corruption

Corruption has been a major factor militating against development in Africa, especially in oil producing countries. Oil revenues have been subjected to serious abuse by government officials. Nigeria, Angola, Egypt and Equatorial Guinea have been identified as countries in which oil revenues found their way into private accounts of corrupt government officials. Nearly two decades after the 1990 Gulf war questions concerning the application of the windfall profits from Nigerian oil remain unanswered. A panel report on the reform of the Central Bank of Nigeria indicated that more than $12.0 billion of oil revenue accruing to the country between 1988 and 1994 was not properly accounted.[7] Transparency International castigated Angolan officials for corrupt practices and indicated that Angola's oil revenues are shrouded in secrecy. A significant proportion of oil revenues is believed to have 'bypassed the budget' (IMF Report 2003) and has gone into private pockets.[8] Egypt is also believed to have over $20.0 billion of the government funds diverted into private accounts and stashed away in foreign banks. Similar problems are prevalent in Equatorial Guinea where the government officials are linked with banks in which public funds are stashed away.[9] The magnitude of corruption in the African oil and gas industry is profound. The situation is a matter of global concern because sub-Saharan Africa is classified as the poorest region in the world. Some countries in Africa are on the path of experiencing economic progress. However, the fact remains that per-capita income is low and half of the sub-Saharan population of 690 million on an individual basis is sustained by less than $1.00/d. It is against this background that the call for accountability and transparency becomes most pertinent.

In recent years the Extractive Industry Transparency Initiative (EITI), advocated by former Prime Minister Tony Blair at the World Summit in 2002, gained momentum. The underlying principle of EITI was to ensure that communities in which mineral resources are tapped in Europe benefit from the activities of the mining companies. EITI derived from the conviction that the citizenry of the mineral producing countries of the world have not reaped the benefits of the naturally endowed resources. The primary aim of EITI is to make the extractive sectors accessible for public scrutiny. It also seeks to ensure that the proceeds from mineral resources are publicly accounted for and distributed in a just and equitable manner. EITI aims at improving natural resources governance in such a manner that would permit full publication and scrutiny of company payments and the government earnings from oil, gas and other mineral resources.[10] Although the concept was initially designed for application in Europe, its intent (poverty reduction) is universally applicable. In view of this, EITI has been widely applied in the Nigerian oil and gas industry. In the Nigerian situation, it was believed that the industry is fraught with sharp practices among officials and IOCs which promote corruption. The implementation of the Nigeria Extractive Industries Transparency Initiative (NEITI) would to a reasonable extent curb the excesses of the government officials in the petroleum industry. Under the NEITI arrangement all contracts are required to follow due process and provide all documentations attesting to compliance with laid down contract award procedures.

Low value addition

Between 1956 and 2005 Nigeria produced about 25 billion barrels of crude oil. Approximately 50 per cent of the volume was exported as equity crude oil to countries in Europe, the US and Asia. Current refining capacity stands at 445,000 b/d. Prior to 1989 refining capacity was less than 300,000 b/d. The exported crude oil is processed in refineries in Europe, the US and Asia. Saudi Arabia, Kuwait, Algeria and Libya have succeeded in acquiring equity interest in refineries and products marketing companies offshore. This was a deliberate step on the part of these countries to achieve offshore downstream integration and create additional value in the crude oil production chain. Other producing countries in Africa including Nigeria are yet to pursue value addition targeted at deriving higher margins in the crude oil production chain. Although Algeria and Libya have achieved value addition in crude oil production their gas production is targeted at off-takers who use the commodity for various industrial activities involving fertilizer plants, petrochemicals, household heating and GTL projects. African gas producers stand to benefit if gas sold to third parties in some cases could be dependent on collaborations leading to the establishment of petrochemical plants or other viable gas based projects in Europe, the US, Japan or Asia. This would create valuable opportunities for exporting countries to acquire equity interest in existing or emerging power plants and petrochemical and fertilizer plants operated by major off-takers. African gas producers should look beyond current revenue streams derived from oil and gas exports and articulate strategies to expand their income base further. This can be achieved by participating in tangible production activities both at home and abroad. It is important for producing countries to hedge their earnings by investing extensively in sectors and markets associated with high returns and linked with stable currencies. Oil and gas resources are susceptible to rapid depletion; therefore it behoves African producers to take adequate measures to create alternative sources of sustainable revenues through diversified investments and technological development in order to safeguard the collapse of regional economies.

INDUSTRY COMMON FACTORS

The oil and gas industries in Africa are principally supervised by NOCs, some of which were formed in the 1960s. The countries under consideration, namely Nigeria, Algeria, Libya, Egypt and Angola, derive over 70 per cent of the government revenues from oil and gas. These countries make deliberate efforts to access the threshold of development through revenues derived from oil and gas. The impetus for achieving development is critically linked to sustainable sources of revenues. As pointed out earlier, agriculture and solid minerals, having suffered neglect from various governments, no longer contribute significant revenues to the national treasuries. Oil and gas now constitute the primary sources of income in African producing countries. For this reason, hydrocarbon exploration is approached with a sense of priority and indeed urgency.

MNCs serve as common denominators in the operations of the industries in the region. Companies such as Shell, ENI, Total, Chevron, Petrobras, ExxonMobil etc. have pragmatised the globalisation concept by featuring and providing world-class services and investment capital to develop the oil and gas fields in Africa. The different terrains without doubt present varying degrees of challenges but it is observed that most of the industries feature onshore and offshore fields. Considering the fact that quota allocation in OPEC is predicated on proven reserves, there is the tendency for countries to vigorously pursue reserve expansion agendas. Operations in the industry are capital intensive; therefore NOCs vigorously pursue exploration, development and production programmes in order to generate revenues to sustain, among others, upstream activities. Oil and gas production in the continent currently accounts for about 10 per cent of the global crude oil and 8 per cent of gas reserves. NOCs consider it necessary to increase Africa's global share of oil and gas reserves; therefore all affected countries actively strategise in order to embark on programmes that will boost oil reserves. It is also important to indicate that plans for achieving economic, social and political goals are vigorously implemented. It is further observed that some fields in the continent, as in the case of Egypt, are experiencing rapid decline in production. In this regard Enhanced Oil Recovery techniques (EOR) are being introduced to improve the performance of the reservoirs and in the process extend the life span of the fields. EOR technique is adjudged beneficial and therefore they are being applied increasingly in countries where reservoir performances show signs of regression.

Impact of oil revenues

The majority of oil producing countries in Africa were basically agrarian economies in the late 1950s and 1960s. With the discovery of oil and gas in these countries the revenue profiles changed for the better. As indicated in Table 1.1, the earnings from oil and gas have steadily progressed: in 1990 oil exports in Algeria accounted for $14.71 billion and in 2001 $19.13 billion was realised. In view of the increased economic activities emanating from the development of China for active global trade, energy consumption in its industrial and domestic sectors expanded. Aggressive economic activities in India have also contributed to the overall spiking of a global oil price. In

Table 1.1 Oil export revenues (US$ billion)

Country	*Year*	*1990*	*1991*	*1992*	*1993*	*1994*	*1995*	*1996*	*1997*	*1998*	*1999*	*2000*	*2001*	*2002*	*2003*	*2004*	*Total*
ALGERIA*		14.71	13.31	12.15	10.86	9.946	11.2	14.07	14.66	11.09	12.52	21.65	19.13	18.82	24.11	32.32	240.53
LIBYA*		13.23	11.23	10.79	8.54	7.875	8.51	10.16	9.58	6.03	7.96	12.72	11.34	11.47	14.05	20.2	163.69
NIGERIA*		13.26	11.79	11.64	10.85	11.04	11.5	14.89	14.39	8.75	12.45	20.04	17.18	17.08	22.18	32.34	229.39
ANGOLA**		2.5	3.2	3.31	2.03	2.3	3.4	4.1	3.1	2.4	3.8	6.1	4.03	5.2	6.15	7.5	59.12

Source: *OPEC Annual Statistical Bulletin 2004. **Estimated values.

consideration of this and other related factors oil revenues of producing countries in Africa have increased considerably. In the case of Algeria, the upward trend attained $32.31 billion in 2004. For the first time in history, oil price in the global market attained the $119/b mark in April 2008, thereby strengthening the economic positions of the producing countries. Oil export earnings in Nigeria showed an identical pattern with Algeria and the similarity in the revenue profile of both countries could be attributed to the identical OPEC quota restrictions for each. In 1990 oil export earnings in Libya amounted to $13.26 billion but dwindled thereafter due to the UN sanctions imposed in 1992. In 2004 efforts were made by the government to gain the confidence of investors through the placement on offer of 135 Blocks for competitive bidding, with the result that some companies have returned to Libya and oil production has increased; in 2004 oil revenues amounted to $20.2 billion. Furthermore, the destabilising 20 year war in Angola notwithstanding, its profile as an oil producing country is on the increase. In 1990 Angola earned $2.5 billion from exports and in 2004 oil revenues increased to about $7.5 billion. Between 1990 and 2004 Algeria and Nigeria in aggregate earned approximately $240 billion and $229 billion respectively. On the other hand Libya earned approximately $163 billion while Angola derived $59 billion from exports. These revenues from oil in relative terms are huge compared to the earnings from the agrarian economies of the various countries in the 1960s and even now.

Oil revenues in the developing countries have been applied in various forms. Some revenue streams have been channelled to key projects such as refineries, fertilizer plants, petrochemicals, LNG and power plants. Revenues from the hydrocarbon sector also facilitated the development of infrastructure which is generally considered compelling in a developing milieu. Most of the countries under reference developed highways and ancillary road networks across major cities and suburban towns. Beyond these efforts the accruing revenues have also promoted the provision of healthcare and educational facilities. The elaborate scholarship programmes in African oil producing countries for manpower development owe their success to oil revenues. Indeed the overall impact in the producing countries is profound and far reaching.

References

1 www.eia.doe.gov/Algeria – 2007
2 www.funding universe.com/company – histories, 2007
3 www.Photius.com/countries/libya 2007 and International Directory of Company Histories, 2004. Vol. 66, St James Press
4 www.eoearth.org and www.eia.doe.gov, Aug. 2006
5 www.eia.doe.gov /Angola – 2007
6 http: //www.Countrywatch.atavista.com, 2007
7 Gerald Acquaah-Gaisie, 2007 Combating Third World Corruption
8 www.mg.co.sa/mail&Guardian online
9 www.globalpolicy.org
10 www.ghanaianobserver.com 2007

2 Nigerian oil and gas industry

Evolution of the industry

The evolution of the Nigerian oil and gas industry has spanned about a century during which several challenges have been encountered and surmounted by the major oil companies. The search for oil was pioneered by the German-owned Nigerian Butimen Company in 1908.[1] Initially several wells were drilled without success. Following this setback the search was discontinued and was not resumed until 1937 when Shell Petroleum Company and British Petroleum Company (BP) were given concessions by the British government to explore for oil in Nigeria.[2] The need to shift to Nigeria as an oil exploration and production base was in part necessitated by the Suez Canal crisis which was at its early stage. There was a general feeling among explorationists that avoiding transportation through the Suez Canal carried a premium political value.[3] The concession secured by the pioneer companies was essentially an exploration licence which was to embrace an estimated 357,000 square miles, covering the entire mainland of Nigeria.[4]

Exploration went on smoothly until 1941 when activities were disrupted by World War II. Under these circumstances all attention was shifted to the war, and oil exploration in Nigeria was suspended until mid-1946. By 1951 Shell BP (Shell D'Arcy) had stabilised and acquired seismic information about the geological features of the concession territory. The seismic and statistical data at its disposal tacitly confirmed that the southern portion (Niger Delta) as opposed to the northern territory was geologically more likely to be oil bearing. In view of this, Shell BP narrowed its original area of operation to 58,000 square miles, mainly in the southern coastal region stretching from the extreme southwestern border of Nigeria to British Cameroon. Shell BP drilled its first hole, which turned out to be dry, in 1951.[5] In an effort to focus its operations properly, in 1957 it further narrowed its concession to an area covering 16,000 square miles comprising 20 Oil Prospecting Licences (OPLs). On 1 January 1961 and 1 January 1962 respectively it converted the OPLs into a total of 46 Oil Mining Leases (OMLs) which covered an area of 15,000 square miles.

At the initial stage, Shell BP enjoyed a complete monopoly of the oil exploration business for a considerably long time (1938–1955). In 1955 Mobil Exploration (Nigeria) Ltd, a subsidiary of American Socony-Mobil Oil Company, obtained a licence to explore oil in areas relinquished by Shell BP covering 281,782 square miles.[6] This was mainly part of the northern region (now designated as Northwestern, North central and Northeastern zones). Around 1957, Mobil Exploration (Nigeria) Ltd reduced its area of concession to 12,700 square miles, mainly in the Sokoto axis. Within the same period, it obtained another oil exploration licence which enabled it to explore 4,000 square miles in the western region (Southwest). Both licences were relinquished to the government due to unsuccessful searches in both the North and the West. However, the reduction of the total area under the control of Shell BP to 15,000 square miles exclusively for oil mining purposes gave other oil companies the opportunity to prospect for oil in the Niger Delta region. Between April 1960 and 30 May 1965, Tennessee Nigeria Limited obtained six OPLs covering an area of 6,863 square miles. Subsequently, it relinquished a large portion of its concession retaining only 64,000 acres mainly for oil mining under its OPL 28.

Other companies followed Tennessee Nigeria Limited in the oil prospecting venture in the coastal regions which had hitherto been monopolised by Shell BP. In December 1961 and June 1962 the Nigerian Gulf Oil Company (affiliate of the American Gulf Oil Company), acquired OPLs covering 3,960 square miles. Nigerian Agip (NAOC), a subsidiary of ENI, was also granted a land prospecting licence covering an area of 2,000 square miles. Within the same period SAFRAP Nigeria Limited also acquired an OPL which entitled it to operate a JV in which SAFREP (Societe Anonyme Française des Recherché de Petroles), RAP (Regie Autome des Petrole) and SOGERAP (Societe de Gestion des Participation de la RAP) had 50 per cent, 40 per cent and 10 per cent equity respectively. In 1965 Phillips Petroleum Company (ConocoPhillips) commenced operations in the Nigerian petroleum industry and obtained an OPL covering 11,400 square miles. This was an area previously controlled by Tennessee Oil. Within the same period Phillips acquired a 50 per cent interest in OPL 34 of NAOC. Finally, on 16 November 1965, the government granted ESSO Exploration two licences to explore for oil in an area covering 58,900 square miles in the Niger Delta. The German owned Nigerian Bitumen Company is no longer operational in the country. However, some of the pioneer companies have metamorphosed into major oil companies through mergers and acquisitions. The Gulf Oil Company and SAFRAP have through this process become ChevronTexaco and Total respectively. It is important to indicate at this juncture that the key players in the industry have increased. In 1971, Pan Ocean Oil Corporation (Nigeria) commenced operations and acquired OPL 71 in the Niger Delta. The operations of the company progressed and on 1 December 1975 OPL 71 was converted into OML 98. Conoco Energy Nigeria Limited also joined the black gold adventure and commenced operations in 1991. Statoil of Norway increased the ranks of the MNCs and was formally incorporated in 1992. The company was awarded OPLs 213, 217 and 218 in the Deep Offshore. Addax acquired Ashland oil and currently operates all the concessions in the Niger Delta region.

Crude oil production

The shift of Shell BP (formerly Shell D'Arcy) was an attempt to provide alternative sources of crude oil derived from the Middle East. Although oil exploration started in 1908, it was not until 1937 that Shell BP applied more advanced technology and techniques in its operational activities in the Niger Delta. These efforts were not rewarded until 1956 when it made its first strike at Oloibiri about 65 miles west of Port Harcourt.[7] By 1957 a total of seven wells had been drilled in the Oloibiri area out of which four were successful. In order to facilitate shipment of the crude oil from Oloibiri, a 10 inch pipeline was constructed to connect a tank farm at Port Harcourt. The oil would then be loaded in barges for final trans-shipment to waiting ocean tankers on the Bonny River.[8] The location of Oloibiri in the rainforest, coupled with the swampy coastal terrain, caused considerable delay in linking the oil well with Port Harcourt through pipeline. At the end of 1956 Shell BP struck another oil well at Afam, a little town located about 20 miles from Port Harcourt. By December, a total of five wells had been drilled at Afam and out of these, three were found to be successful. These wells were subsequently connected to Port Harcourt with a six inch pipeline and the first crude oil was exported from Port Harcourt to refineries in Europe in 1958. Western observers were less optimistic at the time regarding the commercial status of the Nigerian oil industry. They taunted:

> . . . the first oil exports from Port Harcourt will be no more than test shipments to enable the properties of Nigerian crude oil to be evaluated in refineries in Europe; they do not mean that production from Oloibiri and Afam is thought to be commercially attractive.[9]

The industry has experienced remarkable growth since the first shipment of crude oil to refineries in Europe in 1958. Total average daily production of 5,100 b/d in 1958 increased to about 2.4 mmbd in 2006. It is envisaged that production would increase to 2.5 mmbd in 2008 and 4.5 mmbd in 2010. The increase in producibility and indeed the expansion of the crude oil reserves (currently 36 billion barrels) is as a result of huge capital investments in the Nigerian oil and gas industry. Total annual expenditure in the upstream sector is in excess of $8.0 billion annually. On the whole, oil production in the Niger Delta region was far from easy. Occasionally, companies were compelled to carry out their drilling operations from floating barges. In June 1960, for instance, a 14,000 feet well was successfully drilled from a barge.[10]

Stage of development in 1966

In 1966, exactly 10 years after oil was first discovered in commercial quantities at Oloibiri, the Nigerian petroleum industry had recorded significant success. By mid-1966, a total of 277 exploratory test wells had been drilled by the oil companies. Out of these, 136 were found to be successful. During the same period the companies successfully completed 296 of the 352 appraisal and development wells drilled (Table 2.1).[11]

Stage of upstream development in 2005

The petroleum industry experienced phenomenal growth between 1966 and 2005. It can be recalled that in the years preceding 1966 the oil companies involved in the industry drilled a total number of 296 appraisal and development wells. Crude oil production has escalated and this is as a result of the increased number of appraisal and development wells drilled by the MNCs. For instance, in 2002 a total of 187 appraisal and development wells were drilled by the IOCs. The performance of the oil companies in a single year by far exceeded the results achieved by the pioneer companies over a period of 10 years (Tables 2.1 and 2.2). Whereas collaborative efforts between the government and the oil companies were restricted to JV activities in the early years, recent developments in the industry have expanded to include Production Sharing Contracts (PSCs) and Service Contracts (SCs).

The pioneer companies were seven in number but in 2006 about 25 PSCs, most of which are in the Deep Offshore, were operational. The level of production readily serves as an index of the overall development of the industry. It may be recalled that in 1956 aggregate daily production amounted to a modest 5,100 b/d but the escalation of activities, especially in the Deep Offshore, aided by cutting edge technology has significantly elevated total

Table 2.1 Drilling activities of oil companies in Nigeria – 1966

SN	*Name of company*	*Exploratory wells successfully drilled*	*Appraisal/dev. wells successfully drilled*
1.	Tennessee Oil Company Nigeria Ltd.	4	4
2.	Shell BP	102	229
3.	Mobil Exploration Nigeria Inc.	10	2
4.	Phillips Petroleum Inc.	0	0
5.	Nigerian Gulf Oil Co.	11	33
6.	Esso Exploration Inc.	0	0
7.	American Overseas Petroleum Ltd.	2	11
8.	Nigerian Agip Oil Company	3	3
9.	Safrap Nigeria Ltd.	4	14
	Total	136	296

Source: 7th World Petroleum Congress Proceedings, Vol. 2 (Origin of Oil, Geology and Geophysics), April 1967, p. 201.

Table 2.2 Stage of development and level of activities of the industry in 2005

Company	*SPDC*		*Mobil*		*Chevron*		*Texaco*		*NAOC*		*Total*		*Addax*	
Year	**AW*	***DW*	*AW*	*DW*	*AW*	*DW*	*AW*	*DW*	*AW*	*DW*	*AW*	*DW*	*AW*	*DW*
1994	–	8	5	12	4	10	10	1	–	–	1	11	–	–
1995	2	2	1	–	5	–	1	–	–	1	–	9	–	–
1996	–	7	11	73	8	19	1	7	2	2	–	–	–	–
1997	1	7	3	37	–	14	1	3	–	–	–	1	–	–
1998	–	–	–	–	–	–	–	–	–	–	–	–	–	–
1999	1	28	2	24	–	14	–	–	–	1	–	11	–	4
2000	2	17	–	26	2	48	–	–	–	12	–	12	1	5
2001	9	74	12	49	2	69	1	4	–	13	3	4	4	13
2002	4	45	11	30	3	47	1	2	3	14	3	12	1	11
2003	1	31	–	23	–	33	–	–	4	26	6	22	5	–
2004	2	1	0	31	0	32	–	–	–	–	–	–	1	–
Total	22	220	45	305	24	286	15	17	9	69	13	82	12	33

Source: Dept of Petroleum Resources – 2006.
* Appraisal well; **Development well

production capacity to about 2.4 mmbd. Future activities are expected to engender higher performance levels in the industry.

Bomu Dere oil field

Bomu oil field is located close to Dere Town in Gokana Territory of Rivers State. The first well often referred to as (Bomu 1) was drilled on 23 February 1958 and reached a depth of 10,269 feet by 8 April 1958. It was brought to completion by 8 May 1958. Production from this well started nine months after its completion and by mid-1966 a total of 29 wells had been drilled in the field. The wells recorded an average production of 75,000 b/d and in June 1966 the cumulative production from the field reached 113.82 million barrels. Shell operated in Ogoni land for over 20 years. Although the Bomu axis was a potential source of oil, events changed drastically and the area was engulfed in crisis. The Ogoni people expressed dissatisfaction with the operations of Shell and accused the company of exploitation and neglect. The allegation was championed by the late Ken Saro Wiwa. The movement soon involved the youths who accused four of their prominent chiefs of conspiracy and sell out. Sadly, subsequent events led to the murder of the accused chiefs in a mob action. Ken Saro Wiwa and eight other members of the Movement for the Survival of Ogoni People (MOSOP) were arrested and charged with murder. They were hastily tried and convicted of murder by a tribunal which sat in Port Harcourt. On 31 October 1995 a federal government Tribunal sentenced them all to death by hanging, and this was carried out in Port Harcourt on 10 November 1995. This action perpetrated by the Abacha regime attracted global condemnation and led to the suspension of Nigeria from the Commonwealth. The United Nations imposed sanctions on Nigeria which were to last until 1999.

These developments adversely affected Shell operations in the Ogoni territory. The company was accused of complicity in the execution of the illustrious sons of Ogoni land, and consequently it was considered persona non grata in all Ogoni land which led to total suspension of all Shell activities. Initially, an intervention military detachment was dispatched to Ogoni land to provide security cover for Shell operations. The resistance of the Ogoni people increased and all operations were discontinued from about 1996 to 2005. Shell suffered heavy losses as their installations experienced corrosion and outright vandalism. Several attempts have been made by the Rivers State government to resolve the conflict but without success. The injury inflicted by the crisis was profound and invoked deep emotions which could not be easily assuaged. However, in 2005, President Obasanjo constituted a reconciliation committee chaired by Rev. Father Kuka. The primary objective was to find a lasting solution to the crisis.

Collaborative joint ventures

Prior to federal government participation in the upstream sector of the industry, MNCs paid taxes and royalties to the government. In the 1970s and 1980s substantial portions of these revenues were channeled to major development projects such as infrastructure, schools, hospitals etc. Derivation of revenues in the form of royalties, taxes and special levies devoid of direct government participation in the exploration and production activities was seen as not serving the overall national interest. In order to take full advantage of the opportunities in the upstream sector the government, in 1973, endorsed the JV concept first introduced in the Nigerian petroleum industry by Agip. This arrangement created an opportunity for the federal government to acquire equity interest directly in all the activities of the MNCs. The equity holdings of the government in the major oil companies are as follows:

- NNPC 55 per cent; Shell 30 per cent, Total 10 per cent, Agip 5 per cent;
- NNPC 60 per cent, Mobil 40 per cent;
- NNPC 60 per cent, Chevron 40 per cent;
- NNPC 60 per cent, Agip 20 per cent, ConocoPhillips 20 per cent;
- NNPC 60 per cent, Texaco 20 per cent, Chevron 20 per cent;
- NNPC 60 per cent, Panocean 40 per cent.

It is important to indicate that in addition to JVs the government has also introduced PSCs and SCs as avenues for participation in the dynamic upstream sector of the oil and gas industry. Currently about 25 PSCs are in operation. Nine of them are designated as existing while 16 are classified as new contracts. Some of the existing PSCs are Addax, Conoco, Shell Nigeria Exploration, Statoil, Nigeria Agip Exploration and Esso. The JV arrangements required the pooling of funds by both the government and the participating oil companies for purposes of exploration and development of the fields. The federal government is obliged to make periodic contributions towards the funding of exploration and development activities through cash calls. Operations in the upstream sector adhere strictly to schedules; therefore timely response to cash calls is crucial to the successful execution of programmes and adherence to established milestones. This notwithstanding, occasional delays were experienced in the government response to cash calls which distorted the work programmes of the companies. It is essential to note, however, that meeting cash call obligations on the part of the government became increasingly difficult in view of the numerous sectoral demands on it for the provision of various forms of infrastructure and social amenities.

In an effort to circumvent this problem PSCs and SCs, which in the Nigerian context may last for a period of 10 to 30 years, were introduced as alternative funding mechanisms for the smooth execution of programmes in the upstream sector. PSCs and SCs are essentially sole risk operations (i.e. losses are borne solely by contractor) in which a contractor undertakes to carry out on behalf of the government or its NOC, exploration, development and at the option of the government or its agent, production operations. The contractor provides all the funds required for the exploration and development activities. In each case, the contractor is also required to pay signature bonus and other premiums to the government. If oil and gas are found in commercial quantities, contractual provision is made for the contractor (the PSC) to produce the oil and recoup all expenses. In this context therefore, the recoverable oil is for operational convenience partitioned into cost oil, tax oil, royalty oil and profit oil. The cost oil is that portion of recoverable oil that is allocated to the contractor to enable it to recoup all costs incurred in discovering and producing the oil and gas. Tax oil is that quantum of oil allocated to NNPC to generate revenues equivalent to the Petroleum Profit Tax (PPT) due to the federal government. The PPT is paid directly to the government. Furthermore, under the contractual arrangement, NNPC is allocated an additional quantity of oil, the proceeds of which are used to pay for royalties and concession rental fees. Once all these cost components have been deducted or defrayed, the balance of the oil is considered profit oil and is shared between NNPC and the contracting firm in accordance with the proportion set out in the PSC agreement. The SC also has cost recovery mechanisms which are similar to the PSC arrangement. However, in an SC agreement the contractor may not produce the oil and could therefore be compensated in kind by the government. It is also important to note that in the case of PSCs and SCs the contractor is not a co-owner of the concessions and licences under which the fields were explored and developed but was only granted access to the field(s) for the duration of the contract.

Nigerian crude oil export 1969–2004

Crude oil production started at a modest level of 5,100 b/d in 1958 and escalated to 2.4 mmbd in 2006. Cumulative annual production in 1958 amounted to 1.86 million barrels compared to about 844 million barrels in 2005. MNCs continued to invest in research which in most cases led to the invention of cutting edge technologies which are applied in the upstream sector. Introduction of modern technology and chemicals have enhanced the producibility of most reservoirs thereby allowing increasing volumes of hydrocarbon to be recovered. Approximately 24 billion barrels of oil have been produced in the Nigerian oil and gas industry since inception. Each year NNPC exports about 60 per cent of the equity crude oil less the 445,000 barrels daily allocation to the three domestic refineries. In 1969 about 197 million barrels of crude oil were exported and this increased to 557 million barrels in 1989 (Table 2.3). Available records indicate that exports levelled off at about 860 million barrels in 2004. The tempo of activities remains high and NNPC intends to increase crude oil production from the current level of

Table 2.3 Nigerian crude oil export (million barrels)

Year	*Daily export*	*Annual export*	*Cum. annual export*	*% Change*
1969	0.542	197.90	197.90	0.00
1970	1.051	383.54	581.44	193.81
1971	1.486	542.54	1123.98	93.31
1972	1.756	640.98	1764.95	57.03
1973	1.978	722.01	2486.96	40.91
1974	2.179	795.48	3282.44	31.99
1975	1.713	625.35	3907.80	19.05
1976	2.013	734.82	4642.61	18.80
1977	2.030	740.95	5383.56	15.96
1978	1.827	666.89	6050.46	12.39
1979	2.210	806.69	6857.14	13.33
1980	1.960	715.47	7572.62	10.43
1981	1.228	448.37	8020.98	5.92
1982	1.003	366.02	8387.00	4.56
1983	0.935	341.35	8728.35	4.07
1984	1.094	399.35	9127.70	4.58
1985	1.333	486.65	9614.35	5.33
1986	1.221	445.74	10060.09	4.64
1987	1.065	388.80	10448.89	3.86
1988	1.111	405.33	10854.22	3.88
1989	1.526	556.95	11411.17	5.13
1990	1.550	565.75	11976.92	4.96
1991	1.610	587.65	12564.57	4.91
1992	1.585	578.53	13143.10	4.60
1993	1.557	568.31	13711.40	4.32
1994	1.590	580.35	14291.75	4.23
1995	1.665	607.73	14899.48	4.25
1996	1.813	661.71	15561.19	4.44
1997	1.856	677.26	16238.45	4.35
1998	1.833	668.97	16907.42	4.12
1999	1.705	622.36	17529.78	3.68
2000	1.986	725.04	18254.82	4.14
2001	2.009	733.43	18988.25	0.04
2002	1.798	656.34	19644.59	0.03
2003	2.164	789.68	20434.27	0.04
2004	2.356	859.94	21294.21	0.04
2005	2.370	865.05	22159.26	0.04

Production 365 days.
Source: OPEC Annual Statistical Bulletin 2004.

2.4 mmbd in 2006 to about 4.5 mmbd in 2010. Similarly there is a concerted effort by the NNPC to increase the crude oil reserves to 40 billion barrels by the year 2010. Crude oil exports are stimulated by global demand for the commodity and the quantity exported from the crude oil pool of each member country is determined by available national reserves. The projected escalation of production and expansion of national reserves will create an opportunity for annual export volumes to increase in the near future.[12] It is important to indicate at this juncture that gas, although not directly explored for, has become a significant factor in the industry. So far the proven and probable gas reserves in the sector which amount to 187 TCF are mainly associated gas. Gas utilisation in various facets of industrial life has expanded significantly and monetisation has become a major attraction. It is envisaged that in the near future gas will be explored for directly, thereby providing an opportunity for further expansion of the national gas reserves.

Crude oil export revenues

An important counterpart of the exports discussed so far is the accruable revenue from the process. Available data shows that Nigeria has from 1969 earned significant revenues from crude oil exports. In 1969, 197 million barrels of crude oil were exported which earned $0.422 billion. In 1985 exports increased to 487 million barrels and an aggregate export earning of $12.6 billion was realised (Table 2.4). An analysis of the revenue table showed periodic fluctuations in the revenue stream. For instance, it was observed that in 1978 the revenue stream declined and picked up in subsequent years. The same trend was observed between 1986 and 1989 when oil prices crashed on the international market. The drop in revenues from $12.57 billion in 1985 to $4.77 billion in 1986 was rather dramatic. Thereafter export revenues slowly improved and attained $13.27 billion in 1990. Several factors could account for the sudden decline in export revenues. One may partly attribute the drastic decline to a price shock necessitated by a glut in the global oil market. Improved weather conditions could also orchestrate decline in demand and as a result impact negatively on the price of oil. The 1970s and 1980s were the military junta era of Nigeria. There was frequent change of government and such changes, which were against democratic principles, were vehemently opposed by the international community. In this regard various forms of sanctions were imposed by the United Nations in order to deter usurpation of power through undemocratic means. It can be assumed therefore that sanctions might have in part accounted for the drop in export revenues.

Furthermore, one may posit that the communal crisis in host communities in the Niger Delta (the primary oil producing region) also disrupted production activities thereby compelling producing companies to shut down operations. Disruptions arising from community disturbances have often led to closure of flow stations, kidnapping and outright hostage taking of oil company personnel. IOCs hesitate to operate under insecure conditions and the aggregate effect is low production which translates into lower export revenues. The cumulative earnings from oil amounted to a total of $158 billion in 1985 and further escalated to $413 billion in 2004. It can be observed, however, that there was a steady increase in the export volumes in the late 1960s and the 1970s. Thereafter (especially in the 1980s) export volumes stabilised and recorded single digit increase.[13] The same trend is observed in the revenue tables derived from export activities. Crude oil continues to be a

Table 2.4 Revenue from oil 1969–2005 ($ million)

Year	*Revenues*	*Cum. revenues*	*% Change*
1969	422.00	422.00	0.00
1970	718.00	1,140.00	170.14
1971	1,375.00	2,515.00	120.61
1972	1,803.00	4,318.00	71.69
1973	3,049.00	7,367.00	70.61
1974	8,997.00	16,364.00	122.13
1975	7,744.00	24,108.00	47.32
1976	9,444.00	33,552.00	39.17
1977	11,561.00	45,113.00	34.46
1978	9,452.00	54,565.00	20.95
1979	15,624.00	70,189.00	28.63
1980	24,931.00	95,120.00	35.52
1981	17,291.00	112,411.00	18.18
1982	11,883.00	124,294.00	10.57
1983	9,941.00	134,235.00	8.00
1984	11,534.00	145,769.00	8.59
1985	12,568.00	158,337.00	8.62
1986	4,770.00	163,107.00	3.01
1987	7,024.00	170,131.00	4.31
1988	6,267.00	176,398.00	3.68
1989	7,470.00	183,868.00	4.23
1990	13,265.00	197,133.00	7.21
1991	11,792.00	208,925.00	5.98
1992	11,642.00	220,567.00	5.57
1993	10,859.00	231,426.00	4.92
1994	11,040.00	242,466.00	4.77
1995	11,512.00	253,978.00	4.75
1996	14,888.00	268,866.00	5.86
1997	14,391.00	283,257.00	5.35
1998	8,754.00	292,011.00	3.09
1999	12,453.00	304,464.00	4.26
2000	20,040.00	324,504.00	6.58
2001	17,188.00	341,692.00	5.30
2002	17,083.00	358,775.00	5.00
2003	22,184.00	380,959.00	6.18
2004	32,337.00	413,296.00	8.49
2005	47,928.00	461,224.00	11.60

Source: OPEC Annual Statistical Bulletin 2004.

major source of government revenue accounting for about 30 per cent of the GDP and about 85 per cent of foreign exchange earnings.

Impact of oil and gas revenues

Nigeria at independence in 1960 was essentially an agrarian economy which depended on export of cocoa, ground nut, palm oil, tin, bauxite and other precious metals. In 1970 the total revenue derived from crude oil export on an annual basis amounted to $718 million and further increased to $32 billion in 2004. In cumulative terms, a total of $413 billion was derived from oil exports between 1969 and 2004. In the 1960s Nigeria faced serious challenges in the areas of infrastructure, healthcare, education, agriculture and manpower development. The export proceeds from cash crops were modest; therefore development programmes executed in the early 1960s and 1970s were limited. The growth in oil revenues expanded the capacity of the federal government to embark on major projects. During this period roads, hospitals and schools were extensively constructed across the country. Scholarships were awarded by the government to qualified candidates to study engineering, medicine, law, social sciences etc. in order to fill critical vacancies in the government and the private sector. Oil revenues also formed the basis for the construction of the pioneer refinery at Port Harcourt in 1965 and subsequently other refineries at Warri and Kaduna in 1978 and 1980 respectively. In the health sector numerous general hospitals were constructed at the state levels and teaching hospitals were established as affiliates of universities for the training of medical personnel. The establishment of health institutions and provision of medical facilities has significantly reduced infant and maternal mortality. In the educational sector the government relied on oil revenues for the construction of educational institutions at the primary, secondary and tertiary levels. The provision of infrastructure, healthcare, educational institutions as well as refineries and other oil related organisations by the government paved the way for societal restructuring and development in post independence Nigeria. In 1970 three federal government-owned universities were operational. In 2007 the number had increased to well over 50 with an estimated total enrolment of about 400,000. Several polytechnics have been built through oil revenues, which boosted the manpower development process. The oil economy has also stimulated the development of ancillary industries which serve as employment outlets for trained personnel. It is essential to indicate that the fundamental objective of the government in the provision of basic facilities (as obtainable in other developing countries) was to facilitate the emergence of employment-creating industries. In the case of Nigeria this objective was to a reasonable extent achieved and as a result graduates from universities and polytechnics in the 1970s and 1980s readily secured employment in the civil service, Public Sector Enterprises (PSEs) and the evolving organised private sector of the economy. The PSEs did not operate efficiently and as a result depended on the government subvention for survival. From the onset the companies were tacitly established to perform a number of functions; primarily among these were creation of jobs and provision of goods and services at commercial rates for public consumption. They were also expected to serve as outlets for political patronage for close associates of the military class. Essentially the PSEs were structurally deficient (not equipped to compete); therefore their revenues proved inadequate to support the wage bills of the companies. The basic rules of prudent entrepreneurship were violated and the PSEs gradually collapsed leaving behind huge debts for the Federal and State governments to settle.

The establishment of tertiary institutions and the training of skilled manpower is without doubt an important step in societal transformation. However, it is essential to ensure that existing employment outlets have the capacity to absorb trained graduates and technicians. In the case of Nigeria experts are of the view that the supply of trained manpower by far exceeds the aggregate demand for labour in the government and the organised private sector. Oil revenues account for about 80 per cent and 35 per cent of the foreign earnings of the Nigerian economy and the GDP respectively. The high level of returns and seeming guaranteed nature of the flow of revenues from the sector inevitably attracted the full attention of the government much to the detriment of the agricultural and solid mineral sectors. Economists and public policy analysts have repeatedly highlighted the inherent danger in the neglect of other vital sectors of the economy. In response to these professional agitations, the government in 1996 created the Ministry of Solid Minerals to, amongst others, provide alternative sources of revenue and steer the economy away from total dependence on oil and gas. In order to record tangible achievements in the solid minerals sector the government between 2003 and 2006 awarded 4 Bitumen Blocks to potential investors. They are yet to carry out meaningful exploration and exploitation activities. Lack of tangible results from the solid minerals sector in terms of discovery of commercial quantities of Bitumen reserves creates a major impediment in the efforts of the government to create a truly diversified economy.

It is important to note that gas was extensively flared in the industry; therefore prior to 2000 it was not a major source of revenue in the Nigerian economy. In the 1980s the government introduced stringent gas monetisation policies which made gas flaring an unlawful practice and also set 2009 as the starting point for the full enforcement of 'zero flare' in the upstream sector. Nigeria Liquefied Natural Gas Company was established in 1998 and in 1999 the first shipment of LNG was achieved. The pioneer Nigeria LNG plant at Bonny is designed to operate seven Train with a combined capacity of 30 million MT/yr. Similar LNG plants are programmed for operation at Brass and Olokola (OK-LNG) in about 2010. Brass and OK-LNG plants will have 10 million MT and 22 million MT respectively at inception and gradually increase to higher volumes in subsequent years. These plants will serve as sustainable sources of revenue. While being instrumental for increased government revenues, they also impacted negatively on the development of agro-allied industries which are characteristically labour intensive in developing countries. Nigeria has a population of 140 million and an extensive land mass of about 90,770 km^2. More than 70 per cent of the area is suitable for large scale mechanised agriculture. This notwithstanding, the country is a net importer of food. It is often contended that the neglect of the agricultural and solid mineral sectors in the Nigerian economy is inimical to the future of the oil and gas economy considering the fact that overdependence on the sector will inevitably accelerate the depletion of the hydrocarbon reserves. The imbalance in the sectoral development of the economy as evidenced by the poor performance of the agricultural and manufacturing sectors has occasioned a high level of unemployment. Consistent with the basic postulations of sociologists, the majority of the unemployed graduates, technicians and artisans out of sheer frustration now engage in criminal activities, including armed theft of crude oil and petroleum products in the forests as well as the swampy creeks of the Niger Delta. Furthermore, hostage taking for ransom became a new dimension of the criminal in the Niger Delta. The industry suffered a series of hostage takings. The climax of these was the kidnapping of two infants (each three years old) for ransom in July 2007. This ugly act attracted widespread condemnation and as a result the victims were released unharmed. In order to stem this insidious trend it is imperative that equitable development of the oil and gas, agriculture and mineral resources sectors be undertaken so as to engender real growth, employment generation and multisectoral flow of revenues into the economy.

References

1 Frankyl, E. J., and Cordry, E. A. 'The Niger Delta Oil Province; Recent Developments Onshore and Offshore', *World Petroleum Congress Proceedings*, vol 2, April 1967, p. 200.
2 Ibid.
3 Melamid, A. 'Geography of the Nigerian Petroleum Industry', *Economic Geography*, vol. 44, January 1968, p. 43.
4 Schatzl, L. H. 'Petroleum in Nigeria', 1969, p. 1.
5 Egbogah, E. O., and Oronsaye, W. I. *Oil and Gas Journal*, 11 June 1979.
6 Schatzl, L. H. op. cit., p. 1.
7 Although Oloibiri is always referred to as the first site of Nigerian crude oil (1956), Akata-I well produced the first sign of oil in 1953. It was abandoned because it did not contain oil in commercial quantity.
8 'Oil in Nigeria', *Economist*, 21 December 1957, p. 1080. It should be noted, however, that oil production in Nigeria has increased significantly thereby warranting more sophisticated shipment techniques.
9 Ibid.
10 'Oil in Nigeria: Future Indefinite', *Economist*, vol. 197, 1 October 1960, p. 77.
11 Frankyl, E. J. and Cordry, E. A. op. cit., p. 201.
12 OPEC Annual Statistical Bulletin 2004, p. 23.
13 Ibid., p. 13.

3 Petroleum geology of Nigeria

Introduction

Nigeria occupies a land area of about 923,800 km^2 and is located between 14° North latitude and 4°–13° East longitude. It has extensive geological formations. For this reason an assessment of the geomorphology of Nigeria is often taken on a regional basis. In this regard it is common to see geological discussion of North Nigeria (comprising the Chad Basin, Benue Basin, Bida Basin, Sokoto Basin etc.), Southwest Nigeria, Southeast Nigeria and the Niger Delta. This chapter is not intended to provide an elaborate geological account of Nigeria. Rather, it is intended to give an insight into the geological characteristics of the country and perhaps more importantly, to identify the region(s) which are largely associated with oil and gas. Subsequent sections will focus on the brief geology of Northwest, Southwest, and Southeast Nigeria and the Niger Delta. An attempt will be made to provide a more detailed account of the geological characteristics of the Niger Delta. This region currently serves as the hub of all oil and gas exploration and production activities in Nigeria.

GEOLOGY OF NORTHWEST NIGERIA

Northwest Nigeria provides extensive characteristics of Basement geology in the north. It contains all rock types, including Basement complex gneisses and migmatites, quartzites, conglomerates, metasediments, metavolcanics, sheared volcanic rocks, granite intrusive etc.

Basement complex

This formation underlies the entire northwestern region and also embodies rocks superior in age to the Proterozoic metasediments. The metasediments are similar in characteristics with paragenesis which are of high metamorphic grade. The Basement rocks in this region have been subjected to a series of tectonic-metamorphic cycles. They have experienced metamorphism and currently appear as relict rafts and granites.[1] Northwest Nigeria features extensive areas of gneiss and ancient metasediments. The metasediments isolated in the Basement complex are considered residuals of ancient supra crustal cover with linkages of the Birrimian age. Large segments of Basement complex have been transformed to migmatite and granite gneiss. Quartzites are predominant in Northwest Nigeria. They are feldspathic and contain kaolinised plagioclase. Marble is often found intermingled with quartzite proximate to the village of Kwakuti. In addition Calcareous and basic rocks, primarily amphibolites and calc-silicate rocks, feature as narrow bands in the main and adjoining areas.

Younger metasediments

Younger metasediments constitute well defined formations which form belts with a North–South progression. These formations represent remnants of hitherto more extensive super crustal cover. The rock types represent variations which range from psammitic to pelitic sediments, generally referred to as low metamorphic grade. The younger metasediments are believed to have been deposited over 100 years ago and are generally classified as Katangan in age.[2]

Older Granite series

The older Granite series embody rocks which intruded during the Pan-African orogenic cycle. They are divided into three main groups, namely basic and intermediate intrusive, migmatites, and syntectonic to late tectonic granites.

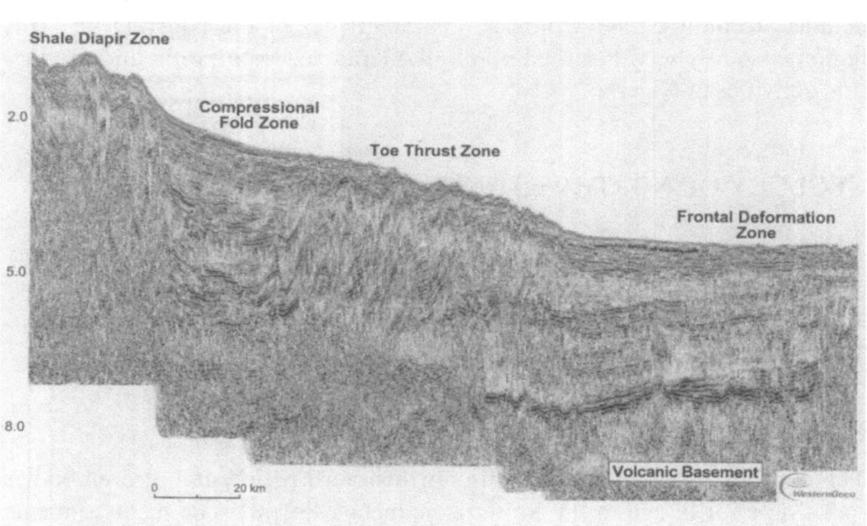

Figure 3.1 Granitic rock type in Northwest Nigeria.

Source: Nigeria – DRSP Joint Development Authority. Guide to the 2004 JDZ Licensing Round Document, Abuja, Nigeria

Basic and intermediate intrusives

These types of rocks are commonly dispersed throughout the Northwestern region. Basic rocks are extensively found in the acid complex which comprises variable and modified acid rocks of igneous origin. They form large intrusions in granitic rocks and are interspersed by acid veins of pegmatite and granite origin. They are characteristically gabbroic and associated with hornblende and chlorite schists.

Migmatisation

Generally speaking migmatisation is presumed to precede intrusion of syntectonic granites. However, in Northeastern Nigeria a constellation of fine grained granites has been classified as of earlier occurrence than the migmatites. These represent a group of inconspicuous discordant intrusions covering a limited area of approximately 200 m as dykes and irregular bodies. They tend to be uniform, equigranular, pale brownish in colour and foliated. Migmatites include muscovite granites, biotites and adamellites which are considered to be in various stages of feldspathisation. Feldspathisation, through various stages of transformation, results in the growth of microcline crystals followed by coarsening of the texture, and culminates in the formation of coarse grained microline porphyritic rocks which share similar characteristics with syntectonic granites.[3] Contact of migmatites with granites is 'intricate and complex'. The syntectonic granites exact gravitational pull on the fine grained granites and have over the period rendered the fine grained granites obscure. Migmatisation has significantly affected the earlier rocks as demonstrated by the occurrence of late microcline in the older metasediments and gneisses. Migmatites can be described as a combination of a group of rocks in which leucocratic granitic components alternate with more basic gneisses.

Granites

Granitic bodies are quite common in Northwestern Nigeria and their sizes range from subecliptical plutons to masses of batholithic scale which stretch over 100 km. These masses, which are concordant and elongated, are often not foliated but feature as intrusive bodies in gneisses and metasediments. Contacts between granites and gneisses are predominantly gravitation, transforming from core granite into metasomatised gneisses. Careful evaluation of the characteristics show a sharp interface between granites and metasediments and no marginal migmatites are observed at the interfacial region. The granites cause contact metamorphism in segments of the metasedimentary belts. Granite rocks in this region are identical in composition comprising mainly quartz, plagioclase and microcline as the primary minerals. The proportion of these minerals in granites varies but cover a spectrum that spans adamellites and granodiorite. Other granite groups, namely porphyritic granite, granodiorite, quartz and syenite are described, based on their petrographic characteristics.

Volcanic rocks

A survey of Northwest Nigeria shows frequent occurrence of subvolcanic complexes which are intrusive in old granite bodies. These rocks are advanced in age and preliminary assessments indicate that they were emplaced during epeirogenic elevation and regional cooling which occurred towards the end of Pan-African orogeny.[4] The region is also associated with secondary sericite, calcite and epidote whose original composition and characteristics are difficult to determine. Carter, in analysing the various complexes, expressed the view that they were associated with faults which may have controlled their emplacement.[5] The subvolcanic complexes are further classified into Maradun group, Gorrusu dyke complex, Dan Gusua dyke complex, Nassarawa dyke complex, Burashika and Kisemi groups.

Maradun group

This group which features a complex of volcanic agglomerates, ryhodacites and ryholites is found at Maradun. The complexes measuring 16 km in length and 2 km in width express a northeast progression. They are bound in the southeastern section by brick-red tuffaceous rocks. The volcanic agglomerates are also associated with ryholotic and ryhodacitic rocks both in the northern and southern segments. The agglomerates feature as a series of low ridges which are continuously exposed and are composed of angular fragments of dark rhyolite, quartz and feldspar which occasionally measure about 40 cm in diameter. Tuffaceous rocks are characteristically low lying and exposed in gullies. They embody micaceous, grits, sand stones and agglomerates in the northern segment and rhyodacite in the southern portion. A cross-section study of the tuffaceous rock confirms that they contain rounded quartz grains, microcline, composite grains and plagioclase which show traces of volcanic and plutonic rocks. The rhyolitic masses tend to form dwarf hills except in the North where they form conspicuous relief features. They are agglomeratic, spherulitic and flow banded. They are also commonly associated with brecciation, silicification and fracturing in their formations.

Gorrusu dyke complex

This complex comprises large feldspar porphyry dykes which form ridges across the Anka–Gummi road a few kilometres east of Gorrusu. The dykes are generally non-foliated and make abrupt contact with granodiorites. Granodiorites are rich in xenoliths which are commonly found in basic igneous rocks. McCurry is of the view that porphyry dykes found in this area contain 'euhedral phenocrysts of anorthoclase, occasional andesine labradorite phenocrysts and strained quartz phenocrysts'. The phenocrysts are known to have sharp outlines while feldspars exhibit sericitic alteration.

Dan Gusua dyke complex

Dan Gusua dyke complex features in the Dan Gusua area and consists of microgranite and microgranodiorite dykes measuring 7 to 10 m in diameter. The microgranites embody large plagioclase anorthoclase and biotite phenocrysts in fine grained groundmass of quartz. Chemical analysis of the rocks from Gorrusu dyke complex and Maradun and Kisemi groups confirm these rocks to be calc-alkaline, alkaline rhyolite and rhyodacite in character. The rocks are also observed to be hypersthene and silica saturated. Major and trace element data indicate that volcanic rocks associated with the northwest region of Nigeria are similar in chemical composition to the high potash shoshonite and latites which are traced to the continental edge of the Pacific orogenic belts.

Nassarawa dyke complex

Microgranite biotite dacite and felsite dykes measuring about 2 m are commonly found in the western segment of Nassarawa. Intermingled with these are boulders which express remarkable flow-banding. McCurry further observed that microgranites display traces of 'plagioclase, anorthoclase and biotite phenocrysts in a medium-grained groundmass of quartz, feldspar and sericite'.[6] The plagioclase with crystals metamorphoses into sericite. The geological characteristics tend to manifest the banding of biotite dacite dykes with numerous zone oligoclase/andesine phenocryst and hornblende phenocrysts in medium-grained terrestrial mass comprising mainly feldspar, sericite and quartz.

Burashika group

This group is an aggregation of porphyries, volcanic lavae, felsites and granites which are embedded in the Gombe stretch. These rocks exhibit characteristics which are also found in rocks at Kisemi and Maradun. They underlie biotite order granite which is in turn covered by cretaceous sediments. There is conclusive evidence arising from further analysis which suggests that the Burashika group is lower cretaceous in age.

Kisemi group

These groups of granites, 0.5 km west of Kisemi, are often geologically characterised as being composed of intrusion breccia which form a vent agglomerate. It is further observed that the breccia consisted of angular and rounded blocks of dacitic origin. These rocks have a linear structure which is associated with a flow phenomenon. Studies indicate that large areas of dacite are usually not brecciated but contain finer grained materials which are traversed by microgranite, pegmatite, amphibolite and epidote. The entire exposure of Kisemi granites is often crossed by a ridge of angulated muscovite quartzite which is believed to be a later quartzite vein or enclave of country rocks. Similar rock features are observed in Gyara, where they form a long flowing profile of low hills demarcated by alluvium which disguise the relationship between outcrops. The rock characteristics also show the existence of porphyritic and non-porphyritic dacites and a regular occurrence of flow banding with north and northeast progression which deepens steeply.

SOUTHEAST NIGERIA

Early studies of the cretaceous and paleogene sediments of South Nigeria show a demarcation into western and eastern positions by the Okitipupa Ridge. Transgression commenced in the east during the Albian age; however, in the west, sedimentation started close to the terminal stages of the cretaceous. The thicker accumulation of sediment in the east has been associated with deltaic build up that is tertiary in nature.[7]

Albian age formation

Abakaliki in Southeastern Nigeria is believed to contain the oldest sediments in the area. The sediments, which are generally unnamed and undifferentiated, collectively constitute the Asa River Group and are poorly bedded and often referred to as Abakaliki shale. These sediments contain sandstone and sandy limestone lenses. The limestone beds vary in thickness with some segments attaining a thickness of 30 m. Paleontological assessment of the shale confirmed compositional characteristics which are linked to the species Elobriceras and Mortinoceras. The beds contain samples of lead and zinc minerals while the shales are deeply eroded and embody traces of echinoids, gastropods, pelecypods and radiolaria.

Cenomanian

Cenomanian age sedimentations are localised in the southeastern portion of the basin proximate to Calabar. These beds are often referred to as the Odukpani formation and their ages span the Cenomanian and Turonian eras. The deposits are primarily composed of limestone and sandstone. Further down the progression of the deposits alternating shales and limestone become more commonly shaly in the uppermost segments. The Odukpani formation which originates from shallow waters features segments which are 600 m thick.

Turonian sediments

The Turonian deposits are by classification the equivalents of the Eze-Aku formation. These formations are found along the Eze-Aku River in the south-eastern segments of Nigeria and are composed of black to hard grey shales which are alternated with sandy shale and sandstones. They vary in thickness, measuring up to 1,000 m in some locations. In some areas the Eze-Aku deposits progress laterally into the Ameseri sandstone facies while other lateral movements in the Nkalagu area terminate in sandy limestone, shelly limestone and calcareous sandstone. The Eze-Aku formation also originates from shallow water.[8]

Coniacian-Santonian

Coniacian-Santonian sediments express a lower degree of thickness than the Turonian and have been equated to the Agwu formation. They measure about 800 m in thickness and embody well-bedded shale which is bluish-grey in colour. They also feature occasional intercalation of shelly limestones and fine grained sandstone.[9]

Campanian

Campanian sediments have close resemblance and characteristics with outcrops of the Nkporo formation. There formations are rare and therefore borehole cores provided a source of petrographic analysis. The formation comprises of mudstone and dark shales occasionally traversed by thin beds of sandstone and sandy shale. The Nkporo formation, like the Campanian, can measure up to a maximum of 1,000 m in thickness. Although no typical Campanian ammonites have been located in Nigeria, it is presumed that the basal segment of the Nkporo sediments is Campanian in age.[10]

Maestrichtian equivalents

The Maestrichtian equivalents in Southeast Nigeria are friable, greyish dark shales which sparingly embody the thin beds of sandstone and limestone. In the Nigerian geological setting the Maestrichtian also has close association with the Nkporo formation which in turn shares some geomorphologic characteristics with the Owelli sandstone, Asata shale and the Enugu shale. These sediments, like the Eze-Aku and Agwu formations, originate from shallow water. Experts are of the opinion that the extensive shallow sea, shallowed more progressively to the extent that coal-accumulating conditions manifested. The bottom part of the 'coal measure sequence', now known as the Mamu formation, is believed to contain marine intercalations which are predominantly composed of ammoniferous shales.[11] Simpson is of the view that the coal bearing segment of the sedimentation of the progression consists of sandy shales, shaly mudstones and sandstone which originate from fresh and low salinity water mass.[12] The Mamu formation is anteriorly overlain by the Ajali formation which manifests its type locality along the Ajali River. The Ajali formation consists of thick white friable segregated sandstone. On the other hand, the Nsukka formation covers the Ajali sandstone and is identical to the Mamu formation. The constituent rocks embody an alternating succession of sandstone, shale, dark shale and occasional intrusion of seams through depositional transgression. Outcrops of the Nsukka formation are noticeable along the Ndu River valley and other road cuttings on the Enugu–Onitsha road. The Mamu formation is proximately situated to the Ajali formation which is well exposed along the 9th mile and the Milikin foot-hills, and features bands of white shale and mudstone, which occur at frequent intervals towards the base line. In some areas the Ajali formation is overlain by a significant thickness of red earth which originates from weathering and ferruginisation activities.

SOUTHWEST NIGERIA

Kogbe, along with other scholars, attributed the southwestern marine formations to the Maestrichtian age. Mudstone, ill sorted ferruginous grit and silt stone are isolated as the primary constituents of the formation.

Marine formation

Maestrichtian–Abeokuta formation

Kogbe, along with other scholars, notably Reyment, posit that the first major marine transgressions associated with the sedimentation which culminated in the Abeokuta formation occurred in the Maestrichtian era. Mudstone, ill sorted ferruginous grit and siltstone are isolated as the primary constituents of the Abeokuta formation. These formations are geomorphologically equivalent to the Nsukka formation in Southeastern Nigeria. The Abeokuta sediments are believed to have been deposited in highly agitated shallow sea under enduring humid conditions. The sediments are derived from fresh and brackish water. The formation is not particularly uniform but it thickens in its transgression toward the Dahomey axis and contains molluscs in the top layer. The formation features thin marine intercallations of about 0.5 m at a depth of 44 m and integrates with the Nkporo formation in a lateral and southward progression. Kogbe further stated that the Nsukka formation transgresses directly on the basement East of Ifon in Edo State. The Ajali sandstone transgressed on the Basement and is capped by the Nsukka formation.[13]

Paleocene age

Ewekoro and Akinbo formations are located in the western part of the basin and are believed to have been deposited in the Paleocene age. The Ewekoro and Imo formations have common characteristics and are therefore considered to have originated concurrently. The Ewekoro formation is primarily of limestone and the type section is about 11–12.5 m thick. The constituent limestone according to Adegoke (Ogbe added the 4th classification) comprises the following:

1 red phosphatic biomicrite (top);
2 algal biosparite;
3 shelly biomicrite;
4 sandy biomicrosparite (bottom).

The base of the Ewekoro formation consists of sandy biomicrosparite which has light brown sandstone and bioclastic fragments as its main constituents. The stratification of the formation is brought into prominence by the lack of uniformity in the quantity and grain size of the intermingled glauconitic and quartz. The shelly biomicrite strata are composed of pure limestone measuring 4.5–6 m in thickness. These strata account for a greater proportion of the Ewekoro formation. The limestone associated with the formation is found to contain microfossils, especially corals, bryozoans, echinoderms, gastropods, crustaceans and scaphopods.[14] The algal biosparite serves as the upper layer covering of the shelly biomicrite stratum. This is a highly exposed section which is water worn and manifests signs of erosion and potholes. Finally, the red phosphatic biomicrite which forms the top layer of the Ewokoro deposits occurs at intervals on the algal biosparite. It also shows extensive characteristics of erosion which degenerate into potholes and scoured surface.[15]

The Akinbo formation is an equivalent of the Imo formation. These shales, which overlie the Ewekoro limestone, are believed to have similar formational characteristics with the type section of the Ewekoro quarry. The Akinbo shales are greenish grey in appearance and the base formation is contoured by a glauconitic rock band which fuses into the Ewekoro formation. The Akinbo formation measures approximately 8–9 m thick in its type locality but progresses to a thickness of about 18 m in some areas. The formation extends westwards to the Dahomey and Togo axis.[16] The formation is extensively fossiliferous and contains molluscs (microforms), ostrapods and foraminifera.

Eocene age

Three primary formations, namely Ilaro, Oshoshun and Ameki, occurred in the Eocene era. Details of these will be discussed in subsequent sections.

Oshosun formation

The Oshosun formation was not easily classifiable; therefore some nomenclatural and superpositional issues concerning the deposits remained contentious for some time. Adegoke among others undertook a detailed analysis of the formation with a view to resolving any set of controversies surrounding it.[17] The Oshosun formation is not completely exposed; therefore characteristics of the sediments were examined based on Akinside borehole extracts (Geological Survey of Nigeria Borehole: No. 1582). The borehole extracts evacuated at different levels are classified as follows:[18]

- 0–36.5 m – Ilaro formation
- 36.5–70.5 m – Oshosun formation
  - 36.5–39.5 m – light greyish-white to purple clay with brown and red mottlings and occasional pockets of grits
  - 39.5–41.5 m greenish clay
  - 41.5–61.5 m – dark grey to greenish-grey clay with some occasional fossil
  - 61.5–64.5 m – finely laminated greyish-green clays
  - 64.5–67.5 m – greyish-white limestone alternating with light grey shale
  - 67.5–70.5 m – light grey, finely laminated calcareous shale with nodules of hard brownish phosphatic material
- 70.5–80.5 m – Ameki formation
  - 70.5–80.5 m – light grey, laminated phosphatic material (shales of the Akinbo formation).

Ilaro formation

This formation was accorded expert evaluation by Jones (1964) who characterised it as '. . . a sequence of predominantly coarse sandy estuarine, deltaic and continental beds which display rapid lateral facies change . . .' His analysis and characterisation were aided by extracts of Borehole No. 1582 pits at Ifo junction and the railway cutting.[20] Borehole studies in Dahomey and Togo confirm the existence of marine and sandy beds (Ilaro formation) close to the shoreline. The formation is sometimes discontinuous in progressions. However, in areas where it is complete, the thickness can range from 37 to 60 m. The type section of the Ilaro formation can be isolated from the Akinside borehole extractions at a depth of 0–36.5 m.

Oshosun formation

This formation is presumed to be derived from marine deposits. This is further supported by the predominant presence of corals, crustaceans, crinoids, pelagic and planktonic foraminifera, fishes, sea snakes and molluscs in the deposits.[19]

Ameki formation

The Ameki formation embodies greyish-green sandy clay which is characteristically fossiliferous. It is also associated with white clayed sandstones and calcareous concretions. A lithological dichotomy shows the existence of a mixture of fine to coarse sandstone and the commingling of calcareous shale, shelly limestone and other sediments. The thickness of Ameki formation was determined to be around 1,400 m in certain sections of its transgression. It is typically found around the 73rd and 87th miles near Ameki and along the eastern railway. Deriving from its thickness, the formation occupies a substantial area of the Eocene strata which covers the Imo formation. The Ameki formation is present in Southwest Nigeria. However, it features as a thin layer over the Oshosun formation and in Southeast Nigeria the formation demonstrates frequent lateral facies changes, coupled with shaly development. The formation is overlain by a thin layer of Ogwashi-Asaba formation and in view of this it is often difficult to establish the interface between the two formations. Further, into the Bende area, the Ameki strata display a frequent alternation of sandy shale, mudstone, clayed sandstone, fine grained sandstone and thin limestone bond.[21]

GEOLOGY OF THE CHAD BASIN

The Chad Basin is an extensive area of inland drainage in the African continent covering approximately 23,000 km^2 with boundaries stretching from the Central Sahara to Southern Sudan. An area estimated to be one tenth of the basin is situated in the northeastern portion of Nigeria. The basin is located at an attitude of 530 m in the west and slopes gently downwards to a height of 300 m within the lake. The distance between the highest and lowest elevation, which culminates in the basin, covers a distance of 240 km. The topography features dunes which cover many kilometres. Some Nigerian rivers in the adjoining areas flow northwards towards the Chad through shallow valleys, but these secondary fluvial movements account for only 5 per cent of the water input into the basin.

Geological history

The Chad Basin in Nigeria is presumed to have originated during the upper Albian era. Experts are of the view that during that period over 1,000 m of continental sediments comprising primarily Bima sandstone were deposited indiscriminately on the Pre-Cambrian Basement.[22] The Turonian age was associated with extensive transgression which triggered the deposition of heterogeneous limestone and shale sequence. These deposits ultimately transformed into the Gongila formation. The Senonian age which followed the Turonian age, was also associated with systematic marine shale deposits 530 m thick and this transgression translated into the Fika formation. Records indicate that during the Maestrichtian age an estuarine-deltaic environment was widespread, thereby aiding the deposition of Gombe sandstone with intercallations of silt stones, shales and ironstones. These sediments, which are about 320 m thick, exhibit limited progression into the basin. At the end of the Maestrichtian era prevailing conditions necessitated the folding of the cretaceous beds into anticlines and synclines.

Stratigraphy of the Basin

Bima sandstone

The Bima formation is believed to have occurred between the upper Albian and lower Turonian age. Outcrops of the formation are manifest in both the Chad Basin and the Benue Basin. It overlies the crystalline basement in a nonconforming pattern and has a thickness which ranges from 100 to 3,000 m. The elaborate variation in thickness is attributed to the irregular relief of the crystalline basement. Experts opine that the Bima sandstone originated from the weathering processes of the granites of the Jos Plateau as the basal beds of the formation are generally feldspathic. However, the upper beds which accumulated when the basin subsided are observed to be less feldspathic. Bima sandstone has diverse lithology which indicates that the accumulation occurred under dynamic and ranging conditions. The composition of the sediments is heterogeneous, varying from poorly sorted and thickly bedded feldspathic sandstone to deltaic sediments.[23]

Gongila formation

The Gongila formation is basically a transitional sequence of the Turonian era situated between the Bima sandstone and the Fika shale. The base of the formation is circumscribed by basal marine limestone which is about 3 m thick. The main Bima formation has a thickness of 500 m.

Fika formation

The Fika formation consists primarily of marine shales which by classification correspond to the upper Turonian and Senonian age. These shales vary in thickness from 100 m at the southern axis proximate to Potiskum to 500 m at the Maiduguri locality. The formation comprises blue-black shales which are occasionally interrupted by limestone intercallations. Research efforts have traced reptile and fish artefacts in the area, thereby confirming the association of the formation to the Turonian and the Maestrichtian era.[24]

Gombe sandstone

The Gombe sandstone has often been characterised as a sequence of estuarine and deltaic sedimentation. This formation features more conspicuously in the southwest portion of the Chad Basin. In the southwest axis it assumes a thickness of about 350 m but diminishes in thickness as it approaches the centre of the basin. The Gombe sandstone has a fragile folding which exhibits east–west transgression trends characteristic of the earlier formations. The lower beds of the formation embody mudstone and ironstone which show thickness of a few centimetres to about 2 m. This notwithstanding, the middle part of the Gombe sandstone is believed to be composed of sandstones which are well bedded. These beds are often attributed to the period between the upper Maestrichtian and the Paleocene age.

Kerri Kerri formation

This formation situates non-conformably on the folded cretaceous sediments and consists of a continental sequence which dips in a northeasterly direction beneath the Chad formation. It has a strong presence in the west as well as southwest of the Chad Basin and is 220 m thick. These sediments have a limited stretch, and borehole drillings in the northeast confirm that they wedge out between Maiduguri and the Damaturu–Gashua road. The area features a submerged ridge which elongates from Biu to Matsena and experts posit that the ridge might have served as a depositional barrier during the Paleocene sedimentation process.[25] The formation is also associated with alternating layers of grit and sandstones in addition to adequately developed cross-bedding which is characteristic of a deltaic environment. It is further claimed that the Kerri Kerri is an extraction of a cretaceous sedimentary rock whose relief experienced more extensive subjugation than the topography of the Bima sandstone. This apart, the formation is often capped with thick laterite which is vesicular in texture.

STRATIGRAPHY OF THE NIGER DELTA

Basic characteristics

The Niger Delta is situated in the Southwestern portion of Nigeria and extends into the Atlantic Ocean. It is bounded by latitude 4°30′–5°20′ North and longitude 3°–9° East. The original delta is believed to have developed in the northern part of the basin during the Campanian age and terminated during the Paleocene transgression. It is also postulated that the modern delta might have evolved during the Eocene era. The Niger Delta displays basin characteristics of deltaic environment encompassing marine, mixed and continental sedimentations which are represented by the Benin, Agbada and Akata formations. The area is associated with dynamic fluvial activities resulting in the deposition of about one million cubic metres of sand in the Mahin Canyon annually.[26]

The geomorphology of the Niger Delta was elaborately analysed by NEDECO (1954), Allen (1964) and Maron (1969) culminating in an extensive elucidation of the characteristics of the formations.[27] Oceanic and tectonic currents are prevalent in the basin especially during the Eocene and Recent transgressions. For descriptive purposes Weber isolated five distinct physiographic provinces in the contemporary Niger Delta, as follows:[28]

Holomarine zone

This zone consists of extensive clay deposition and attains a depth of about 33 m from the outer shelf.

Transition zone

The transition zone is associated with fluvio-marine sedimentation comprising laminated clay, silts and fine sandstone in water depths of 10–33 m. Dynamic currents along the shoreline disperse the fluvial sediments along the coast.

Barrier bars

These feature along the coastal belt and comprise primarily fine medium grained sand. They tend to interfinger with barrier foot sediments at about 10 m depth. Swamps, beach ridges and sand spits are predominant in the barrier bar sedimentation.

Tidal coastal plain

This area, which embodies tidal flats and swamps, is made of sediments which stretch behind the barrier bars. The sediments are heterogeneous, featuring medium coarse grained sand in channel fills. Natural levees are observed to be filled with clayed sand while peaty deposits manifest in swamps and lagoons.

Flood plains

Weber's account of the flood plains suggests that sediments of this environment are primarily composed of medium coarse grained and clayed back swamp deposit. Bathymetric maps provide insight into the configuration of Niger Delta submarine terrain. The Niger Delta extends into the Gulf of Guinea and experts are of the view that the ocean floor of the Gulf of Guinea, which stretches from Escravos River to the Mid-Atlantic Ridge, measures about 1,600 km. It is further observed that the ocean floor has a slope of 1:880 (6 ft per mile) from the shore which progressively increases to about 1:88 (60 ft per mile) at depths of 300–600 ft close to the outer edge of the continental shelf.[29]

Stratigraphic units

Kogbe (edit 1976) opinied that findings arising from geological sutveys and research suggest that rift faulting during the Precambrian era led to the formation of the contemporary Niger Delta. Faunal and sedimentological data further indicate that the Niger Delta is identical in configuration with past transgressions. Further geolological studies in the Niger Delta isolated three distinct stratigraphic units, namely the Benin, Agbada and Akata formations (see Figure 3.2).

Benin formation

The Benin formation stretches from the west and traverses the entire Niger Delta area. It further engaged in a southerly transgression thereby extending its limits beyond the coastland. Over 90 per cent of the constituent parts of the formation is sandstone which is intercalated with shale. Asseez (1976) maintained that the sandstone is coarse grained, gravelly, poorly sorted and subgranular. Some trace of lignite and subdued wood fragments are traced in the formation. He further posited that the formation is a continental deposit which might have occurred in the upper deltaic depositional environment. The formation also features structure units such as channel fills, back swamp deposits, natural levees, oxbow fills and point bars which indicates the heterogeneous nature of the shallow depositional medium. In the subsurface the Benin formation comprises constituent materials in the northern part which are of Oligocene age and progressively display younger sediments towards the deltaic surface. In general, the formation is presumed to range from the Miocene age to Recent.[30]

The formation which is non-hydrocarbon bearing varies in thickness but often exceeds 6,000 ft.

Agbada formation

The Agbada formation (Figure 3.3) is characterised as a sequence of sandstones and slopes. The upper portion, which comprises predominantly sandy units, also features minor shale intercalations. The lower segment comprises shale units which are geomorphologically thicker than the upper sandy deposits. It contains abundance of micro fauna at the base but this pattern decreases upwards. The deposits are generally coarse and poorly sorted which is symptomatic of materials of floviatile origin. The Agbada formation is known to be continuous throughout the subsurface of the entire deltaic region and it is presumed to be continuous with the Ogwashi-Asaba and

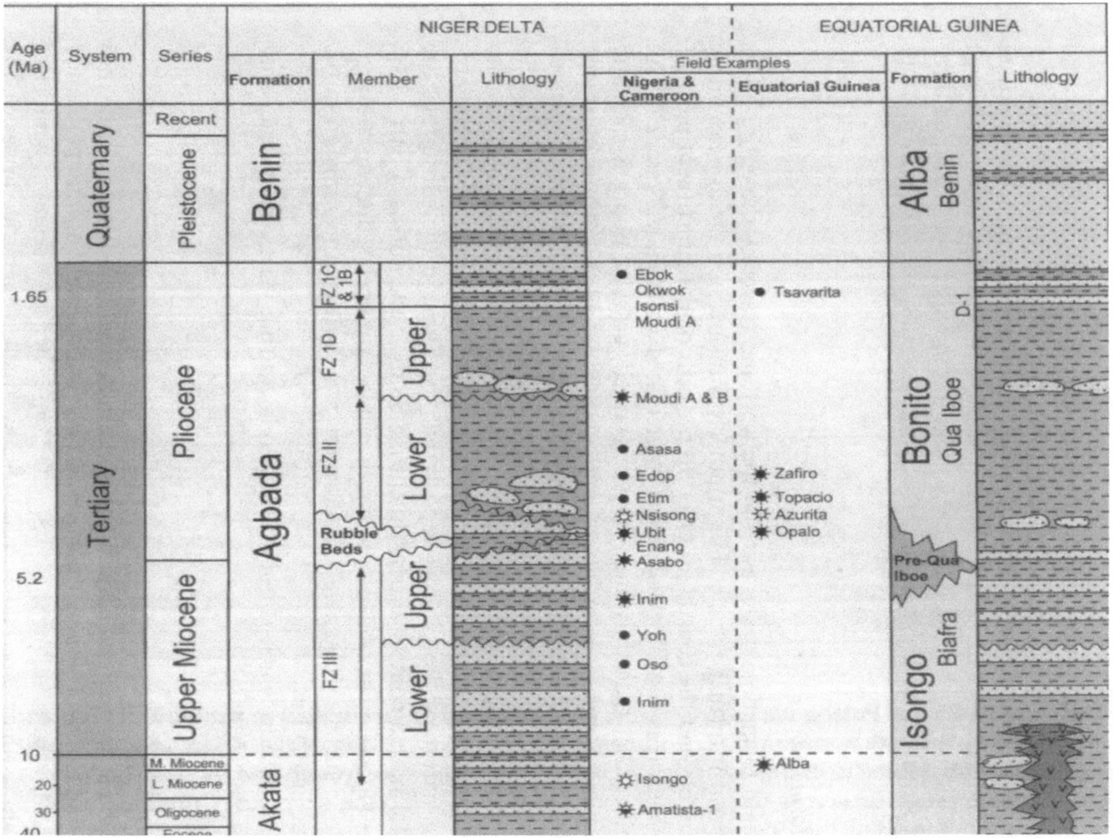

Figure 3.2 Stratigraphic units of the Niger Delta.

Source: Nigeria – DRSP Joint Development Authority. Guide to the 2004 JDZ Licensing Round Document, Abuja, Nigeria

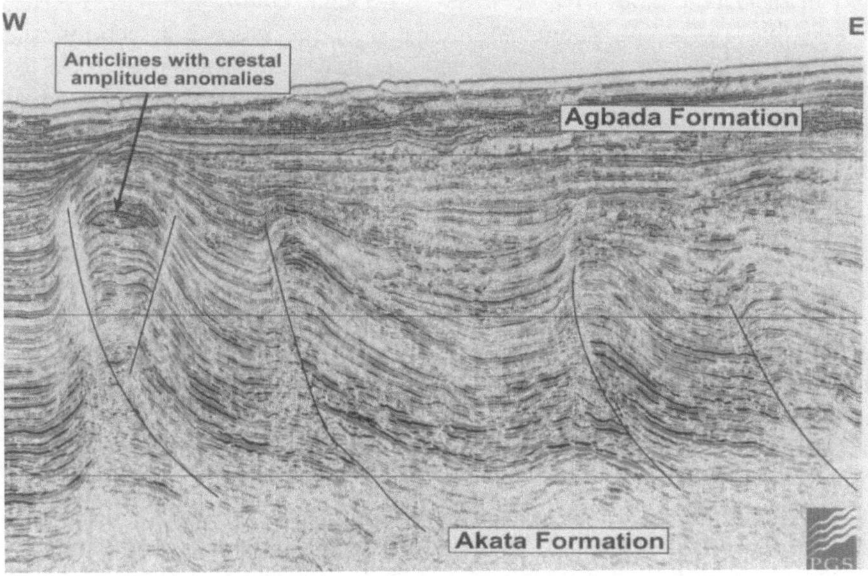

Figure 3.3 Agbada and Akata formations in the Niger Delta.
Source: Nigeria – DRSP Joint Development Authority. Guide to the 2004 JDZ Licensing Round Document, Abuja, Nigeria

Ameki formations of the Eocene–Oligocene transgression.[31] It has a thickness which exceeds 1,000 ft and the age of its constituents range from Eocene in the north to Pliocene/Pleistocene in the south and Recent in the deltaic surface. Hydrocarbon deposits are predominantly found in intervals between the Eocene and Pliocene transgressions which in this instance correspond with the Agbada formations.[32, 33]

Akata formation

The Akata formation is characteristically a uniform shale development which consists of dark grey sandy, silty shale and plant remains at the upper surface. Thin sandstone lenses are observed to occur at its interface with the Agbada formation. The formation is rich in micro fauna, 50 per cent of which is constituted by planktonic foraminifera. The formation is presumed to have occurred in shallow marine shelf depositional environment as evidenced by the benthonic assemblage in the area. Experts are of the view that the Akata formation which measures 4,000 ft in thickness was deposited at the frontal end of an advancing delta and has an age which spans Eocene to Recent.

Hydrocarbon in the Niger Delta

The Niger Delta is considered the primary theatre of exploration and production activities in the Nigerian petroleum industry with activities being carried out in the Delta states of Rivers, Bayelsa, Delta, Ondo, Edo, Cross River, Imo, Abia and Akwa Ibom. As indicated in the preceding sections, the Agbada and Akata formations constitute a world-class source rock for hydrocarbon development and entrapment. Initial exploration and production activities in the oil producing States started on land and gradually shifted to the swampy terrain as a result of technological advancements. Activities have further shifted to the Deep and Ultra-Deep offshore of the Niger Delta. The region has been associated with oil exploration and production for over eight decades. Multinational companies involved in oil and gas exploration and production include Shell, Mobil, Total, Agip, ChevronTexaco, Addax and Statoil. Some indigenous companies engaged in oil and gas exploration and production in the Niger Delta include Consolidated Oil, Dubri Oil, Solgas Petroleum, Atlas Petroleum, Allied Energy Resources and AMNI International. Oil was first struck at Oloibiri in the Niger Delta in 1956 and oil production in the area has progressed from a modest level of 5,100 b/d in 1956 to well above 2.4 mmbd in 2006. Production is envisaged to escalate to 4.5 mmbd in 2010. Current expenditure in the upstream is estimated to be about $8 billion annually. Crude oil and gas reserves are estimated to be 36 billion barrels and 187 TCF respectively. Exploration and production activities in the Deep Offshore are being intensified; consequently it is estimated that crude oil reserves will increase to 40 billion barrels in 2010. The prospects for oil and gas exploration and production in the Niger Delta remain bright especially as more fields are brought on stream in the years ahead.

References

1 McCurry, P. 'The Geology of the Precambrian to Lower Paleozoic Roads of Northern Nigeria'. *Geology of Nigeria*. Elizabeth Publishing Co., 1970, p. 170.
2 Ibid., p. 19.
3 Carter, J. D., Barber, W. and Tait, E.A. 'The Geology of Parts of Adamawa, Bauchi and Borno Provinces in Northern Eastern Nigeria'. *Geological Survey of Nigeria Bulletin* No. 30.
4 McCurry, P. op. cit., 1970, p. 27.
5 Carter, J. D., Barber, W. and Tait, E.A. op. cit., p. 27.
6 McCurry, P. op. cit., 1970, p. 29.
7 Adegoke, O. S. 1969, 'Eocene Stratigraphy of Southern Nigeria'. *Bureau de Récherches Géologiques et Minières* No. 69, pp. 22–48.
8 Simpson, A. 'The Nigerian Coal Field: The Geology of the Parts of Onitsha, Owerri and Benue Provinces'. *Geological Survey of Nigeria Bulletin* No. 24, 85, 1955, p. 5.
9 Reyment, R. A. *Aspects of the Geology of Nigeria*. Ibadan University Press, 133, p. 18 pl.
10 Tattam, C. M. 'A Review of Nigeria Stratigraphy Report'. *Geological Survey of Nigeria Bulletin* 1943, pp. 27–46.
11 Reyment, R. A. 'Review of Nigeria Cretaceous Cenozoic Stratigraphy'. *Journal of Mineral Geology*, Vol. 2 No. 2, 1964, pp. 61–80.
12 Simpson, A. op. cit., 1955.
13 Kogbe, C. A. (ed.) *Geology of Nigeria*. Nigeria: Elizabeth Publishing Co., 1976, p. 273.

14 Ibid., p. 277.
15 Klein, G. de V. 'Intertidal Zone Channel Deposits in Middle Jurassic Great Oolite Series'. *Nature*, 197 (London), 1963, pp. 1060–1062.
16 Antolini, P. 'Eocene Phosphate in the Dahomey Basin'. *Nigerian Journal of Mining and Geology*, Vol. 3, Nos. 1–2, 1968, pp. 17–23.
17 Adegoke, O.S. 'Eocene Stratigraphy of S. Nigeria', 1969, pp. 23–70.
18 Kogbe, C. A. op. cit., 1976, pp. 179–281.
19 Ibid., p. 279.
20 Jones, H. A. and Hockey, R. D. 'The Geology of Parts of South Western Nigeria'. *Geological Survey of Nigeria. Bulletin* No. 31, 87, 1964, p. 8 pl.
21 Kogbe, C. A. 'The Cretaceous and Paleogene Sediments of Southern Nigeria'. *Geology of Nigeria*, 1976, p. 281.
22 Barber, W. 'Pressure Water in the Chad Formation of Borno and Dikwa Emirates, North Eastern Nigeria'. *Geological Survey of Nigeria Bulletin* No. 35, 1965, p. 138.
23 Kogbe, C. A. op. cit., 1976, p. 290.
24 Carter, J. D. Barber, W. and Tait, E. A. op. cit., 1963, No. 30, p. 109.
25 Barber, W. and Jones, D. G. 'The Geology and Hydrology of Maiduguri, Borno Province'. *Rec. Geol. Surv. Nigeria 1958*, 1960, pp. 5–20.
26 NEDECO (Nether Engineering Consultants). 'Western Niger Delta'. The Hague: 1954, p 57.
27 NEDECO op.cit., 1954.
28 Allen, J. R. L. 'Nigerian Continental margin'. *Marine Geology*, Vol. 1, 1969, pp. 289–332.
29 Maron, S. 'Stratigraphical Aspects of the Niger Delta'. *Nigerian Journal of Mining and Geology*, Vol. 4, pp. 1–2.
30 Weber, K. J. 'Sedimentological aspects of oilfields in the Niger Delta'. *Geologie en Mijnbouw*, Vol. 50, 1971, pp. 559–576.
31 Asseez, O. (ed.) 'Review of the Stratigraphy Sedimentation and Structure of the Niger Delta in Kogbe'. *Geology of Nigeria*, 1976, p. 261.
32 Ibid., p. 265.
33 Ibid., p. 266.

4 Nigerian National Petroleum Corporation (NNPC)

Introduction

The Nigerian National Oil Corporation (NNOC) was established in 1971 to ensure the government's participation in oil exploration and production. It joined OPEC in 1971 as the 11th Member country. Its status as a member of the organisation has since grown significantly. In April 1977 NNOC was merged with the Ministry of Petroleum Resources to pave the way for formation of the Nigerian National Petroleum Corporation (NNPC). NNPC was established through an Act CAP 320 (Nigerian Govt. Law) of 1 April 1977. The merging of the Ministry of Petroleum Resources and the Nigerian National Oil Corporation was intended to prevent duplication of functions. The enabling Act empowered NNPC to discharge the following functions:[1]

1 exploring and prospecting for, working, winning or otherwise acquiring, possessing and disposing of petroleum;
2 refining, treating, prospecting and generally engaging in the handling of petroleum for the manufacture and production of petroleum products and its derivatives;
3 purchasing and marketing petroleum, its products and by-products;
4 providing and operating pipelines, tanker-ships or other facilities for the carriage or conveyance of crude oil, natural gas and their products and derivatives, water and any other liquids or other commodities related to the corporation's operations;
5 constructing, equipping and maintaining tank farms and other facilities for the handling and treatment of petroleum and its products and derivatives;
6 carrying out research in connection with petroleum or anything derived from it and promoting activities for the purpose of turning to account the results of such research;
7 doing anything required for the purpose of giving effect to agreement entered into by the federal government with a view to securing participation by the government or the corporation in activities connected with petroleum;

8 generally engaging in activities that would enhance the petroleum industry in the overall interest of Nigeria;
9 undertake such other activities as are necessary or expedient for giving full effects to the provisions of the Act.

The establishment of the NNPC has resulted in greater national participation in oil exploration and production activities, clearly demonstrated by the JVs, PSCs, and SC activities in which the corporation is involved. As at 2000, seven JVs existed in which NNPC holds equity interest. It holds 60 per cent equity interest in all the JVs except the JV between Shell, ElF and Agip JV in which it controls a 55 per cent equity interest.

In 2006 there were nine existing PSCs and an additional 16 new PSCs. However, enhanced exploration and production activities continue to push the number of PSCs upwards. During the same period only one SC existed and it was between NNPC and Agip Energy at the Agbara field.

NNPC structure

NNPC consists of a group of companies which have a central governing body at the corporate headquarters. At the apex of the group structure is the Board of Directors with the Honourable Petroleum Minister as chairman. The management team is led by the group managing director. As at 1998 the top management team comprised four executive directors, responsible for the Exploration and Production (E&P) directorate, Refining and Petrochemical (R&P) directorate, Engineering and Technology (E&T) directorate, Finance and Accounts (F&A) directorate, Investment directorate and the Corporate Services (CS) directorate. It is important to note that a four directorate structure emanated from a major restructuring exercise which was initiated in the Corporation in 2000. Prior to the four directorate structure which came into existence in 2003, and later terminated in 2007, the Corporation operated a six directorate structure.

a The E&P directorate supervises the National Petroleum Investment Management Services (NAPIMS), Nigerian Petroleum Development Company (NPDC), Integrated Data Services Limited (IDSL) and Nigerian Gas Company (NGC).
b The R&P Directorate has in its portfolio Port Harcourt Refining Company Limited (PHRC), Warri Refining and Petrochemical Company Limited (WRPC), Kaduna Refining and Petrochemical Company Limited (KRPC), Eleme Petrochemical Company Limited (EPCL) and PPMC Limited (PPMC). EPCL was privatised by the government in 2005 as part of government reforms in the downstream sector of the industry.
c The E&T directorate is responsible for the supervision and technical input in all engineering-related work of the Corporation. It also supervises the

information technology division. Over the years, staff of the directorate have been involved in the construction of the three refineries, petroleum products pipeline network, tank farms, Nigeria LNG project and other major engineering projects. The primary objectives are to achieve contractor compliance with project goals such as quality, execution, cost control, adherence to completion schedules and environmental protection.

d The F&A directorate supervises all the F&A functions of the Corporation. There are several divisions under the F&A directorate. Although each subsidiary formulates its budget, the ultimate approval of the budget resides with corporate headquarters. All financial requests arising from the subsidiaries supported by duly approved Authorisation For Expenditure (AFE) forms are processed and released subject to budgetary provisions and availability of funds. The directorate seconds accountants to the subsidiaries.

e The Investment directorate is responsible for the supervision of all direct NNPC investments. It has investments in the upstream and downstream sectors. It supervised the 11 OSCs in which NNPC held about 35% equity interest. These companies were privatised by the government as part of the overall privatisation programme. The Directorate also supervises Hyson, Duke Oil, Napoil, Nigermed, NIDAS International and NIKORMA. Hyson, Duke Oil, Napoil and Nigermed are hydrocarbon trading JV companies. On the other hand, NIDAS International and NIKORMA, which are also JV companies, engage in crude oil and LNG transportation respectively.

f The CS directorate supervises administration, human resources, training, medical insurance and pensions departments etc.

NNPC subsidiaries

NNPC has grown over time and nurtured subsidiary companies which engage in different aspects of the petroleum industry. The subsidiaries were created with the primary objective of providing critical products and services in Nigeria and initially in the West African subregion and subsequently at global level. Their operations were designed to generate revenues. Though NNPC spent large sums of money in the establishment of the subsidiaries, their performances have diminished over the years. The subsidiary companies are as follows:

- Port Harcourt Refining Company Limited (PHRC);
- Warri Refining and Petrochemical Company Limited (WRPC);
- Kaduna Refining and Petrochemical Company Ltd (KPPC);
- Eleme Petrochemical Company Limited (EPCL);
- Nigerian Petroleum Development Company Limited (NPDC);
- PPMC Limited (PPMC);

- Nigerian Gas Company Limited (NGC);
- Integrated Data Services Limited (IDSL);
- National Engineering and Technical Company Limited (NETCO);
- Duke Oil; and
- Hyson Nigeria Limited.

Politicisation of the decision making process and failure of past military regimes to appreciate the need for timely approval and provision of funds for the procurement of spare parts and other critical services affected the operations of the plants. A detailed discussion of the subsidiaries and the debilitating problems of their plants and operations will be undertaken in subsequent sections.

Collaboration strategies

Among the various arrangements which exist in the operational relationships between NOCs and IOCs, three types are known to be in common practice, namely JVs, PSCs and SCs. Although often misconstrued, PSCs and SCs are not JVs. In JV arrangements coownership of mutually agreed percentage interests in the OPL, OML concessions, assets and funds engaged in the JV, is a vital element. The basic elements of a JV in business involve:[2]

- a community of interest in the object of the undertaking;
- sharing production proceeds;
- bearing losses proportionate to equity interest if any;
- equal and reciprocal rights to govern the conduct of each other;
- a close and even fiduciary relationship between the parties.

Oil and gas joint ventures

The adoption of JVs in African petroleum industries was primarily influenced by collaborative success between the Italian company ENI and the National Iranian Oil Company (NIOC). ENI modelled an attractive incentive package for developing countries with large oil reserves in the Middle East and Africa. Mughraby in evaluating the ENI model stated:

> . . . by JV formula which is advocated and championed in the external market, ENI demonstrated how National Oil Companies can receive technical help and risk capital from foreign oil companies at minimal loss of income and at no risk whatsoever . . . the ENI formula attracted developing countries which found it a new socio-economic change more than a purely business deal based on financial consideration, as has been the arrangement with major international companies.[3]

Egypt was the first African country to adopt the JV concept in its petroleum industry. It blazed the trail by granting oil leases on small acreages to many IOCs on a short term basis.[4] This approach was a fundamental departure from the practice of granting long-term leases on large acreages to a few oil companies in the Middle East under the traditional concession system. In 1957 the EGPC entered into a JV contract with IEOC, a subsidiary of ENI, for the exploration and exploitation of oil and gas resources. Phillips Petroleum Company and the American UAR Company also finalised JV agreements with EGPC.[5] In view of its success other African countries namely Libya, Algeria and Gabon, adopted the practice. Nigeria embraced the JV and ENI in 1964 granted the federal government an initial offer of 30 per cent equity interest in its subsidiary the Nigerian Agip (NAOC). In subsequent years JV arrangements became dominant in the Nigerian petroleum industry. To date seven major JVs exist between NNPC and IOCs. Nigeria relied on the experience of oil and gas companies in the Middle East to articulate its position and strategy for participation in the industry. The origin and development of JV arrangements in the Middle East and North Africa have attracted elaborate scholarly consideration over the period. OPEC in its philosophy and practice strongly canvassed the need for NOCs to establish JVs with MNCs. Through Resolution XVI Article 90 of 1968 the organisation encouraged all member countries to actively participate in the upstream and downstream sectors of their various oil and gas industries. OPEC advocated an initial minimum participation level of 25 per cent by 1973 and an escalation to 51 per cent by 1982. This advice was heeded by member countries and today all NOCs have equity in their respective oil industries.

JVs

The need to develop capacity, transfer technology as well as attract investment capital required to execute exploration and production activities in the upstream sector, were underlying reasons for the establishment of JV operations between the NNPC and the MNCs. Nigeria commenced active involvement in JV operations with MNCs in the upstream sector in 1973. The move was designed to provide NNPC opportunities for participation in a highly dynamic sector. In the area of funding, NNPC in the majority of cases bears 60 per cent of all funding except in the Shell/Elf and Agip JV in which NNPC equity interest is 55 per cent. At present seven JVs are in operation, as follows:

- NNPC/Shell/Elf/Agip
- NNPC/Mobil
- NNPC/Chevron
- NNPC/Agip/Phillips
- NNPC/Elf
- NNPC/Chevron/Texaco
- NNPC/Panocean

In the years 2001/2002 and 2005 the federal government conducted a licensing round for the allocation of Blocks in the Deep Offshore. A mini-licensing round was conducted in 2006 and a major round was scheduled for 2007 as many other Blocks are yet to be allocated in the Deep Offshore. In this regard therefore, the opportunity exists for forward-looking investors to win oil Blocks and join other MNCs who have reaped attractive returns from their investments.

PSCs

A PSC is an agreement between a NOC (e.g. NNPC) and a foreign oil company. The contractor in this case is not a co-owner (with the NOC) of the petroleum licence or lease. The contractor, guided by preliminary information on the Block, enters into a contractual agreement which covers a period of 10 to 30 years with NNPC and provides funds for exploration, development and production. However, the contract guarantees the contractor the right to recoup all expenses incurred through 'Cost Oil'. In this case cost oil is the quantity of available crude oil allocated to the contractor to enable the company to recover all costs as specified in the contract. The contractor also gets a share of the 'Profit Oil'. In 2007 nine existing and 16 new PSCs respectively were operational.

SCs

SCs are agreements between NNPC and a foreign oil company. A service contractor does not have any working interest or ownership claim in the licence or the lease. In this instance the contractor accepts to carry out on behalf of NNPC exploration and development of the oil field. Contrary to the provisions of a PSC contract, a service contractor does not have an automatic right to produce oil discovered. However, the company may carry out production activities if NNPC grants it the option to do so. A service contractor is responsible for all costs of exploration and development work. Expenses incurred by the contractor are recouped through cost oil and profit oil. At present only Agip Energy has a SC with NNPC.

Oil and gas infrastructure development

In the 1980s and 1990s several projects were embarked upon by the NNPC in an effort to catalyse the industrialisation process. Accordingly, within the period under reference, some oil and gas related projects were executed as follows:

- Old Port Harcourt Refinery – 1965;
- Warri Refining and Petrochemical Company Limited – 1978;
- Pipeline Phase I (Phases I–III link all the Products Depots) – 1978;

- Port Harcourt Refining Company Limited – 1980;
- Kaduna Refining Petrochemical Company Limited – 1980;
- Eleme Petrochemical Company Limited – 1995;
- Pipeline Phase II (1980), phase III – 1995;
- Butanisation Project – 1996;
- Bonny Export Terminal – 1996;
- Nigeria Liquefied Natural Gas Company (NLNG) T1–T2 – 2000;
- Train 3 was constructed in 2002; Trains 4–5 in 2005 and Train 6 in 2007;
- West African Gas Pipeline – 2007;
- Trans-Sahara Gas Pipeline expected to be operational in about 2015;
- Brass LNG expected on stream in about 2010;
- OK-LNG is also estimated to come on stream in about 2010.

It is important to indicate at this juncture that the three refineries (Port Harcourt, Warri and Kaduna), the Butanisation project, Eleme Petrochemical Company, Bonny Export Terminal and Pipeline Phases I–III were funded directly by the federal government with internally generated funds and government guaranteed loans. On the other hand, NLNG, Brass LNG and OK-LNG are JV projects between NNPC and oil and gas MNCs. The dichotomy is essential in view of the sharp contrast in the funding mechanisms of these categories of projects. The West African Gas Pipe Project is jointly owned and funded by Nigeria, Ghana, Benin, Togo, Chevron and Shell. Ghana, Benin and Togo are represented by Ghana National Petroleum Corporation, Société Beninoise du Gaz and Société Togolaise du Gaz respectively.

NNPC transformation programme

NNPC was established in 1977 to operate as an NOC and engage in oil and gas related activities in both upstream and downstream sectors. The Company initiated JV agreements with IOCs and swiftly progressed to develop downstream infrastructure by establishing the refineries and petrochemical companies. Products tank farms were also constructed to serve as storage facilities for refined petroleum products. Furthermore a network of pipelines was constructed to link all the depots. Beyond this, robust management structures were put in place to run the activities of the refineries and other subsidiaries which were established in the NNPC Group. At the early stages of the formation of the Corporation, professionals in the various subsidiaries were given a free hand to run the affairs of the companies. However, in the 1980s and 1990s extraneous factors were brought to bear on the administrative and decision making processes of the Corporation. The military regimes usurped the functions of the professional staff and caused distortions in the operational activities of the companies. Refinery Turnaround Maintenance (TAM) programmes which were hitherto handled by the refineries were politicised, which created unnecessary bottlenecks in the approval processes.

The resultant effect of the politicisation was undue delays in the release of funds for maintenance of key equipment in the refineries and other major plants. A more profound situation set in when maintenance programmes were deferred for upwards of three years thereby seriously jeopardising the operational integrity of the plants. The maintenance periods omitted accumulated and the infrastructure decayed considerably. The result was low capacity utilisation occasioned by frequent breakdown of critical components of the plant especially the Fluid Catalytic Cracker (FCC). The functions of the FCC among others is to crack further the Low and High Pour Fuel Oil (LPFO/HPFO) into additional product groups in order to expand the overall yield of the refining process. The limited processing capacity of the plant translated into lower revenues in the operations of the refineries and indeed the NNPC Group.

It also became apparent that the systems and processes which as a rule must be strictly adhered to in running plants were compromised. This resulted in omission of key steps in plant operations which in some cases led to accidents associated with fire or damage to major equipment in the plants. Corrupt practices also set in and in some cases inferior spare parts were procured by schedule officers.

Furthermore, incidents of contract inflation were uncovered leading to the Aret Adams Commission of Inquiry which investigated the Warri TAM. The findings of the Commission of Inquiry indicted key officials of the Warri Refinery and severe disciplinary action was taken against the affected officers. Similar situations were detected in the Petroleum Importation Programmes which were manipulated by schedule officers to the detriment of the Corporation. It is important to indicate also that the affected officers were subjected to drastic disciplinary actions by the Corporation. Beyond these incidents, several behavioural lapses among the workforce constituted major obstacles in entrenching transparency in the administrative machinery of the Corporation. It is against this background that the transformation programme with the acronym PACE was introduced on 9 July 2004. The acronym PACE translated into (a) Positioning (b) Aligning for higher performance (c) Creating appropriate processes and systems for global competitiveness and (d) Enabling and empowering workforce.

PACE was guided by two world-class consultants – Shell Manufacturing Systems (SMS) and Accenture. In order to comprehend fully the problems of the Corporation the consultants embarked on an elaborate diagnostic exercise aimed at identifying the fundamental problems of the Corporation at corporate headquarters as well as at the subsidiaries of the Corporation. In doing this, various groups of carefully selected staff were identified to work with the consultants. Participation of experienced staff in the exercise facilitated the diagnosis of the basic problems and the root causes. The diagnosis was classified into functional areas which included strategy, personnel, IT, decision making, project management, contracting and procurement, maintenance etc.[6]

Strategy

Under the general domain of strategy the results of the diagnosis confirmed the following:

- Lack of focus in corporate activities which resulted in duplication of roles. Multiplicity of roles in this regard created unhealthy competition between units and departments which inevitably led to utilisation of scarce resources in the achievement of similar objectives. The concept of synergy failed to form part of the modus operandi of most functional units.
- The mission and vision of the organisation were treated as symbolic assertions and therefore not factored into the key decision making processes. Consequently, some programmes initiated and pursued failed to align with the mission and vision of the corporation.
- Over the years NNPC stagnated in the path of professional excellence and growth. Failure to adhere to best practices in the industry deprived the organisation of the opportunity of developing systems, processes and structures required to manage a complex organisation in a highly competitive and dynamic business environment.
- Low level of accountability and work ethic.
- Non-capitalisation of NNPC assets was identified as a key factor responsible for low inflow of investment funds into NNPC through traditional funding agencies. The funding mechanism of the NNPC business was considered weak and this was further aggravated by low returns on investment and lack of financial discipline (budget overruns).

Personnel

In the area of personnel the diagnosis revealed that they, in most cases, lacked world-class skills and competences and portrayed an ageing workforce which could not withstand the rigours and challenges of a high performance enterprise. The training of staff was considered low and haphazard such that the training provided in the majority of cases did not correspond with the needs of the individuals.

Information Technology (IT)

Over the years NNPC procured many computers but the process lacked definite acquisition guidelines and brand specification. As a result, some of the PCs acquired were substandard and lacked redeemable performance guarantees. Consequently, PCs that developed faults could not be traced to the sources of purchase. This arrangement caused unnecessary financial drain on the Corporation. Furthermore it was observed that IT knowledge application was low; therefore critical business areas were not covered.

Decision making

The decision making process was inadvertently bureaucratised and critical processes such as procurement of spare parts for key units of the refining plants suffered long delays.

Project management

NNPC executed many projects on a year to year basis. However, the diagnostic process confirmed that the executions were not guided by a corporate wide project management governance system. It was further observed that the system lacked an accredited quality management system in most segments of the organisation. Lack of a rigorous verification process and challenge during conceptual phase led to the execution of projects that did not conform to the overall business strategy of NNPC.

Contracting and procurement

The contracting and procurement process was observed to lack industry standard compliant policies for sourcing, contracting, acquiring, standardisation and stock management. Failure to adopt best practices led to procurement of substandard spare parts, high-priced items and unnecessary delays in item delivery.

Maintenance and integrity

NNPC established world-class refining and petrochemical companies constructed by globally acclaimed contractors such as Kellogg, SPIBAT, JGC, Marubeni etc. The involvement of these contractors notwithstanding, it was observed that maintenance and inspection were only scheduled for TAM such that availability of funds became the key determinant in its execution. The process lacked reliability, integrity performance evaluation and review. Based on the diagnostic results, senior management staff of the corporation charted a path for the future development of the Corporation within the framework of the following:

- reconfiguration of the NNPC structure and introduction of performance contracts for SBUs;
- introduction of business oriented governance principles to guide the interaction and overall relationship between corporate headquarters and SBUs;
- introduction of world-class processes, IT systems, financial policies and human resource practices in the SBUs;
- full commercialisation of NNPC activities and application of verifiable

performance measurement systems in all business dimensions of the Corporation;
- development of world-class competences in NAPIMS, NPDC, IDSL, Nigeria Gas Company and the refineries.

The transformation programme is geared towards the rejuvenation of all operational and administrative dimensions of the Corporation. NNPC aspires to be an integrated world-class oil and gas company. It is important to note that world-class companies are transnational in their operations. To acquire this status the company must leverage and refine all input variables, namely strategy, personnel, training, IT, decision making, project management, procurement etc. in order to obtain qualitative performance levels. Such performance, if sustained, could propel the Corporation into the required domain of world-class performance. Even with that the Corporation must consistently manifest superior corporate performance indices in order to earn an industry wide rating as a world-class company.

References

1 Nigerian National Petroleum Corporation Act. Ch. 320, 1 April 1977, S.1, SS 6.
2 Olisa, M. M. 'Nigerian Petroleum Law and Practice', Fountain Books, Ibadan, 1987, p. 61.
3 Mughraby, M. 'Permanent Sovereignty over Oil reserves'. Beirut, Middle East Research and Publishing Center, 1996, p. 63.
4 William, H. R. and Meyers, C. J. 'Manual of Oil and Gas Terms'. ® Matthew Bender and Company Inc., 1986, p. 74.
5 Lenczowshi, G. 'Oil and State in the Middle East'. New York, Cornell University, 1966, pp. 10–12.
6 Project PACE. NNPC Publication, 2004, pp. 1–14.

5 Upstream sector

Introduction

Globally the upstream sector of the oil and gas industry has for many years depended on cutting edge technology as a major catalyst of the phenomenal success recorded to date. Advanced technology is extensively applied in 3-D seismic data acquisitions and processing facilitated by high performance PC series, storage capabilities and mobile communication devices. The sector depends on various forms of data and management information, production accounting, well logs and other technological devices which propel the industry. Trends in the industrial and commercial sectors point to the increasing demand for oil and gas as feed stock in the petrochemical industry and for energy in the broader industrial sectors. Areas of application of oil and gas resources continue to expand, but sadly, these resources are finite and depletable. Therefore there is a conscious effort to deploy a gamut of data and information management devices to improve the performance of reservoirs. The devices are also designed to engender cost effective performance in production and elevate the economic threshold of the portfolio of assets.

The upstream activities

The upstream sector of the Nigerian petroleum industry has become one of the most vibrant sectors in Africa. With an initial production level of 5,100 b/d in 1956, daily output in the sector has escalated to well over 2.2 mmbd. The dominant players in the sector, Shell, Mobil, Chevron-Texaco, Agip (NAOC) and Total, have consolidated their positions through acquisition of cutting edge technology and expansion of their crude oil reserves. In recent years the ranks of the players in the sector have increased to include Addax, Statoil, Petrobras, ConocoPhillips etc. It is interesting to note that activities in the upstream have progressively shifted from land, swamp, and shallow continental shelf to the Deep Offshore. There is an aggressive operational campaign targeted at increasing the current production level to 2.4 mmbd in 2006 and further to 2.5 mmbd in 2008. It is also projected that production will attain 4.5 mmbd in 2010. A corollary of the production campaign is an agenda to reduce the 47 per cent gas flare rate to zero in 2009 through aggressive monetisation of gas.

Between 1996 and 2004 major crude oil discoveries were made in the Deep Offshore. The Bonga field discovered in 1996 in a depth of 100 m of water was estimated to have 1 billion barrels of crude oil reserves. Agbami field discovered in 1999 in 1,450 m was projected to have 809 million barrels. Similarly, Erha and Akpo fields were estimated to have 630 million and 700 million barrels reserves respectively. About 16 fields discovered (i.e. 1996–2004) mainly in the Deep Offshore, account for the anticipated addition of 6.31 billion barrels to the national reserves currently estimated to be 36 billion barrels. Associated with these reserves are Bosi and Nnwa/Doro fields which account for 7 TCF and 12 TCF of gas respectively. Some of these fields have gone into production thereby increasing the tempo of oil and gas exploration and production activities in the sector. In 2004 oil majors in the sector discovered 45 TCF of gas consisting of 42 TCF of Associated Gas (AG) and 3 TCF of Non-associated Gas (NAG) respectively.[1] These volumes account for part of the 187 TCF national reserve. The gas reserves are actively utilised in major gas utilisation projects as shown in Table 5.1.

On the whole, ongoing gas projects in the industry have been allocated a total of 96 TCF accounting for 48 per cent of the total gas reserves of 187 TCF. At the domestic level, gas consumption will emanate from the power generation sector, fertilizer plants, aluminum smelting plants and other local industries. Aggregate demand from these areas will expand domestic demand from the current level of 600 million scf/d to about 1.7 trillion scf/d by the year 2010. It is important to note that achievements in the upstream sector

Figure 5.1 Offshore production platform in a Niger Delta creek.

Source: Shell Petroleum Development Company (SPDC), Brief Notes 2, 3rd edn, 2005

Table 5.1 Gas utilisation projects and feed gas requirements

Project	*Feed gas volume*	*Total requirement*	*Start up*
LNG Trains 1–3		12.6 TCF	2002
NLNG Trains 4 and 5	1,340 mm scf/d	11.3 TCF	2005
NLNG Train 6	670 mm scf/d	5.6 TCF	2007
EGP2/WAGP	450 mm scf/d – at peak	1.8 TCF	2007
EGP3/GTL	350 mm scf/d – at peak	3.5 TCF	2005
EAGP-NGL	630 mm scf/d	NA	2009
Brass LNG	1,500 mm scf/d	12.6 TCF	2010
OK-LNG	1,500 mm scf/d	25.2 TCF	2010
ExxonMobil-IPP/LNG	830 mm scf/d	NA	2010
West African Gas Pipeline	200 mm scf/d initial; 500 mm scf/d at peak	NA	2007
Trans-Sahara Gas Pipeline	1 bcf/d at peak	NA	2008
Equatorial Guinea – gas supply	Discussions ongoing	NA	–

Sources: NNPC and PL Consulting, *Nigeria Oil and Gas Industry Outlook*, 2006, p. 180.
All gas allocations to projects are dedicated volumes.

are made possible by JV agreements between NNPC and some of the major oil companies. The JVs are jointly funded and NNPC equity holding is as follows:

- NNPC 55 per cent, Shell 30 per cent, Total 10 per cent, Agip 5 per cent;
- NNPC 60 per cent, Mobil 40 per cent;
- NNPC 60 per cent, Chevron 40 per cent;
- NNPC 60 per cent, Agip 20 per cent, ConocoPhillips 20 per cent;
- NNPC 60 per cent, Texaco 20 per cent, Chevron 20 per cent;
- NNPC 60 per cent, PanOcean 40 per cent.

JV agreements were a common practice in the upstream sector until recently when funding constraints warranted adoption of alternative funding mechanisms. In recent years, the government has opted for PSCs as a way to free funds for other pressing development programmes. NNPC currently has a total of 25 PSCs categorised into nine existing PSCs and 16 new PSCs (Table 5.2). It currently operates one SC with Agip Energy. Existing PSCs are:

- Addax Petroleum (ADPNL);
- Addax (APENL);
- Chevron Nigeria Deep Water Limited;

- Conoco Exploration;
- Elf Petroleum Nigeria Limited;
- Esso Exploration and Production Nigeria Limited (EEPNL);
- Agip Energy and Natural Resources (AENR);
- Shell Nigeria Exploration and Production Company (SNEPCO);
- Statoil Nigeria Limited.

Funding in the upstream sector

The upstream sector (as shown in Table 5.3) is highly capital intensive utilising advanced technology to achieve results in highly challenging terrains. The industry in its 1995 annual budget supervised by NNPC proposed \$2.8 billion to the federal government as cash call requirements for the year. An approval of \$2.04 billion was granted and as a result a shortfall of \$760 million was created. In 2004 a budget of \$4.48 billion was presented for

Table 5.2 The 16 new Production Sharing Contracts (PSCs)

Company	*OPL*	*Year*	*Company*	*OPL*	*Year*
Texaco Nig. Outer Shelf	213	1993	ECL International Ltd	251	2004
Star Deep Water	216	1993	Vintage Oil and Gas	257	2004
Nigeria Agip Exploration	211	1993	Shell Nig. Ultra Deep Ltd	245	2004
Esso E&P Deep Water	214	2002	Shell Nig Explor. Alpha	322	2004
Petrolco Brasileiro Nig. Ltd	324	2001	Chevron Nig. Deep Water	247	2004
Oranto Orandi Petroleum	320	2002	BG Group	332	2006
Phillips Exploration Nig.	318	2002	KNOC	321	2006
Ocean Energy Nig. Ltd	256	2003	KNOC	323	2006

Source: NAPIMS – E&P Directorate, 2006.

Table 5.3 Funding levels of NNPC share of JVs 1995–2004

Year	*Industry proposed (US\$ million)*	*Govt. approved budget (US\$ million)*	*Funding shortfall (US\$ million)*
1995	2,800	2.040	760
1996	2,956	2,005	951
1997	3,373	2,050	1,323
1998	3,600	2,500	1,600
1999	3,474	2,400	1,074
2000	3,472	2,300	1,172
2001	3,501	3,200	301
2002	4,286	3,125	1,161
2003	4,297	3,500	797
2004	4,478	3,200	1,278

Source: E&P Directorate, 2004.

approval but actual approval amounted to $3.2 billion causing a shortfall of $1.28 billion. Government funding continues to diminish thereby warranting the restructuring of projects as well as identifying alternative funding mechanisms.

The prevalence of PSCs in current NNPC upstream contract portfolios is necessitated by dwindling government funding arrangements. In view of the aggressive exploration and production programmes in the Deep Offshore, the funding requirements have escalated significantly. Projections in Table 5.4 indicate that funding requirements in the upstream sector were approximately $8.7 billion in 2006 but slightly declining to $7.2 billion in 2009. The slight decrease in funding requirement is explained by the fact that most of the fields are currently at the exploration and developmental stages. With the exploration and development programmes concluded in subsequent years the actual expenditure on a yearly basis will manifest a slight decrease. In the case of PSCs, funding requirement on an annual basis between 2005 and 2009 is anticipated to range between $3.6 billion and $3.2 billion. In cumulative terms total funding requirements for both JVs and PSCs in 2006 amounted to about $12.7 billion and are expected to decline to $10.4 billion in 2009. The government contribution in JVs in 2006 was about $5.1 billion and declined slightly to $4.1 billion in 2009.[2] In order to guarantee smooth execution of projects in the sector other funding options such as self funding, alternative funding and Special Purpose Vehicles (SPVs) are also relied upon as avenues for the implementation of upstream projects.

It can be conceded that the PSC approach adopted for the development of upstream programmes brings relief and mitigates government risk. It can also be posited that the arrangement maximises PPT and Royalty. This notwithstanding, the Pay Back Period (PBP) hinged on Cost Oil amortisation is unduly elongated, eroding revenues accruable to the government, especially as inflation escalates. Total disengagement from JVs in the upstream sector through the government endorsement of the PSCs arrangements could lead to higher earnings from Royalty and PPT in the short run but will have negative medium and long-term impact on the economy. The preceding sections have evaluated the activities of the upstream sector. It was indicated that the achievements were as a result of the participation of IOCs, through JVs, PSCs and SCs. In consideration of this, subsequent sections of this chapter will explore the profiles of these companies which contributed to the

Table 5.4 Upstream funds requirement 2005–2009 projections

	2005	*2006*	*2007*	*2008*	*2009*
JV	8.7	8.7	8.6	8.3	7.2
PSC	3.6	4.0	3.8	2.6	3.2
Total	12.3	12.7	12.4	10.9	10.4
Govt. share	5.0	5.1	4.9	5.1	4.1

Source: NNPC – E&P Directorate, 2004.

evolution and growth of the sector. The main purpose is to examine their operations and determine their scopes of activities and contributions to the Nigerian economy.

IOCs in the upstream sector

Shell

Shell made an early entry into the Nigerian petroleum industry. The need to shift to Nigeria as an oil exploration base was necessitated by the Suez Canal crisis and a general feeling among explorers that avoiding transport through the Canal carried a political premium value.[3] The company, which was incorporated in 1937, played a pioneer role in the upstream sector and dominated exploration activities between 1938 and 1955. The search for oil was intensified and in 1956 it discovered oil in commercial quantity at Oloibiri in Bayelsa State. At the end of 1956 the company struck another oil well at Afam in Rivers State. A series of discoveries followed, establishing a solid foundation for future exploration and production activities. It maintained its lead in the upstream sector through application of modern technology in the acquisition and interpretation of seismic data. Its drilling and well development campaigns also engaged state of the art technology which in subsequent years distinguished it as the highest producer in the industry.

Although Shell has a pioneer status in the Nigerian petroleum industry, it is associated with a checkered history. In 1973 the federal government, in furtherance of its objective to participate in the upstream sector, acquired 35 per cent equity interest in the company. This was increased to 55 per cent in 1974 and further escalated to 80 per cent due to the government's acquisition of BP's interest in the JV. Observers are of the view that the government action was designed to send strong signals to BP of Nigeria's disapproval of its huge investments in the apartheid regime of pre-independence South Africa. The equity interest in the company was, in 1989, rationalised, thereby creating room for SPDC to adjust its interest to 30 per cent. At this juncture two companies, namely Agip and Elf (now Total), were co-opted to acquire a 5 per cent equity interest each. NNPC further conceded a 5 per cent equity to Total. Shell is currently the operator of the JV in which NNPC holds 55 per cent equity interest, Shell 30 per cent, Total 10 per cent and Agip 5 per cent respectively.[4] In 1977 the upstream sector recorded total production of 2.078 mmbd, of which Shell production accounted for 1.21 million barrels or 41.8 per cent of the total daily production. Its production level in 2006 is estimated to be 1 mmbd. In addition to oil, Shell produces about 700 million standard cubic feet per day (mm scf/d) of gas from 90 fields scattered throughout the Niger Delta region. It has over the years developed a vast network of pipelines covering 6,000 km and this network links about 90 oilfields and 87 flow stations. The pipelines also connect eight gas plants and about 1,000 producing wells through ancillary pipe links.[5]

Subsidiary companies

Shell Nigeria Exploration and Production Company (SNEPCO)

SPDC expanded its activities from land, swamp and the shallow continental shelf to the Deep Offshore in 1993. In order to face the challenges in the sector SNEPCO was created. The company executed aggressive programmes in OPLs 212 and 219 culminating in the discovery of the Bonga reserve (in 1,000 m water depth) in 1996. The discovery is estimated to hold about 600 million barrels of crude oil. It also discovered another reserve southwest of Bonga in 2001. In response to the daunting challenges of the new terrain Shell deployed its global expertise in Deep Water exploration and production to optimise results. The company has since its first engagement in the Deep Offshore recorded steady progress. It currently has participatory interest in OPLs 211, 316 and OML 125 operated by Agip. It also has participatory interest in OPL 209 operated by Esso. In January 2006 Bonga achieved a production level of 187,000 b/d and it is expected that it will attain a production level of 225,000 b/d in due course. SNEPCO has so far engaged a modern FPSO to support the production.

The company inherited from Shell Petroleum Development Company (SPDC) the operations of the EA field which was discovered in 1965. The field, with an estimated reserve of 360 million barrels and potential producibility of 160,000 b/d, commenced production in 2002. The operation would require an FPSO logistics support vessel capable of discharging into proximately positioned tankers. It is part of the SPDC $8.5 investment portfolio earmarked to contribute significantly to the attainment of the 4.5 mmbd aggregate national production level in 2010.

Figure 5.2 An FPSO vessel loaded with materials for offshore operations.
Source: NNPC – Exploration and Production Directorate, 2005

Shell Nigeria Gas Limited (SNG)

Exploration activities in the upstream sector of the Nigerian petroleum industry are primarily aimed at discovering oil. The proven gas reserves of 187 TCF were discovered in association with oil. This notwithstanding, experts are of the opinion that Nigeria is essentially a gas province. Until recently, gas utilisation in Nigeria was extremely low as evidenced by the 43 per cent flare rate (i.e. 230,000 barrels of oil equivalent flared daily). This level of gas flare is a huge economic waste with adverse effects on the environment. The high flare rate earned the country a negative image as the number one gas flaring country in the world, discharging about 70 million MT of CO_2 into the atmosphere annually. There is UN global pressure on all oil producing countries to attain zero flare. Nigeria being a signatory to the Kyoto Protocol has set 2009 as a target for zero flare attainment. More importantly, it has directed all oil companies to put in place gas utilisation programmes. Current estimated domestic consumption of 600 mm scf/d in 2006 is extremely low compared to the proven reserve.

SNG was established in 1998 to promote gas utilisation through supply of the commodity to industrial organisations as well as for domestic consumption. An initial investment of $34 million was executed to put in place infrastructure for gas supply, and about 45 gas sale and purchase agreements have been signed by the Nigeria Gas Company (NGC) with companies at Aba, Agbara and Ota industrial estates.[6] Effective from 2002, SNG commenced gas supply to customers at Agbara and Ota. These companies foresee a future which will be dominated by gas as the preferred energy source for major industrial activities. SNG continues to explore potential gas utilisation opportunities in the industrial axes of Port Harcourt, Lagos and Aba.

Shell Nigeria Oil Products Limited (SNOP)

Shell was until recently heavily involved in the downstream sector of the industry, but low capacity utilisation of the refineries and the protracted era of products scarcity compelled it to re-evaluate its participation. As a matter of strategic business choice, it divested from the downstream sector in which National Oil marketed petroleum products and lubricants. The industrialisation process and expanding commercial activities among a population of about 140 million people suggest a potential increase in demand for petroleum products and industrial chemicals. In consideration of this business opportunity SNOP was incorporated in 2002 and is being positioned to be the largest supplier of refined petroleum products in West Africa. Technically, however, SNOP cannot rely on local refining outlets in view of their low capacity utilisation and frequent breakdown.

So far the downstream sector has been liberalised and licences for private refineries awarded. Investors are, however, worried about the slow pace of deregulation which inhibits market forces to determine the price of Premium

Motor Spirit (PMS) and other products at the pump. Private refineries will come into existence when the price at the pump can guarantee full cost recovery. Such an environment will allow SNOP to pursue its objectives, one of which might be the establishment of a refining outlet that can serve the domestic and international markets. The actualisation of this objective will generate revenues for the organisation as well as contribute toward the stabilisation of the supply of petroleum products for industrial and commercial consumption.

Shell and the NLNG

The Nigeria Liquefied Natural Gas Company (NLNG) was established in 1989 as a JV between NNPC, Shell, Total and ENI. NNPC has a 49 per cent equity interest while Shell has a 25.6 per cent interest. Total and ENI have 15 per cent and 10.4 per cent equity interest respectively. The project is designed to operate about seven Trains with a total annual capacity of 30 million MT of LNG. Train 6, with a capacity of 4.0 million MT per year, came on stream in 2007. Shell and others are actively involved in gas supply to the NLNG project. Trains 1–7 will require about 41 TCF of gas for a period of 20 years. Based on the equity participation of Shell in gas based projects and the volume of its gas reserves, one is inclined to conclude that it would supply a significant proportion of the gas requirements of the projects. Essentially, its involvement will promote gas monetisation and greatly contribute towards the mitigation of the negative impact of gas flaring on the global ozone layer.

ExxonMobil

ExxonMobil has three subsidiary companies, namely:

- Mobil Producing Nigeria Unlimited (MPN);
- Esso Exploration and Production Nigeria Limited (EEPNL);
- Mobil Oil Nigeria Plc (MON).

Mobil Producing Nigeria Unlimited (MPN)

Mobil Producing Nigeria (MPN), incorporated in July 1969, derived its roots from Mobil Exploration Nigeria Incorporated which commenced business in Nigeria in 1955. MPN was granted its first OPL by the federal government in December 1961 and drilled the first well at Ata-1 in 1964. MPN has a daily production capacity of 720,000 barrels and is generally classified as the second largest producer in Nigeria. It is the operator of NNPC/MPN JV in which it has 40 per cent equity interest. The operations of MPN in Nigeria spanning half a century have been entirely offshore. Crude oil production started in 1970 at the Idoho fields located in Akwa Ibom State. It controls 800,000 acres in the shallow swampy waters in Akwa Ibom where production is executed from 90 offshore platforms linked to 353 wells.

It is important to note also that the federal government equity acquisition in Mobil was incremental. The first acquisition of 35 per cent occurred in 1973 and was further increased to 60 per cent in 1979.[7] MPN has remained poised and aggressive in its E&P campaigns. It signed an $800 million loan agreement with a consortium of financiers for the development of the OSO condensate field. The field, which was estimated to have recoverable reserves of 869 million barrels, has been fully developed and commenced production in December 1992. Alongside the OSO condensate project was the OSO–NGL project which converted the associated gas into NGL with effect from November 1998. In 2003 the company fully engaged the Yoho and Awawa fields in OML 104 with recoverable reserves of 400 million barrels and achieved a daily production of 150,000 b/d. These and other production activities increased the total MPN reserve to about 3 billion barrels.[8]

Esso Exploration and Production Nigeria Limited (EEPNL)

Esso Exploration and Production Nigeria Limited was established in 1993 with the primary mandate of executing oil exploration and production activities in the upstream sector. It carried out its first set of operations in Erha field (OPL 209) and currently holds interest in six deepwater blocks covering 3.2 million acres. EEPNL is the operator (56 per cent interest) of OPL 209 in 3,400 feet of water offshore. The field has estimated gross crude oil reserves of 500 million barrels and an estimated producibility level of 150,000 b/d. It also has 20 per cent interest in the prolific Bonga field (OML 118) operated by SNEPCO. The Bonga field has reserve potential of 800 million barrels and initial appraisal estimated a production level of 200,000 b/d of liquids.[9] The company has a similar interest in Bonga South West field as well as Bolia field in which Esso has 20 per cent equity interest. EEPNL (Benue) Limited has 47.5 per cent interests in Chota and 30 per cent in Usan projects operated by ConocoPhillips.

Mobil Oil Nigeria Plc (MON)

Mobil Oil Nigeria Plc has a historical link with Sonocy Vacuum Oil Company which pioneered the sale of kerosene in Nigeria in about 1908. Sonocy metamorphosed into MON in 1951. It has since become a publicly quoted company on the Nigerian Stock Exchange. MON is a major player in the downstream sector of the industry and currently operates over 200 retail outlets in the 36 states of the Federation. MON operates three plants in Apapa, Lagos where it manufactures petroleum lubricants, jelly and insecticides. The 400,000 b/d lube plant operated is one of the largest and possibly the most modern lube plant in Africa.

ChevronTexaco

ChevronTexaco made its entry into the Nigerian hydrocarbon industry in 1913. American Overseas Petroleum Limited which metamorphosed into Texaco Overseas (Nigeria) Petroleum Company (TOPCON) discovered the Koluama field in offshore Niger Delta in 1963. In the same year the company discovered oil at the Okan field near the Escravos River. It operates 11 concessions in an area covering 2.2 million acres mainly in swamps and in the shallow continental shelf.[10] It executed an aggressive exploration and production programme which resulted in the discovery of the prolific Agbami field estimated to hold 800 million barrels of crude oil reserves. This discovery is so far one of the largest discoveries in the history of the Nigerian petroleum industry. In 1999 it also discovered a major gas field with reserve potentials of 12 TCF at Nnwa/Doro, an offshore location. Chevron and Sasol have consummated a mutual interest agreement to monetise gas through the use of Fischer-Tropsch and Chevron Isocracking technologies.[11] ChevronTexaco enjoys 40 per cent equity interest in its JV with NNPC. It is also the operator of NNPC/Chevron Texaco JV and in 2005 produced over 110,000 b/d from 33 fields.

Total

The new Total Group came into existence in May 2003 following two successive mergers. The first part occurred in 1999 when Total joined the Belgium

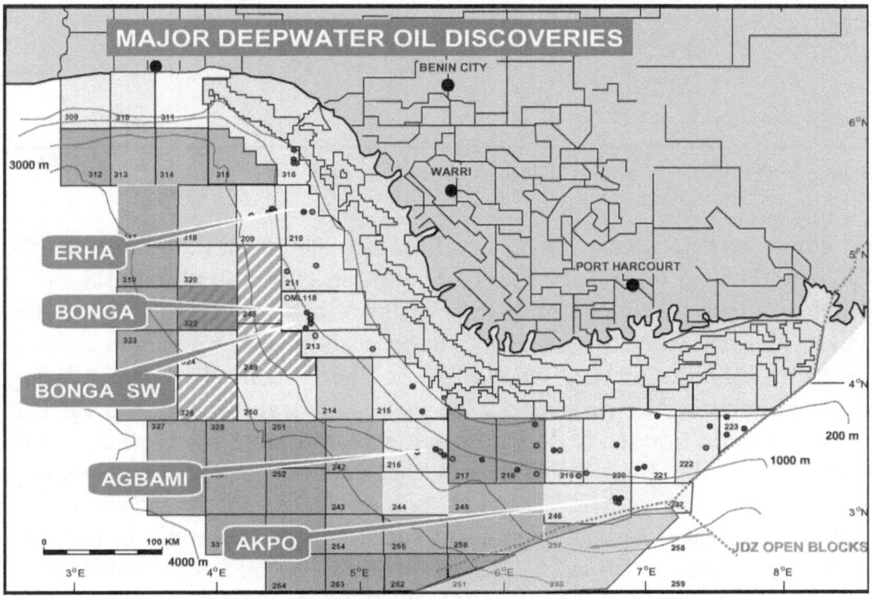

Figure 5.3 Deep Offshore Blocks in the Gulf of Guinea.
Source: NNPC – Exploration and Production Directorate, 2005

Table 5.5 Major Deep Offshore reserves

	Field	*Date*	*Water depth (M)*	*Reserves million bbls*
Oil				
	Bonga	1996	1000	1000
	Ngolo	1997	800	51
	Abo	1997	600	250
	Chota	1998	1050	400
	Ukot	1998	700	300
	Agbami	1999	1450	809
	Erha	1999	1200	630
	Usang/Ukot	1999	750	400
	Akpo	2000	1350	700
	Bonga S.W	2001	1250	500
	Aparo	2001	1250	150
	Bolia	2001	1000	132
	Usan	2002	700	620
	IFO (Bonga In-field opportunity)	2002	1250	78
	Nsiko	2003	1729	132
	Bonga W/N	2004	1141	160
Total				**6312**
	Field	*Date*	*Water depth (M)*	*TCF*
Gas				
	Bosi	1996/7	1450	7
	Nnwa/Doro	1999	1200	12
Total				**19**

Source: NNPC Exploration and Production Directorate, 2006.

oil company Petrofina to form TotalFina. In 2000 TotalFina merged with Elf Aquitaine to form TotalfinaElf. The group currently adopts the name Total in all its operations globally. For historical purposes the discussion will in some cases refer to old names in order to provide a coherent account of the evolution of activities in the group. The French oil company Safrap was registered in Nigeria in 1962. This company transformed into Elf Petroleum Nigeria Limited (EPNL). Its first exploration well in the early 1960s (OB1) led to the discovery of one of the major oil fields in the industry. This field at Obagi (OML 58) has to date produced more than 500 million barrels. Total has recorded significant success in its explorations and production activities which led to the conversion of its OPLs into OML 57 in the 1970s. In the 1980s OMLs 99, 100 and 102 were successfully operated. It put Ofon oil field (OML 102) on stream in December 1997 and reached 60,000 b/d production level.[12] The company has shifted its operations to

Deep Offshore and has discovered Ukot and Akpo fields. It has also increased its exploration portfolio through the acquisition of OPL 223. It further achieved farm-in (being allowed as third party) as operator with 40 per cent equity interest in OPL 215.

Another partner in the OPL 215 development is the concessionaire Northwest Nigeria Limited. It has JV interests in OMLs 99–102 and serves as the operator of OMLs 56 and 58. Another farm-in with 16 per cent equity was also achieved in OPL 247 operated by ChevronTexaco. In 2005, appraisal success was recorded at OPLs 246 (Egina) and 222 (Usan). The company also has 10 per cent equity in NNPC/Shell/Total and Agip JV which has since 1989 progressed to an aggregate production level of about 900,000 b/d. Total also acquired 12.5 per cent participatory interest in Deep Offshore OML 118 (Bonga) field and OPL 219 in which Shell is the operator. It has in its 44 years of operation covered a broad terrain spanning land, swamp, shallow waters and the Deep Offshore. It is an active participant in the NLNG project where it maintains 15 per cent interest and supplies gas as feed stock for Train 1–7 of the project. Its subsidiary company TUPNI serves as technical adviser on the Deep Offshore Block OPL 246 executed in partnership with South Atlantic Petroleum and Petrobras. Amenam/Kpono field, which straddles EPNL's OML 99 and Mobil's OML 70, is expected to add 125,000 b/d to Total's overall production capacity and leverage its reserves.

Nigeria Agip (Agip)

The Exploration and Production Division of ENI commenced activities in the Nigerian petroleum industry in 1962 through its international subsidiary Nigeria Agip Limited (NAOC). Agip made an initial offer of a 35 per cent equity interest to the federal government in 1973. In subsequent years the government exercised its rights to participate in Agip whose operations until recently were located on land, and in the swampy mangrove forests of the Niger Delta. Agip enjoys 5 per cent equity interest in the NNPC JV and is also involved in another JV in which it has 20 per cent equity while NNPC and ConocoPhillips have 60 per cent and 20 per cent respectively. The company controls a concession area of 53,125 km^2 which encompasses OMLs 60, 61, 62, and 63 and its aggressive exploration efforts led to the discovery of Ebocha field in 1965. The OPL of the field was converted into an OML and production from the field commenced in 1970. Agip relies on leading technology in its exploration and production campaigns. This accounts for its success in the acquisition of 21,000 km^2 of 2-D and 4,200 km^2 of 3-D data exclusively in 2003. The company has also successfully drilled over 360 wells out of which 105 were classified as exploratory.[13] An evaluation of the exploration and production campaign in the early years showed a 55 per cent success rate but recent results show a higher performance. It is also important to indicate that a total of 46 fields was discovered since inception and about 32 have been put into production. Agip has through huge investments increased its production assets which include approximately 8 flow stations, 8 gas plants, an export tank farm of 3.6 million barrels capacity and 2 single point mooring buoys for loading tankers.

The company operates a 460 km pipeline network which connects the flow stations and gas plants. In addition, it supplies the Eleme Petroleum Company with NGL and fuel gas through a 180 km pipeline. In 1985 Agip blazed the trail in gas monetisation through the establishment of the Obiafor-Obrikon gas re-injection plant. It took another significant step in 2005 by constructing the Kwale Okpai IPP capable of producing 480 MW in its JV with NNPC. It is important to note also that Agip controls 10.4 per cent equity interest in NLNG and is involved in the supply of gas to Train 1–7 of the project. Through its equity interest and gas supply in the Nigeria LNG, Kwale power plant and the Brass LNG plant, Agip is well positioned to execute additional gas-based projects to significantly boost the gas monetisation effort of the federal government.[14]

Nigerian Agip Exploration Limited (NAE)

Nigerian Agip Exploration Limited, a subsidiary of ENI, was incorporated in Nigeria in 1996. It operated a PSC in Deep Offshore in OPLs 211 and 316 in collaboration with SNEPCO. NAE has developed cutting edge technology in the Deep Offshore resulting in the acquisition of 5,170 km^2 of 2-D and 3,783 km^2 of 3-D seismic data to aid discovery of hydrocarbon reservoirs. It drilled three wells on OPL 316 which led to the discovery of Abo field. It also drilled two wells on OPL 211 which turned out to be a gas field with some trail of oil. Production from the Abo field was executed through an FPSO commenced in February 2003. It set the pace by being one of the first companies to successfully produce oil from Deep Offshore. In pursuance of its upstream strategy, NAE partnered with the Nigerian Petroleum Development Company (NPDC) during the 2000 bid round and won OPL 244 located in the Deep Offshore. A PSC on the Block was signed with the federal government in 2001.

Agip Energy and Natural Resources Limited (AENR)

Agip Energy and Natural Resources Limited is one of the subsidiaries of ENI and was incorporated in 1980. AENR signed a service contract with NNPC on OML 116 and diligently executed a SC which culminated in the discovery of the Agbara field. The field was put into production and has since 1988 produced over 66 million barrels of oil. Additional reserves have been discovered in the Agbara field, which can extend the life span of the field to well beyond 20 years. It acquired over 12,600 km^2 of seismic data to support its aggressive search for oil. In view of the additional discoveries aggregate AENR producibility from the field has increased to 200,000 b/d. As part of the drive to increase national crude oil reserves to 40 billion barrels, AENR signed another SC in December 2000 with the Nigerian Petroleum Development Company (NPDC) to finance and collaboratively develop Okono and Okpoho fields in OML 119. The partnering relationship is designed to bring more fields into production and in the process expand the production capacity of NPDC to 150,000 b/d by year 2007. The partnering arrangement with NPDC has provision for joint conduct of operations, collaborative management through the project management committee, transfer of operatorship and transfer of technology. Okono/Okpoho was executed on a fast track, which made it possible for PSC production to start in December 2001, barely 10 months after the signing of the PSC. The expanded production programme raised the reserve by 300 per cent thereby increasing STOIIP from 56 million barrels to 240 million barrels. AENR has achieved from Okono/Okpoho fields a production level of 50,000 b/d. This by far exceeded the 40,000 b/d anticipated at the inception of the project.

Addax Petroleum Development (Nigeria) Limited

Addax commenced business in Nigeria in 1998 following the acquisition of the assets of Ashland oil. In order to participate in the upstream sector it executed two new PSCs with NNPC to enable it to produce OMLs 123 and 124 located offshore and onshore respectively. It also signed another PSC for the operation of two offshore Blocks, namely OPLs 90 and 225. The OMLs and OPLs were hitherto operated by Ashland Petroleum Development and Ashland Exploration respectively. The company currently produces from OMLs 123 and 124. OML 123, which is located in the swampy offshore, covers 400 km^2 in water depth varying from 10 to 90 m. The Block has six producing fields which are located at Ukpum, Ebughu, Mimbo, Akam, Bogi and Adanga. The first production occurred at OML 124 located approximately 100 km from Port Harcourt. The field which is located onshore covers an area of 3,390 km^2. Two other fields located at Isombe and Ossu came into production in 1975 and 1976 respectively. There was a gradual build up and in 2004 OMLs 123 and 124 reached a production level of 46,000 b/d. Addax Petroleum also operates offshore concessions in OPLs 90 and 225 which have estimated areas of 772 and 1,699 km^2 respectively. Okwori field – one of the fields inherited by Addax – accounts for cumulative production of 38 million barrels and is further anticipated to produce an additional 27 million barrels of oil in the next seven years. The company has so far pursued its exploration and production programmes aggressively and reached a major milestone in December 2004 when it commissioned its FPSO in Singapore. The vessel has a storage capacity of 1.4 million barrels and is capable of producing 38,000 b/d from the Okwori offshore field designated as OPL 90. This field has, with effect from mid-2005, commenced daily production of approximately 21,000 barrels. It was expected to peak at 25,000 b/d in 2006 and record cumulative production of 7.6 million barrels of crude oil in the first year of production.

Indigenous oil companies

Oil exploration and exploitation activities in Nigeria are known to have begun around 1908. Although NNOC was incorporated in 1971 to represent the government's interest, its active participation in the upstream sector started only in 1973 through JV agreements. Indigenous participation in the upstream sector was virtually absent in the early 1970s and the 1980s. In 1990 the federal government made a deliberate attempt to involve indigenous companies in the upstream sector. In furtherance of its objectives 11 local companies were granted exploration licences between 1990 and 1991 in order to kick-start the indigenisation programme. The federal government took cognisance of the low technical capacity of indigenous companies and granted permission for licencees to engage potential off-shore technical partners. Such partners would acquire 40 per cent equity interest in the ventures. The efforts of the indigenous companies have yielded results and as at 2005 Atlas Petroleum International, AMNI International, Cavendish Petroleum, Conoil Producing, Continental Oil, Express Petroleum and Moni Pulo Limited have begun production. The number of indigenous companies has increased and some are making steady progress. While all the companies are important, the discussion will be limited to a few indigenous companies. The aim is to capture some dimensions of their performance and by extension infer the overall performance of this group of companies in the upstream sector of the industry. Specifically the discussion will focus on the following companies using date of incorporation as a basis of classification:

- Consolidated Oil Limited
- Dubri Oil Company Limited (Dubri Oil)
- Solgas Petroleum Limited
- Atlas Petroleum International
- Allied Energy Resources
- AMNI International.

Consolidated Oil Limited (Conoil)

Conoil was registered in Nigeria on August 1984 to engage in petroleum exploration and production. The company was awarded OPL 133 in August 1990 through the federal government discretionary award programme. Conoil executed seismic data acquisition programmes on the Block and based on the information gathered commenced drilling appraisal and developments wells in 1991. The campaign was successful and Conoil has, since the mid-1990s, begun producing approximately 7,000 b/d of crude oil.

Dubri Oil Company Limited (Dubri Oil)

Dubri Oil is Nigeria's pioneer indigenous oil and gas company. It was incorporated in August 1987. Among indigenous companies, Dubri drilled the first well in July 1991 and discovered oil in commercial quantity in the Ovia field in 1995. Although the operational profile indicated that its experience in oil production started with 800 b/d production at the Grilli-Grilli field in August 1987 it is not clear if the field was acquired through a farm-in process. Later production was at OML 96 which is currently at the level of 1,000 b/d.

Solgas Petroleum Limited

Solgas Petroleum Limited, incorporated in September 1990 to explore and exploit oil and gas in the upstream sector, benefited from discretionary allocations. The company was awarded OPL 226 by the federal government in February 1991. The OPL is located in 40–80 m of water about 40 km offshore Delta State. The Block covers an area of 1,553 km^2. Solgas has partnered with a Canadian company, NEXEN, which acquired 40 per cent equity interest in the concession. The company is currently undergoing reorganisation which may affect the equity structure.

Atlas Petroleum International

Atlas Petroleum International was incorporated in 1991 and granted a discretionary offer of OPL 75 in the same year. It executed aggressive exploration and production programmes on the Block. In 1992 Atlas Petroleum partnered with Summit Management Company, Texas to develop the Block. The agreement provided for 30 per cent equity interest for the Summit Management Company in OPL 75.

Allied Energy Resources

Allied Energy Resource was awarded OPL 210 located in offshore Niger Delta in June 1992. In 1993 the company co-opted Statoil Nigeria Limited, BP Exploration Nigeria Limited and Camac International Nigeria Limited. Statoil and BP were each assigned 20 per cent equity interest. Cavendish and Camac International were assigned 57.5 and 2.5 per cent interest respectively. It successfully drilled its first Deep Offshore well in July 1995. Allied Energy has 40 per cent equity in OML 110 where it also acts as technical adviser.

AMNI International

AMNI International was registered as an exploration and production company in June 1993. It currently operates OPLs 469 and 237 which were acquired in 1993 and 1994 respectively. In order to develop the field successfully, AMNI partnered with Liberty Technical Services of Canada. It commenced production in 1996 and output has increased to 20,000 b/d.

References

1 Kupolokun, F. 'The Nigerian Oil and Gas Industry: Examining Current Opportunities and Ensuring Sustainability'. *NNPC Seminar Presentation*. 2004.
2 Ayoola, E. 'Enhancing Investments Through Upstream JVs, Special Purpose Vehicles (SPVs) and others', NNPC E&P, 2004.
3 Melamid, A. 'Geography of the Nigerian Petroleum Industry'. *Economic Geography*, Vol. 44, Jan. 1968, p. 43.
4 'Nigeria: Great Expectations Abundant Opportunities'. Published by Industry Wide E & P Committee for International Conference, 2000, p. 47.
5 'Shell Companies in Nigeria', 9th Offshore West Africa Conference and Exhibition, 2005, p. 32.
6 'Partnering for Sustainable Growth of Nigeria's Oil and Gas Industry'. NNPC Industry Profile for the 18th World Petroleum Congress, South Africa' 2005.
7 9th Offshore West Africa Conference and Exhibition, 2005, p. 33.
8 Ibid. 'Operations in Africa'. Exxon Mobil, www.exxonmobilafrica.com
9 9th Offshore West Africa Conference and Exhibition, 2005, p. 35.
10 ChevronTexaco – Nigeria Fact sheet, www.chevrontexaco.com
11 Chevron and Sasol Launch Global JV., www.Chevrontexaco.com
12 9th Offshore West Africa Conference and Exhibition, 2005.
13 'Partnering for Sustainable Growth of Nigeria's Oil and Gas Industry', op. cit., pp. 82–83.
14 Ibid., pp. 84–85.

6 Marginal field development

Introduction

The Ministry of Petroleum Resources is a statutory agency charged with the responsibility of coordinating all oil and gas related policies, and administering the relevant laws and regulations. In addition the Ministry, through the office of the Minister of Petroleum Resources, supervises the executive organs of the public sector of the petroleum industry and the investments in the industry. It also collaborates with other relevant government ministries to advise the government on issues related to hydrocarbon resources and energy. The executive organs of the public sector are:

- Nigerian National Petroleum Corporation;
- Petroleum Technology Development Fund (PTDF);
- Petroleum Equalisation Fund; and
- Petroleum Training Institute (Effurun).

In the context of the Petroleum Ministry, technical policy issues, royalties and fiscal provisions, regulatory control of the industry and licensing of operations under the Petroleum Act are handled by the DPR. The functions of the department may be summarised as follows:[1]

- advise on policy matters affecting the management of petroleum resources;
- initiating petroleum policies;
- ensuring compliance with petroleum laws and regulations;
- regulating and monitoring the activities of companies operating within the oil and gas industry.

The existing government policies permit private or public interests to participate in the exploration and development of petroleum resources. Interested parties can be indigenous or foreign. Beyond the first discovery of crude oil at Oloibiri in Bayelsa State in 1956, many more discoveries have been reported. Some of the discoveries have remained unexploited for decades thereby jeopardising federal government objectives of increasing oil reserves and generating revenues. The statutory provisions regulating Oil Mining Leases (OML) were, to say the least, narrow in scope. The regulatory procedures, to all intents and purposes, placed more emphasis on active production fields within a concession area. This approach created a situation which led to the neglect of the work programmes associated with exploratory wells within the same concession. These have remained thorny issues, which warranted the amendments of the Petroleum Act 1969.

Petroleum (Amendment) Decree No. 23, 1996

As pointed out in the preceding section, the Petroleum (Amendment) Decree No. 23, 1996 was designed to remove impediments in the paths of indigenous entrepreneurs who seek to participate in the upstream sector for the exploration and exploitation of petroleum resources. The decree sought to release MFs from the control of multinational companies and a few indigenous producers for acquisition by other interested parties. The decree made the following provisions:

- The holder of an Oil Mining Lease (OML) may, with the consent of and on terms and conditions as may be provided by the Head of State, the Commander-in-Chief of the Armed Forces, farm-out any Marginal Field which lies within the leased area;
- The Head of States, the Commander-in-chief of the Armed Forces may cause the farm-out of a Marginal Field if it has been left undeveloped for a period of not less than 10 years from the date of first discovery of the Marginal Field.[2]

Objectives of Decree No. 23

The primary objectives of the above decree are summarised as follows:

- expand the scope of participation in the Nigerian oil and gas industry;
- diversify investments and guarantee generation of sustainable revenues;
- encourage indigenous participation in the oil industry thereby promoting technology transfer;
- expand production output capacity;
- grow the oil and gas reserve base through aggressive exploration and production;
- provide opportunity for portfolio rationalisation;
- create synergy among companies through common use of assets and facilities to ensure optimum utilisation of excess capacity;
- provide opportunity to engage the pool of highly competent Nigerian engineers and technicians in the oil and gas industry;
- create employment opportunities for young professionals and youths.

Understanding marginal fields

For practical purposes a marginal field may be referred to as a field that has proven reserves booked and reported annually to the DPR. Such fields have often remained unproduced for a period of not less than ten years from the date of first discovery of the field. MFs in the Nigeria petroleum industry context have the following characteristics:

- are associated with marginal economics;
- comprise one or more wells which have not been developed by the operating companies due to ranking of the portfolio of hydrocarbon fields;
- are marked by the original operator for farm-out due to portfolio rationalisation;
- have exploratory wells drilled on the structure and reported as oil and gas discovery but have been unproduced for a period of ten years;
- have high gas and low oil;
- have crude oil characteristics different from existing crude stream, which cannot be produced through conventional methods or current technology.

It is important to indicate that farm-out of unproduced, unappraised, abandoned or producing fields on an existing OML to independent operators and indigenous companies, is regulated by legislation. The relevant laws in the Nigerian petroleum industry are as follows:[3]

- Oil Pipe Acts 1956 as amended in 1965;
- Petroleum Profits Tax Act, 1959 with amendments in 1967, 1970, 1973 and 1979;
- Mineral Oil Safety Regulation 1963;
- Petroleum (Drilling and Production) Regulations, 1969 with amendments in 1973, 1979, 1995 and 1996;
- Petroleum Act 1969;
- Nigerian National Petroleum Corporation Decree 1977;
- Associated Gas Re-injection Decree 1979, as amended in 1985;
- Petroleum (Amendment) Decree 1996;
- Deep Offshore and Inland Basin Production Sharing Contract Decree 1999.

Technical and economic considerations

MFs generally refer to development fields which (in economic terms) are borderline cases. Such fields are traditionally small in size. They can also benefit from less conventional hardware and are amenable to application of discretionary fiscal incentives to attract independent producers. MFs are not considered sufficiently cost effective; therefore they rank low in the portfolio rationalisation of companies and may be considered unviable for a number of reasons including:

Technical considerations

- reservoir parameters;
- water depth;
- distance from existing infrastructure;
- high development cost;
- field size.

Economic considerations

- fiscal regime;
- oil and gas price (production cost);
- cost of finance;
- internal Rate of Return (IRR);
- net Present Value (NPV) of project;
- pay-back period;
- rate of inflation.

MNCs based on technical and economic reasons raised strong arguments in support of their desire to retain the fields. On the other hand, the federal government and its agents (DPR, Ministry of Petroleum Resources and NNPC) view the problem from a different perspective. Each viewpoint raises an array of issues, which require careful analysis. Subsequent sections will dwell on the fundamental issues in the general domain of MF development. The agitations of the multinationals and indigenous entrepreneurs will be examined in tandem with the government position on the matter.

Concerns of MNCs

The MNCs are of the view that development of MFs at this time is undesirable. This argument is premised on a number of reasons, which include the following.

Efficiency in field development

Economics

Oil and gas field development is traditionally guided by a ranking system such that only those which meet the minimum commercial viability criteria are considered for development. MFs fail to satisfy the basic profitability criteria.

OPEC quota

MNCs also contend that under the prevailing OPEC quota system development of MFs would amount to developing high cost fields at the expense of low cost fields. This approach (i.e. marginal field development under the prevailing circumstances) in their view negates prudent economic analysis and choice in project execution.

Funding

Cash-call has been a recurrent issue in the upstream sector of the oil and gas industry. MNCs are concerned that cash-calls (financial contributions from the government to fund JV operations) emanating from the federal government are often delayed. It is contended therefore that delayed response of the government to programme funding requirements accounts for the non-development of MFs.

Fiscal issues

The government has mapped out incentive packages for indigenous companies for MF development. Such incentives are not extended to MNCs. They are therefore of the view that some MFs could be developed if similar incentive packages are extended to them.

Legal issues

Legally the MNC is the concessionaire of the entire block in spite of the fact that a third party (in this case the indigenous entrepreneurs) is the operator of the field(s) under the farm-out arrangement. Although all obligations of the concessionaire under existing laws remain with the MNCs, it lacks the authority to monitor the activities of the operator stringently.

Operational issues

The technical proficiency of minor players is a point of concern to MNCs. The operations of an incompetent operator in a licence held by a MNC is capable of creating problems on some aspects of the operations especially in the areas of environmental protection and remediation, community involvement and project abandonment. These are serious issues of concern which call for an indemnity from the small operator. However, the ability of the small players to indemnify the MNCs and the scope of the indemnity is often not clearly stated. MNCs are of the view therefore that the non-development of MFs is necessitated by extraneous factors as outlined in the preceding sections. In this regard they contend that the categorisation of fields as marginal and allocating them to indigenous companies without due reference to their concerns and agitations is not justifiable. Furthermore they assert that the practice amounts to replacing the developments of efficient fields with non-efficient ones. The net effect of these actions is not in the overall interest of either the government or MNCs.

Need for MF development

In the preceding sections MNCs' contentions in favour of their retention of MFs were closely examined. This notwithstanding, the government feels strongly about the assignment of the MFs to indigenous operators. The underlying reasons for the decision are stated as follows:

Technology transfer

Active participation of Nigerian entrepreneurs in the upstream sector of the industry started in 1990. These incursions have yielded technical and financial success. Execution of oil and gas projects has provided more opportunities for the employment of Nigerians who, through direct participation, gain technical experience. The execution processes provide opportunities for technology assimilation which catalyses the technology transfer process. The initial success of indigenous operators stimulated the whole industry to provide more opportunities for local entrepreneurs.

The government is of the view that over a period of 40 years the exploration and production business has nurtured a pool of highly competent engineers and technicians. These individuals can operate as consultants, managing elaborate engineering and other related projects. In this regard the government is under enormous pressure from local entrepreneurs who demand opportunities for direct participation in the sector. Under these circumstances the farm-out of MFs without doubt provides a suitable political solution and an opportunity for the actualisation of the Nigerian Content Development objectives.

Sole risk operations

In the past MNCs have complained about poor government response for cash-calls that have been attributed to delays in the pace of work programme execution. Under the current policy arrangement MFs will be developed on a sole risk basis. Third party operators will not have recourse to the government thereby eliminating the perennial problems of under-funding and cash-call arrears.

Oil producing communities

In recent times, especially under the democratic dispensation, agitations for resource control by oil producing communities continue to gain momentum.

The States in the Niger Delta region, and indeed the oil producing communities, claim that they have been excluded from the oil exploration and exploitation processes. This in their view is an act of marginalisation. In this context, however, MF development provides opportunities for the inclusion of oil producing communities and other interested Nigerians in the ownership structure of the scheme. The foregoing position of the government shows clear justification for the development of the fields. The Nigerian oil and gas industry is sufficiently robust and capable of producing more business opportunities for MNCs. Consequently, relinquishing the MFs could be considered an appropriate step needed to promote the NCD policy. So far, the areas of disagreement between government and the MNCs have been harmonised and on the basis of mutual agreement 24 MFs were awarded to Nigerian companies in 2003. This singular action is expected to yield benefits which among others include:

- opportunities to gainfully engage technically competent Nigerians in oil and gas operations;
- opportunities for technology assimilation and transfer;
- increased employment opportunities;
- provision of a springboard for indigenous company development;
- increased production capacity;
- creation of appropriate multiplier effect on the economy.

Excluding MFs from existing JVs

Conventionally, JV agreements provide that any of the parties to a JV agreement is at liberty to transfer all or part of its individual interest provided other parties are duly informed about the beneficiaries of the transaction. This provision is essential for purposes of smooth consummation of the JV farm-out negotiations. Existing JV partners were unwilling initially to cooperate with the government in its efforts to farm-out specific MFs. The concern of the government derived from the fact that existing discoveries remain undeveloped, not as a result of marginality, but due mainly to existence of better investment opportunities in other fields. For purposes of equity and advancement of national interest the government adopted some strategies as follows:

Sole risk operation under JOA

Olisa (1997) opined that the petroleum JOA states in clear terms the legal relationship between concurrent owners of licences, leases or concessions. One of the concurrent owners is often designated as an operator who is entitled to conduct, manage and control all JV operations while adhering to the basic provisions of the JOA. It also outlines the fundamental rules and procedures governing the joint development of the licences, leases or concessions pursuant to the overall interests of the concurrent owners. In any JV agreement, disagreements over drilling of wells or project development occur occasionally. Such disagreements often result from differing technical views within the group as it relates to the interpretation of data of the field. Political considerations may also compel a company to support a well or field development programme. In view of the disagreements which may arise in making an investment decision on a particular field, the JOA embodies provisions which permit some members of the group to proceed with certain types of work without the dissenters. It is important to indicate at this juncture that in the case of the Nigerian oil and gas industry NNPC has the right to file a notice of sole risk development. However, some technical dimensions associated with the process require clarification.[4]

Non-consenting parties

In practice non-consenting parties (i.e. parties choosing not to participate) have the right to join the project at a later date provided an agreed premium has been paid. It is, however, important to point out that such a late entry premium is deliberately high in order to discourage non-consenting parties from participating at a future date.

Tariff allocation

A corollary to the issue of late entry of non-consenting parties is the allocation of an agreed tariff (usually in $/barrel or per cent of net production) to the farmee (i.e. the small operator) for the services it is providing as a compromise position.

Part surrender of licensed area

The terms and conditions of licensing in the Nigerian oil and gas industry require the licencees to surrender a certain percentage of the licensed Block. The enforcement of the provision (or clause) could compel MNCs to relinquish a reasonable number of MFs, thereby making it possible for the fields to be farmed out to interested parties.

Enabling Act

After over 40 years of oil and gas exploitation and production, indigenous participation in the sector has been rather low. The Petroleum (Amendment) Decree No. 23, 1996 was therefore designed to pave the way for indigenous entrepreneurs to participate in the production of MF. For about a decade after the amendments, the law remained ineffective largely because the DPR on the one hand and NNPC on the other, lacked any fundamental machinery to enforce the law. The major producing companies, for purposes of self-interest, grossly misinterpreted the motives of the law. These obstacles apart, the federal government in 2002 successfully rolled out the MF development programme.

Farm-out arrangement

The decisions of the MNCs in oil field development are governed by empirical data delineating the economic potentials of each project. In a typical project portfolio ranking process, an MF stands a remote chance of choice for development at the point of discovery. In this regard, farm-out presents a logical means of encouraging the development of existing MFs. In oil industry practice, consideration would entail carrying out a specific work obligation (known as earning obligation) which would involve drilling of a number of wells or development of an entire field. This approach calls for a mutual agreement on a spending ceiling between the farmor and the farmee following which the farmee executes the programme on behalf of the parties. Farm-outs provide the required flexibility to improve or refine a farmor's acreage portfolio without recourse to relinquishment and new licence awards. The Emerald field in the UK North Sea is a typical example of a field development on the basis of farm-out arrangement and agreed tariff structure. Farm-outs are predominantly motivated by a desire on the part of the farmor to:

- prudently manage a company's acreage portfolio;
- finance heavy expenditures; and
- benefit from different options of risk and rewards.

Farm-in agreement

In practice various forms of farm-in agreements are adopted for development, as outlined below.

Acquisition of interest

This is the conventional type of farm-in in which the farmee executes a mutually agreed work programme in return for a percentage interest in the field. In a typical MF development programme the farmee would be responsible for all dimensions of development including drilling and financing.

Overriding royalty payment to farmor

In order for this condition to apply the farmor would assign its interest ab initio to the farmee thereby paving the way for the development and financing of the field. In exchange, the farmor is entitled to earn an overriding royalty interest. Once the field has been developed and the farmee has earned a mutually agreed Rate of Return (ROR) on the investment, the farmor can invoke its rights and lay claim to a proportion of its initial interest. Typically if company 'X' were to have 100 per cent interest in a MF, it might decide to farm-out the development of the field to a third party. The third party would carry out development and financing of the field. Furthermore, it would operate the field and pay all relevant costs of operation, fiscal obligations and in return earn profits from the project. However, the farmee is obligated, under this arrangement, to pay company 'X' in the early years a small percentage of the gross revenue, known as overriding royalty.

Tariff payment

Under this option, the farmee receives a $/barrel tariff as compensation for the funding and development of the field. Payment for the farmee can also be calculated as a percentage of the net production.

Parameters of equitable consideration

For purpose of harmony it is essential for each party to a farm-out arrangement to determine the worth of the farm-out interest on offer, in order to establish equitable consideration. This provides the platform for determining equitable interest, overriding royalty or tariff payment for carrying out a specific work obligation (the earning obligation). It is important to indicate, however, that the oil and gas market is not amenable to swift and easy analysis. Comparisons hinged on a dollar per barrel basis can indeed be misleading and perilous. It is important that both farmor and farmee carry out an accurate assessment of the worth of their interests. For these reasons, subsequent sections will delineate 'parameters' to which highest values should be attributed by each party in the farm-out negotiation. Some of the parameters are as follows:

Financial worth

In evaluating the financial worth of an arrangement, factors to be considered include Net Present Value (NPV), tax synergy, corporate offsets and exploration effects. Cash value refers to the cash flow emanating from the economic evaluation of interest involved. Such cash flow is based on reserve estimates, production etc.

Historical fiscal benefits

The farm-out arrangement has a pragmatic dimension which allows the farmee to allocate a portion of outstanding fiscal benefit related to a particular interest. The interest earned in each farm-out arrangement is directly related to the value each party attributes to the relief. Capital allowance and investment tax credit are common fiscal relief available to a farmee.

Tax on work programme costs

These are tax reliefs derived or earned as a result of expenditure incurred by the farmee in executing a work programme. The value or quantum of the tax relief can be mutually determined.

Tax synergy

Farm-out arrangements also allow new entrants to enjoy tax relief in respect of unsuccessful exploration cost incurred on other blocks. Such costs are for practical purposes offset against PPT in an assigned interest.

Farmor expected returns

In the oil and gas industry, acquisition of an interest in a predictable field is considered a low risk investment; therefore such investment can only attract a Return on Investment (ROI) of 15 per cent. On the other hand, on a farm-out arrangement in which the farmee bears the risk of development on behalf of both parties an ROI of 21 per cent would be expected.

Marginal field allocation

For many years indigenous companies in the upstream sector craved the government for the allocation of MFs and finally the government took concrete steps to fulfil the objective. It is important to indicate, however, that the issue of MFs was sensitive and required comprehensive dialogue with the MNCs. As at December 1999, these discussions had identified a total of 116 fields located in the Niger Delta which had the characteristics of MFs. Out of these fields, which had reserves ranging from 5 million barrels to 50 million barrels of oil or more, 24 MFs were put on offer for competitive bids on 31 August 2001, and 150 companies expressed interest in the MFs programme through purchase of the application forms. Out of these, 142 companies submitted duly filled forms with pre-qualification bid packages.

Dr. R. Lukman, the Presidential Adviser on Petroleum and Energy, in his press briefing confirmed that the 142 companies submitted 408 pre-qualification bid packages for the 24 available MFs. The preceding analysis indicated that the exercise generated a high level of enthusiasm among indigenous companies. The bids were screened and evaluated by a government committee. On 24 February 2003 the federal government awarded 24 MFs to 31 indigenous companies (Table 6.1). Some of the fields put on offer were discovered in the early 1960s but remained unproduced or abandoned for over three decades. In this regard the decision of the government to transfer the fields to small indigenous players is not only justified but opened up an avenue of revenue generation for the Nigerian economy. All beneficiaries of

Table 6.1 Marginal field allocation – 2003

S/N	*Field*	*JV Owner*	*Operator*	*STOIIP*	*Booked reserves*	*Year discovered*	*Location*
1	Assaramatoru	SPDC	Prime Energy Ltd. (51%)	30.8 mm stb	4.7 mm stb	1973	Rivers
2	Ibigwe	SPDC	Walter Smith Petroleum (70%)	84.4 mm stb	17.2 mm stb	1965	Imo
3	Tom Shot Bank	SPDC	Associated Oil and Gas Services (51%)	53.1 mm stb	8.6 mm stb	1980	Akwa Ibom
4	Tsekelewu	SPDC	Sahara Energy (51%)	15.3 mm stb	2.2 mm stb	1979	Ondo
5	Atala	SPDC	Bayelsa Oil Co. Ltd.	23.1 mm stb	24 mm stb	1982	Bayelsa
6	Asuokpu	SPDC	–	10.4 mm stb	2.7 mm stb	1989	Delta
7	Uquo	SPDC	Frontier Oil Ltd.	94.9 mm stb	14.2 mm stb	1971	Akwa Ibom
8	Okwok	SPDC	–	15 mm stb	–	1968	Akwa Ibom
9	Oza	SPDC	Millennium Oil and Gas	26.3 mm stb	7.3 mm stb	1959	Rivers
10	Ofa	SPDC	Independent Energy Nig. Ltd.	56.9 mm stb	5.2 mm stb	1970	Delta
11	Eremo	SPDC	Excel Exploration and Prod. Services	15.7 mm stb	3.9 mm stb	1978	Bayelsa
12	Asuokqu/Umutu	SPDC	Platform Petroleum	36.9 mm stb	13.3 mm stb	1980	Delta
13	Stubb Creek	SPDC	Universal Energy	69.3 mm stb	18.4 mm stb	1971	Akwa Ibom
14	Qua Iboe	SPDC	Network E&P	54.8 mm stb	5.2 mm stb	1960	Akwa Ibom
15	Ajapa	Chev/Tex.	Britania–U Nigeria	4.64 mm Bo	13.0 bill. CF gas	1987	Delta
16	Dawes Island	Chev/Tex.	Eurafiic Ltd.	5.26 mm bbls	1.39 mm bbls	1979	Rivers
17	Akepo	Chev/Tex	Sogenal Ltd.	13.04 mm bbls	3.86 mm bbls	1993	Delta
18	Ekeh	Chev/Tex.	Movido E&P	12.0 mm stbo	4 mm stbo	1986	Bayelsa

Table 6.1 Continued

S/N	*Field*	*JV Owner*	*Operator*	*STOIIP*	*Booked reserves*	*Year discovered*	*Location*
19	Ke	Chev/Tex.	Del.sigma	–	–	1965	Rivers
20	Ogedeh	Chev/Tex.	Bicta Energy	25.66 mm bbls	8.92 mm bbls	1993	Delta
21	Oriri	Chev/Tex.	Goland Petroleum	12 mm stbo	3 mm stbo	1991	Bayelsa
22	Ororo	Chev/Tex.	Guarantee Petr. Ltd. (55%)	25.74 mm bbls	5.6 mm bbls	1986	Ondo
23	Umusadege	TotalFinaElf	Midwestern Oil and Gas (70%)	50 mm bbls	10 mm bbls	1974	Delta
24	Umusati/Igbuku	TotalFinalElf	Pillar Oil Ltd.	25.8 mm bbls	7.7 mm bbls	1981	Delta
25	Obodogwa/ Obodeti	TotalFinalElf	Energy Oil & Unipetro (50%)	14.6 mm bbls	4.7 mm bbls	1980	Delta
26	Amoji/Igbolo	TotalFinaElf	Chorus Energy Ltd.	16.7 mm bbls	4.6 mm bbls	1971	Delta

Source: *Partnering for Sustainable Growth of Nigerian Oil and Gas Industry*. NNPC publication for World Petroleum Congress, Johannesburg, September 2005.

the MF allocation have signed farm-out agreements with the major operators.

Challenges of MF development

Allocation of MFs to indigenous entrepreneurs is a welcome development. However, those awarded MFs currently face serious financial challenges. The development of an average field requires capital outlay of $30 million. Such a financial package is beyond the reach of small-scale operators in the upstream sector. This is because they lack the collateral to attract such loans. It is interesting to note that Nigerian banks have, over the years, not shown significant interest in the funding of oil and gas projects. Some of these banks, with a capital base of about $200 million, consider the gestation period of upstream projects (approximately 3 years) too long for financing. They are seen to be more favourably disposed to financing trading activities which involve short-term loans. In the case of the NLNG, which has become a success story, a consortium of Nigerian banks financed equivalent to about $160 million, while offshore banks in a given instance financed over $600 million. The dearth of investment funds has turned out to be the major obstacle in the implementation of the MF development programme.

Another hurdle which companies have to overcome is the issue of identifying willing technical partners. Most of the companies lack international exposure; therefore, convincing offshore companies to participate in MF development programmes in Nigeria is a difficult task. Whereas the majority of those who received a concession are facing financial difficulties, a few others have mobilised and commenced development activities in the field. The plight of the beneficiaries can rightly be attributed to the weak financial infrastructure of the Nigerian economy. The primary concern among keen observers is that with each passing moment the work programmes remain unimplemented and may at some point lapse. The allocation of MFs to small companies should bring with it some joy and a sense of fulfilment. Unfortunately, events following the award of the MFs clearly indicate that the beneficiaries have an uphill task ahead of them. The strong desire to achieve robust NCD amongst others is hinged on the successful execution of the MF development programmes. Successful implementation of this laudable programme calls for a comprehensive strategy, which should have as one of the ingredients a federal government guaranteed loan scheme. The risk element in MF programmes is considerably lower because each field has proven and booked crude oil reserves. This suggests that if the fields are successfully developed through special loan schemes, revenue accruing from the development of the MFs could easily be dedicated to amortise the loan. One may recall that the government in the past awarded some oil blocks to Nigerians on concessionary grounds. Some of the fields were developed; however, a majority remained undeveloped due to financial constraints and were subsequently taken over by the government. The inability of indigenous companies to participate in the mid-1980s caused a setback to government objectives of empowering Nigerian entrepreneurs. The MF development programmes should not be allowed to suffer similar setbacks.

Significance of MF development

Political expediency

The pressure for the award of MFs to indigenous entrepreneurs spanned one and a half decades. In order to create a pool of them the DPR entered into long negotiations with major producing companies, as for many years MFs were capped and did not feature as priority areas in the portfolios of the major companies. It was also observed that the upstream sector had total IOC domination. Indigenous companies found it increasingly difficult to penetrate the sector in view of the capital intensive nature of the industry, and delay in the production of MFs was seen by the government as a loss of revenue into the National Treasury. The limited reserves associated with the fields did not serve as significant incentives for IOCs to develop them. It became obvious that unless appropriate actions were taken, the fields would not be developed in the foreseeable future and it was on this premise that the federal government insisted that major oil producing companies in the Nigerian upstream sector must relinquish fields which remained unproduced after ten years. Through this process the government recovered 24 MFs for allocation to suitably qualified indigenous companies.

The decision of the government to award MFs had political and economic undertones. It was the desire of the government to create opportunities for the local companies to participate in the upstream sector and essentially MFs were to serve as preparatory grounds for more serious challenges in the sector. The federal government was perceived by indigenous entrepreneurs, including members of the Petroleum Technology Association of Nigeria (PETAN), as being insensitive to their aspirations. More importantly the point was made that the idea of leaving all upstream activities in the hands of foreign companies portends danger, especially as it relates to the achievement of the national aspirations of the country. For this reason the government forged ahead with the allocation of 24 MFs to 31 local companies to assuage the intense demands for indigenous participation. Some of the companies have started production in the fields.

Economic considerations

The 24 MFs awarded to local entrepreneurs had proven oil and gas reserves: the fields were estimated to hold 600 million barrels of crude oil. Some fields also hold significant volumes of natural gas. The primary concern of the government was the empowerment of local entrepreneurs who would engage in activities that could lead to technology acquisition and job creation. The award of the fields and subsequent commencement of production has achieved the fundamental objective of job creation among highly competent professionals. The new jobs created will increase the earning capacity of workers and will create spin-off in the local economy. One may assume that 35 per cent of the proven reserves are recoverable and in this regard it could be posited that at least 210 million barrels of oil would be recovered from the fields located in the Niger Delta. In July 2008 the price of oil for the very first time attained $147/b in the international market. It can be further contended that in the event that oil prices recede to the $50/b level, revenues accruing from the MFs in the Niger Delta will account for about $15 billion over a period of 10 years. This is a reasonable revenue stream for not only the local entrepreneurs but for the local economy. The increase in the earnings of local companies through participation in MF development will pave the way for future engagement in the broader and more dynamic offshore operations in the upstream sector. Through the award of MFs the government has put in motion a process that will generate sustainable economic and political solutions critical for the empowerment of local entrepreneurs and the rejuvenation of the economy.

References

1 'Guidelines for Farm-Out and Operation of MFs'. DPR, 2000.
2 Ibid. 'MFs Bid Guidelines Development Strategy Options Papers'.
3 'Laws of the Government of the Federal Republic of Nigeria. Petroleum Amendment Decree, 1996', paragraph 16A, subparagraphs 1 and 2.
4 Olisa, M.M. 'Nigerian Petroleum Law and Practice', 1997, p. 74.

Further reading

Jones, G.C.L, Jacobs, F.A. and Toh, C.M. (1984) 'The Economics of Marginal Offshore Oil Discoveries in China', *Oil and Gas*, 82(1).

Soetedja, V., Suyana, D. and Kontha, N.H. (1998) 'Case History of Marginal Oil Field Developmen', Asia Pacific Oil and Gas Conference and Exhibition, Perth, pp. 175-183.

'Industry Minnows Make Most of Marginal Oil and Gas Fields' (2007) www.goliath.ecnext.com

Vincent, J.R. (2001) 'Resource Depletion and Economic Sustainability in Malaysia', *Journal of Environment and Development Economics*. (Cambridge University Press (on line) www.journals.Cambridge.org).

Frynas, J.G. (2000) *Oil in Nigeria: Conflict and Litigation Between Oil Companies and Communities*, www.book.google.com

Nkwoh, B. (2004) 'International Market Research: Marginal Fields Hold 300 million barrels in Nigeria', www.strategis.ic.gc.net/marginal/fields

'Marginal and Stripper Well Revitalisation', www.fe.doe.gov/program/oilgas/marginalwells

Hannesson, Rognvaldar (1998) *Petroleum Economics: Issues and Strategies of Oil and Natural Gas Production*, Quorum Books, West Point CT.

7 Oil field service companies

Introduction

The Nigerian petroleum industry derived its roots from early exploration and production activities pioneered by the Nigerian German Butimen Company in 1908. Exploration and production entail various functions which require collaborative efforts of different companies. In practical terms oil and gas production requires the technical input of Oil Field Service Companies (OSCs) in the provision of an array of services in order to successfully develop and produce an oil field. In the early years of oil exploration and production the upstream sector may not have been distinctly defined. This is understandable because the industry was at the early stage of development and therefore unsophisticated. The tempo and scope of exploration and production activities have changed drastically thereby warranting advanced technological input from OSCs. In the case of Nigeria, one can posit that OSCs evolved at about the same time as the early exploration and production activities. The historical origins of these companies in Nigeria are not adequately captured. However, it can be assumed that some of the early OSCs in the Nigerian petroleum industry may have operated as units or affiliates of the pioneer oil companies. The rapid expansion of the industry and the escalation in the demand of petroleum products globally has warranted the introduction of high precision technology to produce oil and gas in a cost effective manner. In view of the various dimensions of oil and gas production, companies have segregated into areas of specialisation. The Nigerian oil and gas industry currently has multinational and indigenous OSCs providing different kinds of services to the major oil producing companies.

Although the oil and gas industry started around 1908, government participation started only in 1971. This implies that in 1970 (i.e. 10 years after independence) for instance, the industry was devoid of indigenous participation in both the upstream and downstream sectors. This situation had little to cheer about. To reverse this trend, therefore, the government promulgated the Enterprises Promotion Act 1977.[1] The primary objective of the Act was to create opportunities for indigenous participation in various commercial activities in the economy. Based on the provisions of the Act, the federal government, through NNPC, acquired 35 per cent equity interest in the OSCs. The reasons for the acquisition were to promote technology transfer, increase Nigerian involvement in development and promote indigenous participation in the oil field services subsector. In this regard OSCs were considered strategic for the realisation of national objectives. Some of the core business areas of these companies are as listed in Tables 7.1 and 7.2.

Multinational OSCs

There are several multinational OSCs operating in Nigeria. NNPC, up until 2006, held equity interest in 11 of these companies. They have existed in the upstream sector for upwards of five decades, operate alongside the major producing companies and have made huge profits over the years. This notwithstanding, none of the companies has made any deliberate effort to establish a befitting office complex or expand the assets base. Understandably, oil rigs are traditionally leased in view of the high cost and the dynamic nature of technological innovations. This is, however, not the case for other fixed assets and office equipment. Most multinational OSCs lease office space and other equipment used in the office. Over the years, the turnover of their operations in Nigeria has grown significantly yet their assets base remains stagnant. This practice does not demonstrate any form of commitment to the economic well-being of the country. OSCs in the upstream sector on a yearly basis declare low profit of about 2 per cent of the turnover which runs in excess of ₦40 billion (40 billion Nigerian Naira).[2] NNPC as a strategic shareholder has, on a number of occasions, drawn the attention of the companies to the low profits and dividend declared in their annual accounts especially in the context of their variance from industry wide profit levels in the oil services subsector. The OSCs argue, however, that the operations are financed solely by their parent companies that have to unilaterally bear the risk of securing loans from the financial institutions. The cost of funds, they argue, accounts for the low level of profit. Under these circumstances the companies argue that the low level of profit does not create opportunity for capital accumulation which is essential for the expansion of the assets base.

Technology transfer

One of the primary objectives of the government in the acquisition of shares in the OSCs was to create avenues for indigenes to have relevant exposure to the operations of the oil and gas industry. It was the thinking of the government that the acquisition of equity interest in the companies would provide the leverage for Nigerians to hold key managerial positions in the companies. More importantly, there was need for indigenes to acquire critical technical skills which would allow them to establish companies in the oil services subsector. Although these companies have operated for upwards of five decades, the level of transfer of technology is still at the rudimentary stage. Experts in the industry argue that technology transfer has not been achieved because the primary inputs in the sector are imported. OSCs currently operating in Nigeria do not have standard fabrication yards. This is to ensure that all fabrication contracts are handled by the home offices. The home offices in most cases procure the materials from sister companies which quote high and non-competitive prices. The practice turns out to be a profit depletion mechanism. It has also been observed that OSCs have made little effort to procure chemicals such as baryte and bentonite locally although a survey by the Ministry of Solid Minerals indicates that baryte and bentonite are available in large quantities at various locations across the country. Meanwhile the companies imported raw baryte and bentonite into Nigeria for processing. The government considers this a waste of scarce foreign exchange and placed an embargo on the importation of these chemicals. Rather than put in place facilities that will promote local sourcing of baryte and bentonite the companies resorted to lobby government officials to reverse the policy. An informal survey carried out indicated that at the end of 2006 no MNC had assembled a scheme to source such raw materials locally.

Indigenous OSCs

Emergence of indigenous OSCs in the upstream sector started in the mid-1970s. Prior to 1970 Nigerians were engaged in less technical areas which offered little opportunity for technology assimilation and transfer; most of the key operatives of the industry in the 1970s were expatriates. The introduction of the Enterprises Promotion Act in 1977 created avenues for active participation of Nigerians in the oil services industry. As pointed out earlier the acquisition of equity interest in the 11 oil services paved the way for Nigerian engineers and senior technicians to secure employment in the companies. In the 1980s the first set of Nigerian engineers employed in the sector rose to functional management positions and got exposed to the technology and techniques of providing services in the upstream sector. It is significant to note that between 1989 and 1995 Nigerian engineers who had worked in some of the MNCs floated their own companies in the oil field services subsector. The trend continued until 1997 when more indigenous OSCs came into operation. These companies have gained experience and as a result executed challenging contracts in the upstream sector.

Indigenous companies in the oil and gas industry have grown in number. The growth of the local companies is in line with federal government objectives of domesticating the industry in the foreseeable future. Indigenous companies have demonstrated a high level of competence by executing key projects for SNEPCO, Mobil, Addax, ChevronTexaco and other major oil companies in Nigeria. Indigenous OSCs and multinational OSCs provide services which cover a broad spectrum. These companies have in recent years consolidated their relationships and organised shared ideas to explore avenues for creating a formidable front for aggressive participation in the industry. Interested companies have constituted themselves into an association (Petroleum Technology Association of Nigeria) – see details of OSCs in Tables 7.1 and 7.2. In 2001, the federal government allocated MFs to some indigenous companies. It is important to indicate that beneficiaries are companies owned by a group

Table 7.1 Oil field service companies

Name of service company	*Category of service*
BJ Service Company Nigeria Limited	Cementing
Baker Nigeria Limited	Well test Drilling mud
Baroid of Nigeria Limited	Drilling mud
Camco Limited, Nigeria	Drill bits Completions
Cooper Cameron Corporation	Wellhead supply Maintenance
Deutag Drilling Nigeria Limited	Drilling
Ecodrill Nigeria Limited	Well test
Global Marine Nigeria Limited	Drilling rigs
Halliburton Energy Services (Nig.) Limited	Mud supply cementing
Homan Engineering Company Limited	Rig site construction
Intel Service Limited	Integrated logistic service
Mallard Bay Drilling Nigeria Limited	Drilling rigs
NRB Drilling Service Limited	Drilling rigs
Nexus Drilling Nigeria Limited	Down hole dehydration technology, secondary recovery technology, production optimisation techniques
Noble Drilling Nigeria Limited	Drilling rigs
Optimized Technology Limited	Oil field software development Internet services
Remm Oil Services Limited	Coring Machine shop service Drill bits
Schlumberger Oilfield Services	Reservoir evaluation, wire line services, cementing and stimulation, well completion and productivity
Global Petroleum Resources Limited	Engineering project management, global technical manpower support services, quality management services, oil field procurement and drilling, management services

Source: *Partnering for Sustainable Growth of Nigerian Oil and Gas Industry*. NNPC publication for World Petroleum Congress, Johannesburg, September 2005.

Table 7.2 Member companies of the Petroleum Technology Association of Nigeria (PETAN)

Name of member	*Category of service*
Lonestar Nigeria Limited	Drilling and workover, Drilling and petroleum engineering
National Marine Authority	Shipping and marine
Nestoil Limited	Corrosion, engineering, environmental procurement
Oildata Wireline Service Limited	Cased hole electric-line logging Petrophysical and reservoir data service Tubing Conveyed Perforating (TCP) Production logging Services
Oilscan Limited	Cased hole electric-line logging TCP
Oiltest Services Limited	Laboratory services Mechanical wireline/completion Production logging services PVT services Surface and bottom hole sampling Well analysis – well production testing
Pan-African Int. Limited	Subseas wellhead systems Floating production Systems – FPSO Engineering and design Pipe recovery, pipe coating Mooring systems
Reservoir Fluids Lab. Inc. (Sart Nigeria Limited)	Laboratory service PVT services
Sowsco Well Services Limited	Cementing and high pressure pumping Wellhead maintenance service
Supergas Limited	Oilfield pipeline engineering Civil engineering Procurement for the petrochemical industry
Weafri Well Service Company limited	Cementing and high pressure pumping
Weltek Limited	Manufacture of wellhead/flow Station/compressor fail-safe control panels Production maintenance service
Zumax Nigeria Limited	Mechanical wireline using lift barges Mechanical wireline/completions Cementing and high pressure pumping
Ciscon Nigeria limited	Completion services Casing and tubular services Rental tools Machine shop services

Source: *Partnering for Sustainable Growth of Nigerian Oil and Gas Industry*. NNPC publication for World Petroleum Congress, Johannesburg, September 2005.

of individuals who developed their careers through major oil companies and OSCs in the upstream sector. Some of the beneficiaries have partnered with foreign companies to produce the MFs.

Challenges of indigenous OSCs

Indigenous companies face serious challenges in the upstream sector. Some of them lack experience and as a result find it difficult to engage suitable foreign partners for project execution. In addition, they experience funding constraints which limit their capacity to bid for medium to large-scale projects in the industry. The oil producing companies capitalised on the inadequate technical skills of the indigenous companies as well as their funding constraints and it became a common practice for oil producing companies to award contracts to locally registered multinational OSCs at the expense of the indigenous companies. With the growth of the upstream sector, the expenditure level increased to about $8 billion. This level of expenditure became attractive to foreign based OSCs and other equipment and service vendors. Most of them have no intention of establishing branches in Nigeria but were simply interested in winning lucrative contracts for equipment supply or execution of major contracts. To achieve this object foreign companies recruited indigenous companies to serve as agents or representatives. These companies were used as fronts in quoting for jobs in Nigeria and were given token amounts as commission in the event of a successful bid. The problems of indigenous companies were further compounded by both the low capital base and risk aversity of Nigerian banks. Prior to 2005, most Nigerian banks operated with a capital base of less than $100 million. This being the case, local banks found it difficult to finance upstream projects which individually quite easily go beyond $20 million. In the light of the low capital base, banks perceive upstream projects which have execution periods ranging from two to three years as having long gestation periods. In the alternative, banks prefer financing commercial activities which have short pay-back periods.

Financing oil and gas projects

Consistent with the experiences of most developing countries, Nigeria is constantly in need of investment capital. The earnings from oil over the past five decades notwithstanding, investment capital is in seriously short supply. This in part accounts for the deceleration of oil and gas infrastructure development and indigenous participation in the capital intensive upstream sector. It is important to note at this point that the NCD policy is designed to grow incrementally indigenous participation in the upstream sector through the execution of less technical jobs (e.g. conceptual designs, front end engineering, fabrication etc.) by indigenous contractors on a competitive basis. The envisaged growth requires reliable sources of funding. It also calls for relaxation of collateral requirements for loans and a paradigm shift on the part of banks from short-term funding of commercial activities to high revenue yielding activities in the oil and gas sector. There is need for banking institutions to recognise the expediency of the local content development policy and support oil and gas investments through robust syndicated loans. This will provide a suitable avenue for Nigerian banks to capture a significant proportion of the investment opportunities in the upstream sector.

Project finance options

Funding of oil and gas investments is often achieved through a variety of financing options. These include bank loans, stocks, bonds, promissory notes, JV contracts, PSCs, SCs and Foreign Direct Investment (FDI).

Collateralised loans option

The Nigerian banking sector has registered a high growth rate in the past two decades. This success apart, the capitalisation of the banking industry is relatively low in relation to over $8 billion funding requirements of the oil and gas sector. The federal government funds 60 per cent of this amount through cash calls while the balance of 40 per cent is funded by the JV companies. Oil and gas investments often have long maturity periods spanning about three years or more. Most banks hesitate to finance such projects in view of the so-called long gestation period and the perceived risk elements associated with the projects. Only recently, the federal government awarded 24 MFs to Nigerian entrepreneurs. These fields will require funding, and the participation of local banks in the granting of loans to interested companies will usher in a new set of investment activities in the industry. It is important to caution, however, that participation of local banks should be without prejudice to the well-known constraints of collateral on the part of indigenous OSCs who are the main sponsors of the MF projects. Total expenditure on the NLNG project to date has exceeded $7.5 billion. So far six Nigerian banks have syndicated $160 million which accounts for 2.1 per cent of the total expenditure on the NLNG project. This project and others which will come on stream in the future provide valuable opportunities which have not been exploited by the local banks. There is therefore need for Nigerian financial institutions to support oil and gas projects in order to stem the huge capital flight associated with the offshore financing of these investments.

Stocks option

The capital market in developed economies is a major source of funds for oil and gas projects. Although the current capitalisation of the capital market is estimated to be approximately ₦12 trillion (about $70 billion) the Nigerian Stock Exchange (NSE) is yet to float the stocks of upstream sector companies. This makes it difficult for companies to raise capital locally for oil and gas projects. At present, indigenous contractors rely heavily on internal cash flow and overdraft facilities to fund oil and gas investments. Considering the high interest rates (approximately 21 per cent) associated with overdrafts, this funding option dilutes the profitability index of oil and gas investments. The limitations of the capital market in this regard inhibit the execution of oil and gas projects, which ultimately affects economic expansion and the overall industrialisation process. The absence of the upstream sector in the capital market is a critical problem and the expert attention of bankers is required in order to generate innovative solutions that will allow upstream sector companies to be quoted on the Nigerian stock market.

Bond issuance

The bond market in Nigeria is localised and certainly at the infant stage. Although the federal government and some State governments have floated bonds, the financing option attracts low patronage from the general public and the organised private sector. Investment capital derived from the process is low; therefore bond issuance may for some time to come not be classified as a viable funding option for oil and gas projects.

Nigerian Content Development

The issue of NCD in the past was treated with levity by the MNCs. This without doubt accounted for the 5–8 per cent NCD recorded in the petroleum industry in 2005. It is important to indicate at this point that in recent times the federal government approach to the issue has taken a positive direction. Taking into account the achievements of Brazil, Indonesia, Malaysia and Mexico, a new policy posture has been introduced. The above countries have achieved local content development exceeding 50 per cent. Nigeria intends to achieve 75 per cent local contents development in 2010. NNPC has for this reason created a division in the corporate headquarters to drive the NCD strategy. More specifically, the following jobs and others are to be executed locally to promote technology assimilation and value addition in the execution of jobs in the oil and industry:[3]

- front end engineering;
- fixed platform weighing 5,000 MT;
- top side of FPSO;
- third party service (NDT, Mechanical Tests and PWHT) and certification of welding;
- all low voltage cables;
- subsea systems, risers, flow lines etc.;
- 75 mm carbon-steel pressure valves;
- waste management;

- IT software;
- seismic data processing.

The designation of certain jobs to be executed locally will create opportunities for indigenous companies to participate in competitive bids. MNCs would be required to comply with the policy and establish appropriate machine shops, manufacturing outfits and fabrication yards to produce various inputs for use in the upstream sector. Commencement of such activities will create jobs and also impart skills which support the activities of indigenous companies. Essentially, the full implementation of the NCD policy will strengthen indigenous OSCs, which should lead to increased revenues and rejuvenation of the Nigerian economy.

References

1 Nigerian Enterprises Promotion Act 1977.
2 Ariweriokuma, S. 2001 'Impact of Multi-NOC Practices on Local Content Development'. NNPC COMD Project, 2001, pp. 25–31.
3 'Enhancement of Local Content in the Upstream Oil and Gas Industry in Nigeria'. INTSOK Study 2003, Sections 1–6.

8 Nigerian Content Development (NCD)

Introduction

Active participation of the federal government in the upstream sector started in 1971 through the acquisition of equity interest in major oil producing companies. The government, through its various actions and policy initiatives, demonstrated the need for indigenous entrepreneurs to participate in both the upstream and downstream sectors of the petroleum industry. This was a deliberate attempt to infuse Nigerian participation in the sector. Although NNPC participated in the sector through JVs, revenue derived was mainly from the disposal of equity crude oil, royalties and taxes. It is important to indicate, however, that total expenditure in the upstream sector is approximately $12 billion. The JV partners benefited directly from the expenditures in the sector because some major contracts for services and equipment for oil exploration and production were awarded to their affiliate companies. Through active participation the MNCs as well as OSCs were able to channel a significant proportion of the upstream expenditures to their parent companies overseas. Nigerian participation was extremely low and this was attributed to low experience, low level of technical skills and inadequate investment capital. Over the years, professional skills in core engineering, geology, reservoir engineering, geophysics, instrumentation etc. among Nigerian professionals have increased. This notwithstanding, MNCs continued to award contracts for basic procurements, Front End Engineering Design (FEED), conceptual design, fabrication etc. to companies offshore. This practice without doubt created an unfavourable business climate for Nigerian entrepreneurs. Consequently, an estimated $12 billion expenditure in the upstream sector failed to stimulate and impact on the local economy. In essence, the huge repatriation of earnings from the upstream sector stifled the employment objectives of the government.

Constraints of NCD

The Nigerian petroleum industry has recorded impressive growth since 1956 when oil was first discovered at Oloibiri. However, it is sad to note that the expansion of the industry has failed to achieve significant NCD and indigenous participation. Dr. R. Lukman in a keynote address at the national workshop on NCD indicated that over 90 per cent of an estimated $12 billion annual expenditure in the upstream sector 'escapes the domestic economy'.[1] He further stated that in Algeria, Malaysia, Indonesia, Brazil, Oman, Venezuela and Norway local content utilisation has reached advanced stages of development. At present, there are several indigenous companies striving to penetrate the core business areas of the upstream sector to supply materials or services or execute projects. Such efforts have not yielded reasonable success because of the propensity of MNCs to procure inputs from overseas vendors. This practice, among others, accounts for the low 5 per cent retention of the $12 billion annual expenditure in the Nigerian economy. Oladele,[2] however, posited that the low level of NCD derives from:

- deficient capitalisation arising from the tendency of Nigerian entrepreneurs to operate as 'one man' businesses;
- capital and structural deficiencies associated with poor training and low managerial ability; and
- inability to attract funds due to lack of suitable collateral and positive corporate image.

He further stated that some indigenous companies have been transformed into commission agents. In this instance, irrespective of the value of the procurement, the agent earns a small commission while the bulk of the revenue and the corresponding 'value addition' remain offshore. Oladele also pointed out that indigenous companies could strengthen their organisation through partnering and alliancing with international companies. Identifying such partners, he opined, should be facilitated by the oil producing companies and the government. Gaius-Obaseki in a keynote address at the Conference on Partnering and Alliancing in the Nigerian Oil and Gas Industry, strongly advocated the formation of strategic alliances between indigenous companies or with foreign companies in order to guarantee NCD.[3] This view was also well articulated by Ogiemwonyi in a similar paper on NCD at the Offshore Petroleum Conference in Houston.[4] However, Olurunfemi in his analysis attributed the problem to the inability of commercial banks to provide long-term loans for the execution of projects in the oil industry.[5] Ofuhrie contended that systemic and institutional problems are the primary obstacles to the actualisation of policy objectives. He indicated that systemic problems embody not only deficient infrastructural facilities but also lack of essential raw materials. Deficient infrastructural facilities such as erratic power supply and communication do not provide an enabling environment that permits adherence to project completion targets. Ofuhrie further posited that the lack of these basic facilities diminishes the competitive edge of the indigenous contractor. For instance, the absence of a local iron and steel industry inevitably compounds the problem because many inputs such as drilling tubular casings, wellheads, platforms etc. are derived from steel.[6] Omene undertook an analysis of the subject and advanced criteria for the determination of local content component in the MNCs' contract award processes.[7] It is important to note that both Ofuhrie and Omene in examining the problem expressed a common view that the government should introduce enabling policies to promote NCD.

Although local content development in Nigeria has been illusive, some OPEC member countries such as Kuwait, Saudi Arabia and Venezuela have made substantial progress in local content development in their respective petroleum industries.[8] Norway is a classic success story of local content development. Hagen[9] stated that Norway, starting in 1972, undertook a systematic approach to promote local content development. At present Norway has achieved local content utilisation ranging from 65 to 90 per cent in various categories of inputs. The programme was channelled through the industrial base comprising mining, maritime and the processing industries as springboards for the domestication of inputs in the petroleum industry. The Norwegian government, Ministry of Petroleum and Energy, Petroleum Directorate and other relevant research institutions also facilitated the process. Nigeria urgently needs to devise a means to stem the high rate of capital flight arising from the unfavourable procurement and contract award practices of the MNCs.

Realising the consequences of the practice, NNPC reiterated to the MNCs the need to incorporate Nigerian content in all segments of their operations. Over the past few years Nigerian content has been brought to prominence through workshops, seminars and other industry interactive fora organised by NNPC, the MNCs and the National Assembly. The workshops and seminars were primarily aimed at sensitising all stakeholders and to articulate all related issues with a view to developing robust and sustainable solutions.[10] Initially, the Nigerian content issue was widely misunderstood, especially among the MNCs. The policy was interpreted to imply that only indigenous companies can bid for and win contracts in the upstream sector. This has been clarified through workshop presentations and the provision of an appropriate definition of the term 'Nigerian content'. For the purpose of this discussion the concept would be defined as follows:

> Nigerian content is the aggregate of goods and services rendered, manufactured or fabricated which derive from the local environment significant proportions of raw materials, skills and technical know-how for the execution of projects for the major oil companies or OSCs operating in the Nigerian Oil and Gas Industry.[11]

NCD therefore promotes the utilisation of locally manufactured goods, raw materials, skills and services in the project execution processes of the oil and gas industry. The definition implies that any company established and operating in the upstream or downstream sector that mobilises locally manufactured goods, raw materials and skills to execute a job, satisfies the Nigerian content policy and can therefore be an active beneficiary of contracts in the petroleum industry.

The level of participation in the various workshops and the consensus expressed clearly confirmed the sensitivity of the subject. In consideration of this, a committee on NCD was set up in October 2001 with the primary objective of reviewing all documents on the issue. This was aimed at producing a policy statement as well as a draft bill for consideration by the National Assembly. This task was successfully executed by the committee and a report submitted three months after its inauguration. In order to broaden the perspective and parameters of the study the Presidential Adviser on Petroleum (later Minister of Energy) Dr Edmund Daukoru further commissioned Intsok of Norway to conduct a study on NCD based on the antecedents of Norway in successfully implementing a local content development programme. The expanded study was concluded in August 2003. Following the conclusion of the two studies a committee was set up to harmonise the findings. The report of the studies confirmed that whereas countries like Norway, Brazil, Malaysia and Indonesia had achieved local content development in excess of 45 per cent in their various economies, Nigeria was struggling to scale the 5 per cent mark. The study in addition to identifying the key success factors also delineated primary impediments to NCD, as follows:[12]

- low technological capacity;
- lack of funding from financial institutions;
- inadequate and incoherent policies/legislations;
- inadequate infrastructure;
- unfavourable business climate;
- lack of partnering between indigenous contractors and technically competent foreign companies.

The conclusions drawn from the studies conform with the views expressed by experts in the various workshops and seminars held in the industry. These findings and opinions clearly confirm the need for introduction of a definite policy framework to actualise the NCD objectives. In this regard the NCD aspirations are as delineated in Table 8.1.

One primary index for the evaluation of national economic well-being is the GDP. A comparative analysis of Indonesia and Nigeria shows a vast contrast. Whereas Indonesia and Nigeria had similar per capita income in 1960, the GDP of Indonesia doubled in 1999 while Nigeria stagnated. Revenues from the oil and gas sector were not adequately channelled to create broad-based industrial growth. A World Bank study also showed that at a broader level Nigeria's growth in GDP at the sub-Saharan African level has been slow. It was further observed that GDP growth rates in non-oil-producing neighbouring countries during the period under reference were higher. Nigeria, as an oil-producing country, has challenges like other producing nations, namely to

Table 8.1 Projected Nigerian content value contributions

Sector	*Average annual spend*	*Value contribution*	*Value contribution*	*Value contribution*	*Value contribution*	*Value contribution*
	$m	*$m 2006*	*$m 2007*	*$m 2008*	*$m 2009*	*$m 2010*
Engineering	900	270	340	495	605	720
Installation	1,100	220	233	263	295	330
Construction	1,100	330	465	528	625	770
Fabrication	1,500	500	610	705	850	1,000
Procurement	5,400	475	600	1,000	1,700	2,500
Others incl. gog hub*		105	320	650	1,150	1,480
Total	10,000	1,900	2,568	3,631	5,225	6,800

* Gulf of Guinea hub

Source: Nigerian National Petroleum Corporation – NCD 2006.

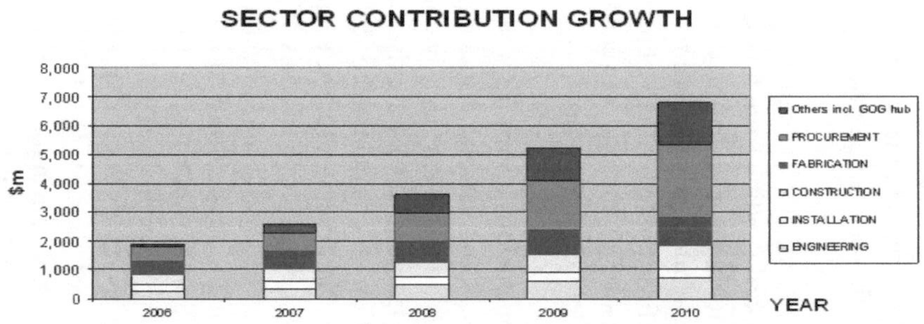

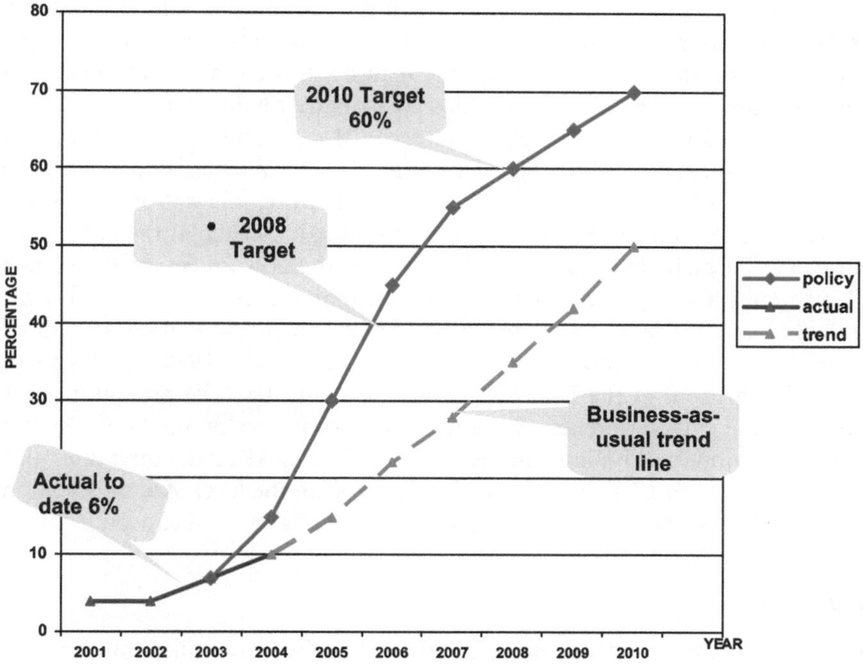

Figure 8.1 Nigerian Content Development targets.

Source: Nigerian Content Division – NNPC 2006

secure long-term welfare for the citizens by transforming the non-renewable resources into far more broad-based industrial wealth. The contribution of the manufacturing sector to the GDP remained at slightly between 4–8 per cent. In contrast, however, the performance of the industrial sectors in Malaysia and Indonesia escalated to approximately 32 per cent and 26 per cent respectively.[13] The low performance of the Nigerian industrial sector clearly shows that in the short term it may not be a major source of material inputs for the execution of projects in the upstream sector. The INTSOK study, in analysing the level of NCD, evaluated the experiences of other NOCs. The aim was to determine the policy thrust of the various countries in their efforts to engender local content development. In consideration of this, therefore, subsequent sections of the chapter will examine the local content development strategies of Brazil, Indonesia, Malaysia, Mexico and Norway.

Brazil

Historically Brazil has been involved in land and shallow water oil and gas exploration and production since 1930. However, offshore production started only in 1977. Despite the rapid increase in production over the past decade, Brazil remains a net importer of oil and gas. The upstream sector of the industry from the onset was under the firm control of the government and only indigenous companies were allowed to mine for minerals. In order to consolidate its gains in the upstream sector, the government introduced an Import Substitution Policy (ISP). This policy to a great extent encouraged the development of local technology for the manufacture, fabrication and provision of services in the organised private sector and in particular the oil and gas industry. Through the manufacturing and fabrication processes Brazil was able to meet a substantial proportion of the requirements of the oil and gas industry. The ISP and associated programmes encouraged rapid growth of Local Content Development (LCD) to a 90 per cent level in 1980. LCD is now a major component of the Brazilian oil and gas industry. The presentation of an articulate and comprehensive local content staff development strategy is a major criterion for winning a licence in the industry. Affected companies also suggest a range of LCD options, which depends on the level of sophistication of technology to be applied in the project and available local capacity.

Indonesia

Indonesia commenced oil and gas production in 1890. The industry was controlled by MNCs until 1950 when a new set of legislation was introduced. The new laws transferred ownership right to the government and government owned companies. To this end Pertamina, the NOC, was formed in 1968 to regulate the oil and gas industry and to negotiate and manage PSCs executed by MNCs. PSCs which are commonly used by NOCs for exploration and production activities in the industry were essentially invented by Indonesia. Pertamina invested heavily in shipping, petrochemicals, fertilizers, steel and hotels. These investments were beset by ethnicity and corrupt practices, which caused dismal performance, and so attention is currently refocused on the oil industry as the main income earner. LCD is an important component of the oil and gas industry policy with the current level approximately 25 per cent and expected to increase to 35 per cent through the execution of new PSCs.

Malaysia

Malaysia's entry into the oil industry can be traced back to 1910 but it did not become a significant oil and gas producer until 1980. Petronas, the NOC, was established in 1973 and given the exclusive right to own and mine petroleum resources. It was assigned the responsibility to negotiate and manage PSCs with MNCs. All JV agreements with IOCs have provisions for technology transfer as a key process for LCD. Each contract embodies an LCD strategy both in terms of procurement and employment. A clause in the PSC agreements states that companies with *Bumiputra* (local ownership) management or owners should be preferred when competitive, but otherwise local content is defined as value added in Malaysia irrespective of owners. Malaysia has achieved 70 per cent LCD in the upstream sector of the industry and the country has enjoyed a stable macro-economic environment which has resulted in stable exchange rates and cost levels.[14]

Mexico

Mexico is a long time player in the oil industry having been one of the highest oil and gas explorers as far back as 1920. It was during that period second only to the US in oil production. The titles to the oil resources in the 1920s were owned by foreign companies which repatriated large amounts of the oil revenues and left little revenue for development in Mexico. The situation was considered exploitative, triggering the nationalisation of the oil and gas resources in 1917. This process transferred ownership of the oil and gas resources and mining rights to the State. Complete nationalisation of the industry occurred in 1938, leading to the establishment of PEMEX to inherit and produce the existing oil and gas producing fields. PEMEX gained monopoly in both the upstream and downstream sectors when foreign companies were later excluded from participating in the Mexican petroleum industry. The oil and gas workers unions in Mexico are influential and they control four out of nine directorship positions in PEMEX. LCD policy is also vigorously pursued with the majority of the inputs in the upstream sector procured locally. The Mexican government introduced a policy which makes it mandatory for PEMEX to procure all materials locally if the cost differential between local materials and imported material is less than 15 per cent. This policy has virtually guaranteed utilisation of local inputs and the protection of the supply chain in favour of the indigenous companies.

Norway

Oil and gas production in Norway started in 1971, when Statoil, the NOC, and its affiliates produced 3.4 mmbd mainly for the export market. IOCs played a major role in the development of the industry by pioneering the exploitation, development and operation of the major oil fields. As a precondition for winning oil licences, major IOCs operating the fields were required to execute an agreement with their Norwegian partners. The agreement requires the IOCs to implement indigenous capacity building and technology transfer programmes. The IOCs were in some cases required to draw up and implement the transfer of operating schedules to Statoil. The programme presented a 'steep learning curve' for the NOC, which paved the way for the smooth take-over of operations of more than half the Norwegian fields within a relatively short period.[15] The complete nationalisation of the industry also compelled multinational engineering companies to partner with local entrepreneurs to develop engineering capacity. These policies consolidated the LCD policy, making it possible for Norway to achieve a 50 per cent local content utilisation in the upstream sector.

An assessment of the parameters in Table 8.1 indicates that Nigeria commands a large resource base compared to all the countries evaluated in the analysis of the LCD programmes. The analysis further shows that all the NOCs, including Nigeria, have similar policy measures. All five countries have introduced policies that encourage indigenous local content utilisation, staff training and technology transfer. However, it was observed that policy design, policy transparency and the resolve to enforce the regulations vary significantly among the countries. In consideration of the preceding, the INTSOK study recommended the following as necessary steps for the achievement of LCD beyond the present 5 per cent mark in Nigeria:

- a legal framework to implement local content policy should be put in place;
- an office should be established to coordinate the implementation of the policy. Adequate personnel and tools need to be provided to enable the office to perform its functions;
- the NCD co-coordinating office should articulate a vision as well as clear objectives which will guide and drive the implementation process;
- the NCD policy should establish distinct criteria to measure growth and improvement of local content development;
- oil and gas companies should be made to identify with the policy and accept some obligation for the realisation of LCD;
- a reliable database of major inputs (materials and services) should be established;
- the coordinating office should set standards and communicate best practice in various local content categories to every player in the industry (oil and gas companies, service companies etc.); and
- oil and gas companies should present half-yearly reports outlining their level of performance in achieving local content objectives.

It is of interest to note that a recent study of selected OSCs operating in Nigeria showed that, whereas the companies were progressively declaring high turnovers during the period under review (1994–2005), the profits declared remained low at between 1 per cent and 6 per cent of turnover, which in each case was in excess of ₦4 billion. The companies operated on 'Cost Plus' basis such that only a 5 per cent mark up is added to the total cost of goods and services. This practice serves the interest of the MNCs because in most cases the procurement costs are inflated. Potential profits of the companies are also reduced through 'transfer pricing' mechanisms. Most of them import raw materials from their home offices or sister companies. The cost of imports from affiliated companies is usually higher because procurements are not made on a competitive basis. Furthermore, head offices of the companies also charge administrative costs for arranging procurements and other documentation processes. These are all profit depletion mechanisms which would adversely affect the NCD objectives. Beyond these practices, NCD experienced a setback because OSCs and other service providers were reluctant to source raw materials locally. This is because affiliate offshore companies are sustained through the procurements and provision of services for the companies operating in Nigeria.

A case in point is the recent government policy which banned importation of baryte and bentonite, both of which are available in Nigeria. These raw materials can be beneficiated in local factories and used as drilling mud in the upstream sector. The OSCs argued that the grade is low (i.e. below 4.2 specific gravity) and that use of the products would not be conducive for drilling. The process of beneficiation involves refining and upgrading the raw material to attain the required minimum standard for input in the drilling process. No effort was made to beneficiate bentonite locally and rather strong petitions and appeals were forwarded to the government agencies demanding the abandonment of this policy.

The results of the study further confirmed that the low profits declared by the companies are unrealistic and totally inconsistent with Nigerian and indeed acceptable global oil and gas industry practice. This implies that the federal government, NNPC and other Nigerian stakeholders cannot enjoy capital appreciation on their investments. Essentially, the low profits declared cannot generate a sufficient multiplier effect to stimulate the private sector to produce goods and services for consumption in the upstream sector of the Nigerian petroleum industry. Furthermore, the low fixed assets base of the companies partially contributes to the unemployment problems of Nigerian society. For instance, the practice of fabricating oil rig platforms offshore and towing them in on barges for installation in Nigeria strongly negates the NCD policy. It is a ploy for transferring fabrication jobs to affiliate companies thereby boosting employment offshore at the expense of the host communities which are overloaded with unemployment related problems.

Local participation strategy

In order to clear possible barriers to indigenous participation in the upstream sector, the NCD Committee report ranked existing operating and service companies on the basis of ownership on scale A–E. 'A' in this case represents a wholly owned indigenous company while 'E' represents a wholly owned foreign company. Other companies depending on the degree of ownership fall between the 'A' and 'E' spectrum. According to Kupolokun,[10] this categorisation became necessary in order to ensure that low technical impact jobs are reserved for Nigerians. The Committee report further identified 17 technology areas which are avenues for indigenous participation to achieve the NCD. Nigerian content achievement target dates are graphically portrayed in Figure 8.1. Similarly, areas amenable for rapid NCD are also depicted in

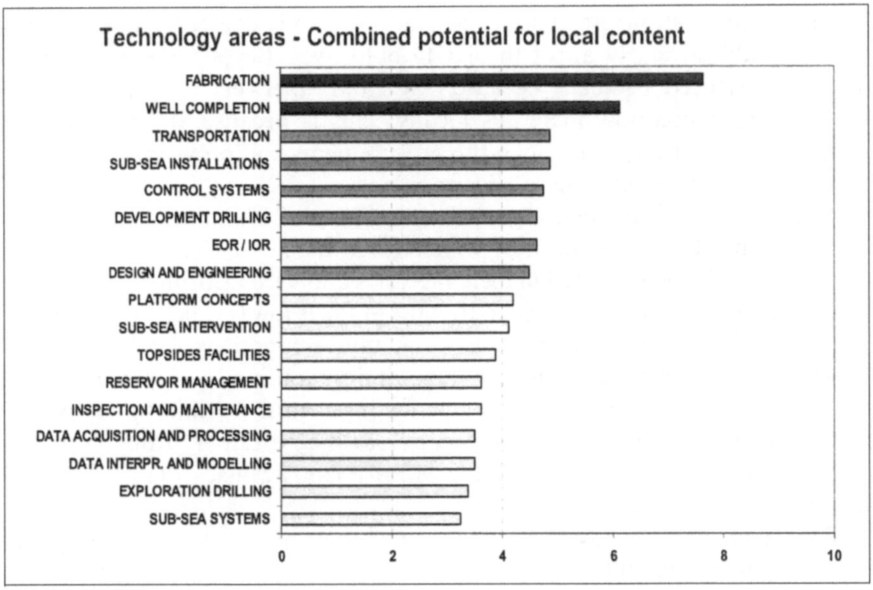

Figure 8.2 INSTOK Study 2003 – Nigerian Content Development (service area achievability index).

Figure 8.2. A close assessment of the bar graphs plotted on the Nigerian content achievability matrix shows that fabrication has a high achievability rating. Following closely are well completion, transportation, subsea installations etc., whereas the lower portion of the matrix features subsea systems, deep-sea exploration and drilling, seismic data interpretation and modelling etc. Fabrication and other related job groupings are less capital intensive and for this reason they can easily be engaged.

Subsea systems, deep-sea exploration and drilling require cutting edge technology. These are highly technical and require huge capital outlays to mobilise in the upstream sector. This area in the short term offers a low potential for NCD. It is important to indicate, however, that the 17 technology areas identified by Intsok consultants are not totally inclusive. It is simply an indication of the vast investment opportunities in the upstream sector. For instance MF development, although not listed in the report, can be identified as a high potential area for NCD. It is for this reason that 24 MFs were awarded to 31 Nigerian companies in order to boost indigenous participation in the upstream and also to promote NCD.

NCD policy directives

The issue of Nigerian content remained lucid for many years. For this reason oil companies and OSCs shifted high revenue-generating activities, namely fabrication, FEED, procurements, software development etc., to their sister companies or affiliates. This practice negates NCD objectives and requires reversal through appropriate policy measures. In furtherance of this and to achieve a 45 per cent and 75 per cent NCD in 2007 and 2010 respectively (Figure 8.1) NNPC directed all operators industry-wide to comply with NCD objectives as follows:

- FEED and detailed engineering design for all projects are to be executed in Nigeria effective from December 2005.
- Project management teams and procurement centres for all projects in the oil and gas industry are to be located in Nigeria effective from the first quarter of 2006.
- All operators and project promoters are to forecast procurement items for inclusion in a Master Plan (MPP). Such a plan is to be submitted to NNPC every January.
- Topsides of fixed platforms weighing more than 5,000 MT are to be fabricated in Nigeria.
- Fabrication of all piles, decks, anchors, tanks are to take place in Nigeria.
- A minimum of 50 per cent of the total tonnage of FPSO topside modules is to be fabricated locally.
- All third party services (NDT, mechanical tests, PWHT), certification of welding procedures and welders must be executed in Nigeria. The Nigerian Institute of Welding is to certify all such items in collaboration with international accreditation bodies.
- All FPSO contract packages are to be bid on the basis of carrying out topside integration in Nigeria.
- All operators and project promoters are to ensure that recommendations for all major contract awards for all major projects are forwarded to NNPC constituted Boards of such oil and gas companies for approval. All approvals must include evidence of a binding agreement by the main contractor with a Nigerian content subcontractor. Such agreements shall indicate the cost and detailed scope including total man-hours for engineering, tonnage of fabrication and relevant defining parameters for materials to be procured locally as well as other services.
- All fabrication of subsea systems including risers and flow lines, subsea assemblies and ancillary facilities including system integration tests are to be performed in Nigeria.
- Low voltage earthing cables of 450/750V grade and control, power, lighting cables of 600/1,000V grade are to be purchased in Nigeria.
- All equipment and materials manufactured in Nigeria must be fully utilised and clauses that create impediments or exclude participation of local companies should not be included in any Invitation to Tender (ITT).
- All carbon steel pressure vessels of not more than 75 mm shell thickness shall be fabricated locally.

- All seismic data processing projects and all reservoir management studies are to be undertaken in Nigeria.
- All waste management, onshore and swamp integrated completions, onshore fluid and mud solids control, onshore Measurements While Drilling (MWD), Logging While Drilling (LWD) and directional drilling activities are to be performed by indigenous companies having genuine alliances with multinational companies.
- All international codes and standards used in the industry are to be harmonised to support utilisation of locally manufactured products such as paints, cables, steel pipes, rods, sections, ropes etc. and to improve capacity utilisation in local industries.

The directive of the federal government and NNPC to oil producing companies and OSCs in the industry to embrace the local content development policy has begun to yield tangible results. In the 1980s the government established Niger Dock to undertake dry docking of offshore rigs and other marine vessels. The company did not live up to the challenge and expectations of the government. For this reason offshore rigs and marine vessels were towed to companies overseas for dry docking. Interestingly, however, the government privatised Niger Dock and the new management (Jagal Group) has repositioned the company to effectively perform those functions for which it was established. SNEPCO awarded the Bonga Buoy project to SBM which partnered with Niger Dock to execute it. The fabrication of the Buoy provided employment opportunities for highly skilled welders and engineers. Alongside the Bonga Buoy fabrication are other major fabrication projects. The Chevron Meren X wellhead platform was also fabricated at Niger Dock. The NPDC Okpoho field platform in a similar manner was fabricated at the SAIPEM facilities in Port Harcourt. Furthermore Solt, a Nigerian based company, fabricated the drilling jackets, bridges and buoy for the Total Amenam-Kpono field.[15]

The NCD drive has also led to domestic fabrication/supply of

- Shell 120,000 b/d Forcados Yokri and 45,000 b/d Tunu flow stations by Globe Star/Suffolk and Warri based Natomi;
- Shell sponsored pressure vessels and scrubbers for Forcados Yokri, executed by Dorman Long;
- collaborative supply of drilling rig Ensco 100 SF by SDF, Sewops; and
- provision of integrated drilling services by indigenous companies Tecon and Ciscon.

The fabrication of these facilities in Nigeria confirms that a change of attitude on the part of MNCs and progressive compliance with the policy when fully operational, will create opportunities for elaborate execution of fabrication projects locally. The aggregate impact of domestic fabrication of buoys, platforms, FPSO topsides, piles, decks, anchors, jackets etc. would be felt in the areas of job creation and the economic empowerment of local entrepreneurs. This will engender savings and expenditure which will spin-off an elaborate multiplier effect in the domestic economy. Furthermore, the fabrication process and other programmes will strengthen the skills of technicians and enrich the experiences of individuals. Local content-related activities will stimulate employment opportunities, correspondingly depopulating the ranks of the unemployed restive youths in the Niger Delta and other states of the country. Gainful employment of youths through these processes especially in the Niger Delta will drastically reduce communal disturbances which have often interrupted oil production activities. NCD, if fully supported by the major oil companies, OSCs and other companies in the organised private sector will yield results capable of promoting relative harmony and stability in the oil producing areas. The enactment of an appropriate law by the National Assembly will suffice to entrench the practices properly in the oil and gas industry. Such action on the part of the government will send clear signals to all concerned that the domestication of raw materials and other inputs in the execution of jobs in the upstream sector is an irreversible government action.

Penetrating the supply chain

The Nigeria Content Enhancement Study conducted by INTSOK of Norway in collaboration with the NCD Committee and other collaborating organisations, as part of its findings observed that major oil companies have numerous suppliers who offer a wide range of goods and services. The contractual arrangements between the oil companies and the suppliers often involve a chain of subcontractors. This supply chain can be considered a unidirectional sequence of activities involving flow of materials from the supply source, finished products and after-sales. The supply chain in the oil industry also has additional activities such as finance, communication, transportation and other specialised services. This chain of activities is usually closely guarded in order to prevent entry of potential participants. From the supply chain and the NCD perspective the source of the supply is only justified if the end-product or service procured promotes local value addition. The primary objective of the Nigerian content initiative therefore would be to penetrate the supply chain in the upstream sector through a set of enabling legislations.

References

1 Lukman, R. 'Keynote address at the NNPC National Workshop on Improvement of Local Content and Indigenous Participation', 2001, pp. 2–4.
2 Oladele, O. R. 'Opportunities for Indigenous Participation'. Paper presented at the NNPC Workshop on Improvement of Local Content, 2001, p. 7.
3 Gaius-Obaseki, J. E. 'Partnering and Alliancing in the Oil and Gas Industry'. *NNPC News* Vol. 22, No. 7, 2001, p. 9.
4 Ogiemwonyi, C. O. 'Alliancing as a Tool to Increase Local Content in the

Nigerian Oil Industry'. Paper presented at the Offshore Petroleum Conference, Houston, April/May 2001, pp. 2–6.

5 Olurunfemi, M. A., 'The Role of Financial Institutions in Promoting Local Content and Indigenous Participation in the Oil Industry'. Paper presented at the NNPC National Workshop on Local Content, Aug. 2–3, 2001, pp. 1–10.

6 Ofurhie, M. 'Fiscal Policies to Support Local Content Development and Indigenous Participation'. ibid. pp. 4–5.

7 Omene, G. E. 'Definition of the Scope of Local Content and Indigenous Companies in the Upstream Sector of the Petroleum Industry'. *NNPC News*, Vol. 22, No. 7, 2001, pp. 5–6.

8 Lukman, R., 2001, op. cit. p. 3.

9 Hagen, P. 'Norway's Experience in Local Content Development', op. cit. 2001, pp. 3–8.

10 Kupolokun, F. 'Opportunities for Local Participation in the Oil and Gas Industry'. Extract of selected speeches 2003–2004 and paper delivered at the stakeholders seminar on the Nigerian Content Development Bill in the upstream sector of the Petroleum Industry, 2004/2005.

11 Ariweriokuma, S, 2001 'Impact of Multi-NOC Practices on Local Content Development'. *NNPC, COMD Programmes Project* pp. 41–42.

12 INTSOK, 2003 'Enhancement of Local Content in the Upstream Oil and Gas Industry in Nigeria'. *Nigeria and Norway Collaborative Study*, 2003.

13 Ibid.

14 Ibid.

15 Kupolokun, F. 2004 'An Address on the Occasion of the Commissioning of the Bonga Field Buoy at Niger Dock Yard, Snake Island, Lagos'.

9 The Joint Development Zone (JDZ)

Introduction

The Gulf of Guinea lies in the concave recess of West and Central Africa which is filled by water mass from the Atlantic Ocean. The area is constituted by Nigeria, Equatorial Guinea, Cameroon, Angola, Gabon, Democratic Republic of Congo and Democratic Republic of São Tomé and Príncipe (DRSTP). Although the Gulf of Guinea is made up of eight countries, subsequent discussions will focus on Nigeria and DRSTP as the principal parties in the JDZ. The Gulf of Guinea contains one of the richest and most prolific fields in an area of about 1.2 million acres offshore. This prolific area, often referred to as the 'Golden Rectangle', is proximately located to the offshore intersections of the maritime boundaries of Nigeria, Cameroon and Equatorial Guinea. The region is highly endowed with hydrocarbon and fishery resources. It is believed to have proven crude oil reserves estimated to be between 5 to 14 billion barrels with producibility potential of 1.0 mmbd.

The Middle East, which is known for its high oil and gas reserves, has for some decades been embroiled in one conflict or another. There is growing concern that owing to the escalation of conflict in the region and the increasing antagonism against the United States, its allies and other major consumers, oil might be applied as a political weapon against certain developed countries. The emergence of the Gulf of Guinea as a potentially high oil and gas bearing region serves as a relief for some of the major oil consuming countries entangled in conflicts in Iraq, Afghanistan, Lebanon and Israel. This chapter is not intended to provide an elaborate account of the various conflicts in the Middle East. Rather it is devoted to analysing the collaborative efforts of the member states of the Gulf of Guinea region especially as it relates to their economic empowerment. As pointed out in the preceding chapter, eight countries are recognised as the constituent sovereign states of the region. The rapid rise in its profile of, and its strategic importance to Africa was responsible for the decision of the member countries to form a Joint Commission to adjudicate over the region. Although a treaty has been signed by eight countries, only Nigeria, Equatorial Guinea and DRSTP have ratified the document.

Search for oil in the Gulf of Guinea

The search for oil in the Gulf of Guinea has spanned 15 years which provided authentic seismic insight into the geological configuration of the region. Exploration and exploitation activities in the area are dominated by Shell, ExxonMobil and Total. These companies are responsible for the discovery of major fields in the Niger Delta. Ubit with 2 billion barrels and Edop with 1 billion barrels are giant fields operated by ExxonMobil in its OML 67–70 concession area. In 1994 the two fields accounted for a significant proportion of its 700,000 b/d production. These resources have inevitably attracted global interest in the West African subregion as an alternative source of oil for the USA, OECD and other consuming countries. Nigeria and São Tomé and Príncipe have initiated a bilateral agreement which provides for the collaborative administration of the JDZ.

Oil exploration and production in Nigeria shifted from land and swampy terrain to the shallow continental shelf and in recent years has advanced into the Deep and Ultradeep Offshore portions of the Niger Delta. In 1990 the federal government, in an effort to encourage indigenous entrepreneurs, awarded ten Blocks to ten indigenous companies. Oriental Oil was allocated OPL 224 which was converted to OML for crude oil production purposes. Between 1991 and 1992 United Meridian Corporation (UMC) and the Walter Oil and Gas Corporation acquired acreages north of Bioko Island in the territory of Equatorial Guinea. The acquired acreage is sandwiched between the oil producing areas of Nigeria and Cameroon. The tempo of oil exploration in the region increased in 1993 with the entry of the independents. Perenco SA acquired 30 per cent participatory equity interest in the 5,000 b/d Moudi Permit located in Cameroon. Ashland discontinued its operations in Nigeria and Addax acquired OPL 98 – a 6,000 b/d field hitherto operated by Ashland. In 1991 Oriental partnered with Conoco Inc. to explore OPL 224. Conoco, having drilled three dry wells, discontinued the partnership in 1994. In contrast, Walter Oil and Gas Inc. focused on natural gas condensate production in the Alba field. This field, from expert studies, was estimated to have a proven condensate reserve of 50 million barrels capable of producing 6,000 b/d. In 1997 CMS Energy acquired Walter Oil and Gas and expanded its investment by building a 700,000 MT/year methanol plant on Bioko Island. The investment was a strategic move to monetise the growing gas reserves in Alba while contributing to the zero flare policy. The level of activities in the region is extremely dynamic as several acquisitions took place within a short space of time. CMS's acquisition of Walter Oil lasted for about 5 years after which Marathon took over its operations. In 1992 UMC, an independent upstream company, made an incursion into the Niger Delta Deep Offshore and leased OPL 222 for $40 million. It later partnered with Mobil to produce the Zafiro field. The field, with a reserve potential of 100 million barrels, came on stream with an initial production of about 300,000 b/d in August 1996. Thereafter, several other wells were drilled and developed at Zafiro thereby expanding recoverable reserves to 1.2 billion barrels and producibility to well over 350,000 b/d. It is important to indicate that Marathon Oil has also made significant progress by expanding the initial condensate reserves to 600 million barrels. The campaign in the Gulf of Guinea has escalated. Consequently, Mobil Producing Unlimited concluded plans to raise its production in OMLs 67–70 from the 1994 production level of 300,000 b/d to about 700,000 b/d in 2007.[1]

The joint development initiative

Nigeria has common boundaries with the Republic of Benin, Cameroon, Chad, Niger and São Tomé and Príncipe. The boundaries with these countries are, to say the least, ill defined. This has led to periodic conflicts between border communities and the outcome of such conflicts has often turned out to be sour, causing unnecessary diplomatic strain. Nigeria is recognised globally as a leader in Africa and for this reason is expected to champion the promotion of peace and harmony in the continent. In recognition of its role of advancing peaceful co-existence, the federal government took the initiative to establish a National Boundary Commission (NBC) in 1987.[2] The NBC was constituted primarily to negotiate the delimitation of land and sea boundaries with the neighbouring countries mentioned above. In order to achieve this objective, international workshops were organised for representatives of the constituent states to attend. These workshops involved the Republic of Benin, Cameroon, Chad, Equatorial Guinea and Niger. São Tomé and Príncipe as an island state was initially not listed as having tangible delimitable boundaries with Nigeria. However, a critical evaluation of the maritime boundaries between Nigeria and DRSTP confirmed significant overlap. Under the provisions of the United Nations Convention on the Laws of the Sea (UNCLOS) of 1982 DRSTP has definite boundaries which have to be negotiated and delineated. In consideration of these facts, the National Boundary Commission included DRSTP as a legitimate participant in future workshops and deliberations. Although UNCLOS came into effect in 1982, the principles of limits of sovereign states derived from the Truman Declaration of 1940. UNCLOS internationalised the laws with the participation of member states in the negotiation and adoption of the principles. Prior to UNCLOS, delimitation processes lacked uniform adjudication.[3]

The negotiations leading to the establishment of the Nigeria–DRSTP JDZ was led by the National Boundary Commission. Discussion for boundary demarcation which started in 1992 took place initially between Nigeria, Republic of Benin, Cameroon, Niger, Equatorial Guinea and Chad. Some of the workshops made far reaching recommendations to the government. One of the main recommendations was the joint exploration and exploitation of cross-border mineral resources. The discussions with Equatorial Guinea led to the amicable unitisation of the Zafiro/Ekanga field which straddled the boundary between Nigeria and Equatorial Guinea. In recognising the potential benefit which could accrue to both countries, the NBC at the instance of the federal government entered into boundary negotiations with DRSTP in 1999. The negotiations were inconclusive thereby temporarily halting the possibility of cross-border exploration and exploitation of oil, gas and other resources. Presidents Obasanjo and Trovoada took the initiative and resolved the outstanding issues on 28 August 2000. This accord, reached in São Tomé and Príncipe, paved the way for the formation of a JDZ between Nigeria and DRSTP (hereafter the Treaty). This was signed on 21 February 2001. It provides for joint control of exploration and exploitation of oil, gas and other natural resources. It further provides that Nigeria and DRSTP shall enjoy 60 per cent and 40 per cent respectively of all benefits or obligations emanating from all development activities carried out within the zone. It further provides that the affairs of the JDZ shall be administered by the Joint Development Authority (JDA) under the supervision of the Ministerial Council. The Treaty shall last for four years in the first instance and is renewable for another 30 years.[4]

Nigeria and DRSTP are signatories to the United Nations Convention of the Laws of the Sea (UNCLOS) Montego Bay Treaty of 1982. This Treaty allows sovereign states to assert claims on their territories including territorial waters, inland waters, contiguous zones and Exclusive Economic Zones (EEZs) up to a maximum of 200 nautical miles. The Continental Shelf, stretching a distance of 350 nautical miles, is also used as a delimitation yardstick. The purpose of this chapter therefore, is to examine the evolution of the Nigeria–DRSTP JDZ. Additionally the discussions will focus on the configuration of the JDZ and JDA with a view to determining the mutual benefits derived by the parties involved. JDZs are widely used by many sovereign states across the globe and in view of this, an attempt will also be made in subsequent sections to examine various JDZ models adopted by some of these countries. The evaluation of the models would provide a suitable platform for the discussion of the Nigeria–DRSTP model.

JDZ models

The concept of JDZ has been widely adopted in agreements involving Bahrain/Saudi Arabia (1958), Germany/Netherlands in the Ems Estuary, Austria/Czechoslovakia (1960), and Kuwait/Saudi Arabia (1965) in their overlapping zones. The Bay of Biscay Convention regulates the situation between France/Spain (1974), Norway/United Kingdom (1976) over the Frigg gas field, Abu-Dhabi/Qatar (1979), Iceland/Norway in respect of Jan Mayen (1981) and Australia/Indonesia (1989) over the East Timor Continental Shelf etc. Intersection of territorial boundaries of sovereign states is a common occurrence which has been known to exist for many years. JDZs generally vary in complexity and geographical scope. The purpose of this section therefore is to examine existing Joint Development Models (JDMs) with a view to having an insight into the various approaches deployed to arrive at mutually acceptable boundary delimitation solutions. Such an insight will be contrasted with the Nigeria–DRSTP (JDZ) model. The ultimate aim would be to determine the adequacy of the provisions in the Nigeria–DRSTP Treaty in forging an enduring collaborative relationship.

One of the early JDZ agreements was recorded in 1960 between Austria and Czechoslovakia. The agreement was initiated in order to jointly develop oil and gas deposits located in segments of their overlapping boundaries and did not refer to any particular geographical area, but to certain deposits which representatives would classify as of common interest. A mixed Technical Committee was established and empowered to internally determine the volume of the resources.[5] The North Sea is a typical example of a terrain in which hydrocarbon resources cross international borders. Some international oil boundary disputes were solved through the law of capture as in the case of the border oil fields of Iraq and Kuwait. In the offshore setting the cost of production is much higher and the challenging terrain warrants the use of sophisticated technology. Perhaps of greater concern is the fact that the drilling of closely spaced fields is extremely difficult and in most cases impossible in the offshore setting. In such situations various forms of agreements have to be fashioned between the governments and licence holders. The North Sea unitisation agreements involved the United Kingdom, Norway and the Netherlands. The experiences from these treaties paved the way for the emergence of contemporary joint development treaties. The Frigg treaty was the first cross-border field development agreement in the North Sea. The field was discovered in 1971 on the Netherlands side of the median line. However, seismic data showed that it straddled some segments of the UK territorial waters. Early discussions over the Frigg field were fraught with rancour and as a result the negotiations proceeded over an extended period. In the second instance, the Stratfjord field with an estimated reserve of 4 billion barrels of oil, discovered in 1974 on the Norwegian side of the median, also straddled UK territory. Based on the experiences from the Frigg fields, both the UK and Norway resolved to go ahead with exploration and development pending the final determination of the extent of overlap of the field on the adjoining territory. Finally, the Markham field with gas reserves of 0.5 TCF straddled the boundaries of both the UK and the Netherlands. With the benefit of hindsight, especially as it relates to the Frigg and Stradjford fields, the team of experts involved in the Markham negotiations acted proactively and exhibited greater cooperation than had been experienced in earlier cross-border negotiations. The Norwegian–UK Agreement of 1976 governing the Frigg gas field distinctly defined the area in question relying on geographical and geological factors. The initial area demarcated was subsequently extended to other gas bearing fields which constituted the territories in which resources could be explored and exploited.[6] Abu-Dhabi and Qatar in delimiting their boundaries relied on geographical coordinates. They also adopted a principle of equal sharing of royalties, profits and fees. The equal rights principle of the JDZ notwithstanding, Abu-Dhabi had the prerogative of granting exploitation rights to concessionaires. In some cases a complex unitisation process was undertaken and in this case the resources in the straddled area were brought into a common pool. The licensees, granted a concession by each licensor, would enter into an agreement with the licensee of the other party and both would mutually appoint an operator. The agreement also regulated the movement of persons, materials, inspectors, safety, transfer rights and taxation. The rights and privileges of each State remained unencumbered by the collaboration.

In 1960 the Ems–Dollar Treaty was signed between Germany and the Netherlands to establish a JDZ in the Ems Estuary. The overlapping zone was delimited into two zones – one for each country by a provisional line. Both Germany and the Netherlands exercise jurisdiction over their respective subzones. They may, under their respective domestic laws, grant concessions for exploration and exploitation of the resources in the joint development zones. The concessionaires rights and obligations are usually spelt out and they are entitled to an equal share of the oil and gas being recovered from the area. The expenses are also proportionately borne by the parties.[7] Kuwait and Saudi Arabia have intersecting boundaries and the overlapping area was partitioned into a neutral zone in 1965. Both countries mutually agreed to coordinate the joint exploitation of the oil and gas resources in the area. In this regard Kuwait and Saudi Arabia retained their rights over the natural resources in the partitioned zone. The joint exploration and exploitation in the neutral zone, notwithstanding prior concessions granted in the area, proceeded unencumbered. The agreement also provided for freedom of movement, access, safety and tax regimes for all concessionaires operating in the neutral zone.[8]

As pointed out earlier, several models of JDZs exist but the agreement between Indonesia and Australia over the Timor Gap is perhaps the most elaborate scheme. The zone of corporation was in 1972 divided into three areas designated X, Y and Z. In the first area 'X', a Ministerial Council was established to exercise the rights and responsibilities of the two states as it relates to policies of exploration and exploitation of oil and gas resources. On the other hand, the JDA is the body responsible for the day to day management of the affairs of the JDZ. The agreement embodied specific clauses relating to surveillance, security, marine, environmental, search and rescue, air traffic, research, and other related matters. The accord also included unitisation clauses in the case of resources extending beyond X. The operations were principally carried out through PSCs. In areas Y and Z each state was entitled to engage third parties to explore and exploit oil and gas on its behalf. It is, however, the responsibility of the executing party to inform the other party of the agreement initiated and also pay 10 per cent of the income tax collected. In the case of unitised production, the criminal laws of each state apply to its nationals or residents although the activities of the PSC are governed by a contract.[9]

The treaty between Australia and Indonesia over the Timor Gap has taken a different dimension because East Timor became an internationally recognised independent country on 19 May 2002. The new status put it in a position to renegotiate the agreement between Australia and Indonesia over the East Timor Gap. The Gap covers an area of about 135 nautical miles of sea-bed which was hitherto undelimited by Australia and Indonesia. East Timor has since its independence engaged Australia with a view to amending, and in some cases abrogating, certain provisions of the agreement entered into by Australia and Indonesia. Prominent among the new East Timor conditions is the change of 50–50 share of the production inherited from the 1989 Timor Gap Treaty. East Timor demanded that the provision be changed to 90–10 split in favour of East Timor. Under the early treaty, Australia was entitled to 85 per cent of the Continental Shelf or natural prolongation. This claim is strongly contested by East Timor. Several areas of contention existed and these issues have delayed exploration of resources especially in the Greater Sun Rise area. The aggregate effect is the delay in the unlocking of the resources of the Timor Sea for the benefit of the East Timorese people.[10]

Nigeria–DRSTP JDZ

The Nigeria–DRSTP JDZ is an area constituted by overlapping maritime boundaries. The area lies between latitudes 1° and 3° North and longitudes 4° and 8° East in the Gulf of Guinea (Figure 9.1). The JDZ is estimated to cover an area of 34,548 km^2 in water depths of 1,500 m in the northern segment to 4,000 m in the south western sector. The area lies proximate to the Golden Rectangle which has a reserve of about 5 billion barrels of oil. The UNCLOS of 10 December 1982 (Montego Bay) was primarily designed to provide a forum through which guidelines for the demarcation of sea boundaries between neighbouring countries could be discussed and adopted. The difficult and extensive negotiations associated with sea boundary delimitation are well known. In recognition of this, therefore, the convention as an interim measure acceded to the fact that states with opposite coasts can in the spirit of cooperation and understanding, and pending the final delimitation, enter into collaborative arrangements.

Nigeria and DRSTP took cognisance of this provision and in furtherance of their mutual interest to explore and exploit the oil, gas and other natural resources jointly in the overlapping maritime zone initiated the Treaty. The purpose of the Treaty was to establish legal premises of an international nature which lend support to the establishment of a JDZ. In developing countries, economic and political stability are two major preconditions for the inflow of Direct Foreign Investment (DFI). Ordinarily, oil companies would hesitate before risking investing huge sums of money in exploration and exploitation of oil and gas resources unless the political as well as legal risks have been drastically mitigated by the host country. Wars, terrorism and political instability constitute huge business risks. On the other hand, disputed boundaries also present serious obstacles in the oil and gas business environment.

In situations where international boundaries are entangled in economic and political problems, companies would only be willing to undertake oil and gas related activities at the boundaries if both governments make a commitment to establishing a business-friendly environment. Essentially, international boundaries between countries constitute political issues which can only be resolved by the affected parties. Boundary demarcation and negotiation of the EEZ under Article 74 of UNCLOS is contentious. In this regard, reaching an amicable solution is often time consuming. The alternative to prolonged negotiations, however, is the conversion of the overlapping area into a JDZ. Intersecting boundaries are occasionally bedeviled by oil fields which straddle the lines and such fields may be unitised and the benefits shared equitably among the party states.[11] Nigeria and DRSTP opted for joint development and the Treaty consummated by the Presidents of the two countries has ingrained in it basic provisions which include the following:[12]

- a Joint Ministerial Council;
- a Joint Development Authority;
- a Regime for Petroleum in the Zone;
- health, safety and environment;
- resolution of disputes.

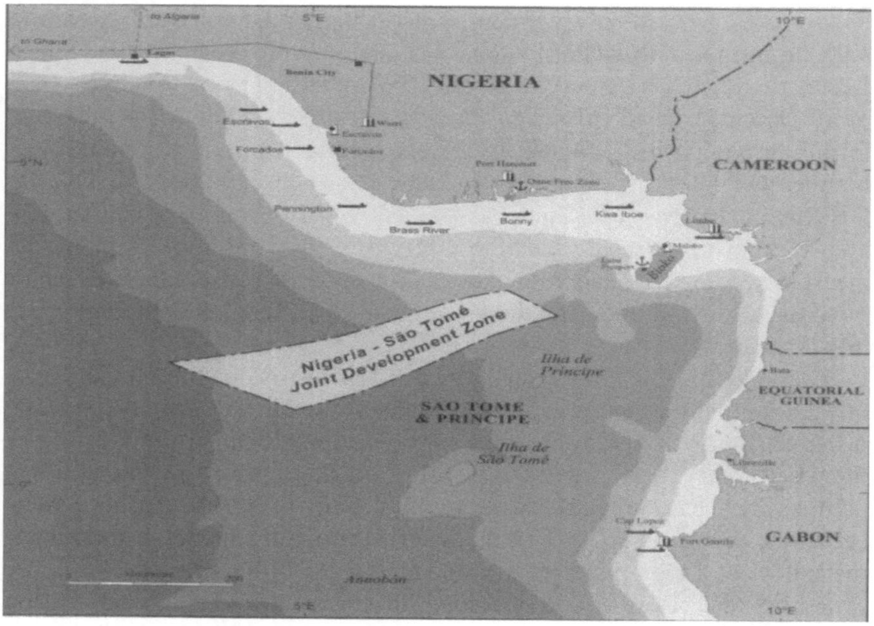

Figure 9.1 Gulf of Guinea JDZ.

Source: Nigeria – DRSTP Joint Development Authority: Guide to the 2004 JDZ Licensing Round Document

Joint Ministerial Council

The Joint Ministerial Council which derives from the provisions of the Treaty was established to perform specific functions that are necessary to advance and achieve many of the basic objectives of the Treaty. It provided that the Council shall at each given time be comprised of a minimum of two Ministers or a maximum of four Ministers or persons of equivalent rank appointed by the Head of State of each party. The Council is also empowered to deliberate on matters concerning the activities of the Joint Development Authority (JDA) and shall in addition perform the following functions:[13]

- give directives to the Authority on the discharge of its functions;
- approve rules, regulations and procedures for the effective function of the Authority;
- consider and approve audited accounts and audit reports of the authority;
- consider and approve the annual report of the Authority;
- review the operations of the Treaty and make recommendations to the State parties on any matter concerning the functioning or amendment of the Treaty as may be appropriate;
- approve a development contract which the Authority may propose to enter into with any contractor;
- approve the termination of development contracts entered into between the authority and contractors;
- consider and approve the annual budget and the opening of bank accounts by the Authority;
- settle disputes in the Authority;
- approve the appointment and remuneration of auditors.

Joint Development Authority (JDA)

In establishing the JDA the Treaty provides that it shall report to the Council, have a judicial personality in international law and will enjoy similar status and capacity under the laws of each of the party States. Such capacity is to enable it to perform its legitimate duties. In practical terms it is the administrative organ established to administer the activities of the JDZ. It is accountable to the Council and has several functions, some of which are as follows:[14]

- demarcate the zone into Blocks, prepare bid documents, and execute the bid process;
- execute development contracts with contractors subject to Council approval;
- supervise the activities of contractors;
- recommend to Council the termination of activities of contractors;

- allocate to the party states the financial or other benefits in proportion to their participatory interest;
- control the movement of vessels, aircraft, equipment and structures into, within and out of the JDZ;
- regulate and direct on all matters related to the supervision and control of operations in addition to health, safety and environmental issues;
- regulate marine activities in the zone;
- preservation of the marine environment bearing in mind the relevant provisions and international laws governing the zone;
- collaborate with other agencies in the generation, collation and exchange of scientific, technical and other data concerning the zone and its resources;
- prepare annual reports for submission to Council;
- advise and recommend to states parties on issues concerning material changes to the law which may be necessary to promote the development of the resources in the zone.

Boundary dispute negotiation

International boundary disputes arise as a result of claims and counter-claims by sovereign states over land, space or territory which lie within a contiguous geographical location. In this regard therefore the desire to negotiate would be necessitated or guided by many factors among which are:

- the economic, cultural or occupational value attached to the disputed territory;
- the expansionist philosophy and perception of an individual state;
- perceived threat to national security;
- commercial value of the disputed land;
- encumbrance of logistics and accessibility.

The boundary disputes are often very extensive in nature; therefore several methods may be adopted to resolve them. Tumsah[15] opined that boundary disputes may be resolved through the following:

- Negotiation – this involves dialogue between the affected parties and may culminate in establishing a mutually accepted boundary line.
- Arbitration – this involves the use of third parties or experts who understand the issues and deep feelings involved in the dispute such that they are able to strike a compromise which is acceptable to both parties.
- Litigation – often states involved in boundary disputes may resort to litigation in order for the judiciary to examine the facts and deliver an unbiased judgment. The ultimate objective in this case is the demarcation of the disputed area.
- Wars – territorial disputes create tension and such tension, if not properly

managed, could lead to war. The conflict between Iran and Iraq originated from territorial disputes over oil fields. The same is true of the dispute between Iraq and Kuwait. One may also indicate that the conflict between Israel, Palestine and Lebanon derived from disputes linked with unresolved boundary delimitation issues. In the case of the countries under reference the situations deteriorated into bloody wars and terrorist attacks.

- Establishment of JDZs – JDZs are usually a product of extensive negotiations between sovereign states which have overlapping boundaries. The purpose of the establishment of such a zone is to create an opportunity for both parties to benefit by jointly developing the affected area.

JDZ oil and gas regulations

It is important for oil and gas activities in the region to be carried out in an orderly fashion. For this reason the Treaty provided for oil and gas regulations which will guide the conduct and activities of concessionaires. The JDA serves as the custodian of all acreages in the JDZ as well as OPLs and OMLs originating from the JDZ. In this regard the JDA is empowered to act as follows:[16]

- grant, subject to approval of Council, Exploration Licences (ELs), Oil Prospecting Licences (OPLs), OMLs and PSCs to companies operating in the JDZ registered in Nigeria and DRSTP respectively;
- supervise all operations carried out under OPLs, OMLs and other contracts consummated by the Authority;
- enjoy at all times unlimited access to all areas covered by the OPLs, OMLs and contracts and other installations which are operated or maintained in support of these licences, leases and contracts in order to carry out inspections or related activities;
- summon in writing the holders of OPLs, OMLs, contractors or their subcontractors to appear before the Authority at a specified time and place for purposes of providing information pertaining to its operations;
- order the suspension of operations carried out under the OPLs, OMLs and contracts in order to provide an environment which is safe and conducive to carry out exploration and production or other related activities.

Partitioning the zone into Blocks

One of the functions of the JDA is the demarcation of the JDZ into distinct Blocks of appropriate dimensions for the purpose of granting OPLs. Each Block so demarcated is for practical purposes assigned a distinct identifying reference number. The boundaries of the Blocks, if found necessary, are further defined by meridians of longitude of integral units of 5 minutes and in the same manner by parallels of latitude which are in units of 5 minutes.

Exploration Licence (EL)

An exploration licence accords the licencee the right to undertake speculative geophysical surveys in a specified area. The holder of an EL may drill holes not exceeding 300 feet below the ground provided such drilling is for the purposes of facilitating geological and geophysical work. The grant of an exploration licence in a particular area does not necessarily preclude further grant of licence or lease in oil and gas exploration and production. An EL may not automatically convert into an OPL or OML and enjoys a life span specified in the agreement. In accepting the grant of an EL the licensee undertakes to at all times carry out its activities under close supervision of qualified personnel and such supervisory activities shall be conducted to the satisfaction of the JDA.

Oil Prospecting Licence (OPL)

An OPL grants a licensee the right to explore and exploit for oil and gas within an assigned Block. It may upon the discovery of oil in commercial quantity be converted into an OML. In the case of an OPL the licensee is entitled to drill holes to any depth of choice in furtherance of exploration and production objectives. The licensee may also evacuate all or part of all earned equity oil upon fulfilment of all obligations placed on it by the terms of the OPL – royalties, rents, petroleum tax regulations and other applicable laws. The JDA provides that an OPL shall enjoy an initial life span of about four years which upon formal application in writing may be extended for an additional mutually agreed period.

Oil Mining Lease (OML)

The OML confers on the lessee the exclusive right to '. . . search for, win, work, carry away and dispose of all petroleum in, under or throughout the Block described in the schedule'.[17] The Treaty provides for a life span of 20 years for the OML, with a renewal clause as contained in the PSC. The nature of the lease in the Nigerian context is not quite clear, especially with regard to the fact that prior to 2006 no categorical decisions have been made in law courts concerning the nature of the lease. In 2006 SAPETRO applied to the government to convert portions of OPL 246 into additional OMLs but this was denied. In pursuance of this decision the federal government took steps to recover the unutilised portions of OPL 246 following the conversion of the OPL into an OML. The recovered portion of the Block was later awarded to the Oil and Natural Gas Corporation (ONGC) of India. The lessee took steps in the Court of Appeal to restrain the federal government in its effort to repossess the unconverted portion of the Block. In consideration of this the court gave a ruling that the Minister of Petroleum, guided by the Petroleum Act 1969, acted properly in not granting SAPETRO's request to convert the OPL into additional OMLs. Under these circumstances SAPETRO appealed the judgment to the Supreme Court. This was perhaps the first time such a case would be decided by a court of competent jurisdiction. In examining related issues Olisa[16] expressed the view that the nature of the grant of an OML in the Nigerian context is inexplicit as to whether or not the lessee can enjoy a freehold, a tenancy at will, personal property or an interest in the land. It is important to note, however, that in countries where oil and gas laws have developed over a long period substantial litigation exists concerning the juridical nature of the lessee's interest in petroleum matters.

Production Sharing Contract (PSCs)

The activities in the JDZ are executed through PSCs. Furthermore, the Authority is empowered by the Ministerial Council to enter into contractual agreements with companies incorporated in Nigeria or DRSTP. With respect to PSCs, the Authority may exercise additional functions as follows:[18]

- execute a PSC with a contractor with a view to granting the contractor exclusive right to undertake petroleum activities in the assigned OPL;
- ensure that the contractor is obligated to subject its activities to the supervision of the Authority and observe all regulations in respect of the PSC;
- ensure that every PSC embodies specific undertakings by the contractor to the Authority to observe and perform all obligations enunciated in the contract;
- take appropriate steps against a licencee, lessee or contractor in relation to an event which pertains to terms of the contracts;
- place on offer on all available Blocks for competitive bid;
- develop a model PSC for Ministerial Council approval and adopt the same as a standard PSC for exploration and production activities.

JDZ licensing round

The Gulf of Guinea remained unnoticed for many years but recent events in the region have brought it into prominence. The extensive seismic activities spanning several years have shown that the region is oil bearing. Initial estimates indicate that the region has reserves ranging between 5 and 14 billion barrels of oil. Some gas reserves have also been discovered. As pointed out earlier in this chapter, Nigeria and DRSTP agreed in principle and further initiated a Treaty that, pending the final demarcation of the overlapping maritime boundaries of the respective states, the overlapping zone could be developed jointly. In furtherance of this mutual understanding a JDA was created. For purposes of allocation and contract execution part of the JDZ was initially demarcated into nine Blocks. These Blocks (1–9) were placed on offer for competitive bidding on 23 August 2003. All licensing round tenders undertaken by the JDA are subject to competitive bidding by interested parties. The tendering is rigorous in nature and involves several processes which culminate in the award of Blocks to successful candidates.

Available records indicate that 20 indigenous and MNCs submitted 33 bids. The number of bids notwithstanding, the Authority was constrained by the quality of the bids and technical and commercial competence of the companies to award only one Block to a consortium of ChevronTexaco (51 per cent), Esso (40 per cent) and Dangote Equity Energy Resources (DEER) 9 per cent. In furtherance of the developmental objectives of the zone, Blocks 2–6 were put on offer in 2003. The PSC for Block 1 has been signed and negotiations on Block 2–6 have been concluded paving the way for the signing of the PSCs.

Oil Block bid process

The tendering process administered by the JDA as a first step invites tenders from interested parties. Information provided includes:

- Block(s) earmarked for allocation,
- the bidding system to be adopted,
- details of the tender agreement to be executed,
- rights and responsibilities of potential concessionaires,
- the period within which applications are to be submitted.

Applications for competitive tenders

The intent to participate in the operations of any of the Blocks on offer is registered with the JDA through a written application. Such applications would be filed with a set of documents or information as follows.

- proof of payment of prescribed processing fee;
- proof of good financial standing and technical capability of applicant;
- details of work programme to be executed on the Block;
- the signature bonus offered by the applicant;
- detailed outline of the proposed annual expenditure;
- specific date when operations will start following the granting of a tender agreement by the JDA;
- detailed recruitment programme of staff of both party states and schedule of regular performance report.

Formation of the JDZ has strengthened the relationship between the two countries. The level of interest in the region continues to grow. It is expected that the areas of cooperation would increase, thereby paving the way for fisheries and other resources to be exploited in the region. One major consideration is the safety provided by the region and the potential for steady supply of oil and gas from the region. Nigeria and the US have already set in motion a Gulf of Guinea energy security initiative. This initiative will create an enabling environment that will promote the safe execution of oil related activities in the region.

Benefits of the JDZ

Political benefits

Prior to the formation of the JDZ by Nigeria and the DRSTP no tangible bilateral activities existed between the two countries. The determination on the part of the collaborating States led to a mutual agreement to explore and exploit natural resources jointly in the overlapping maritime boundaries. Formal demarcation of the maritime boundaries was deferred to occur some time in the future. The creation of the JDZ and the JDA has paved the way for officials of the Nigerian Boundary Commission and the Ministry of Petroleum Resources to visit DRSTP regularly to pursue the objectives of the JDA. Offices have been opened in Nigeria and DRSTP in order to administer the activities of the authority. The two nations have become closer to one another and areas of collaboration have extended to commerce. In July 2003 a military coup d'etat was staged in DRSTP by disgruntled military officers of the island state. One of the demands was the abrogation of the Treaty between Nigeria and DRSTP establishing the JDZ and the JDA. Sadly, the incident occurred during an official visit of President Fradique De Menezes to Nigeria. In spite of the embarrassment inflicted on Nigeria by the ill advised action of the military, President Obasanjo intervened and established dialogue with the recalcitrant officers. The issues at stake were resolved and the military coup was reversed. As a mark of solidarity President Obasanjo accompanied President Fradique De Menezes to DRSTP to ensure that his return to the island state was uninterrupted. Both countries have been cooperating in other mutually beneficial areas. The successful execution of the Treaty by both countries has demonstrated the capacity of countries in the region to amicably resolve boundary related problems. Other political benefits derived from the completion of the Treaty between Nigeria and DRSTP are as follows:

- amicable resolution of boundary disputes devoid of prolonged litigations;
- strengthening of bilateral cooperation and peaceful coexistence in the subregion;
- promotion of regional integration especially in the context of the enunciation by the New Partnership for African Development (NEPAD) and the African Peer Review Mechanism (APRM);
- promotion of democratic culture;

- promotion of Extractive Industry Transparency Initiative (EITI);
- exchange of Nigerian Industry Technocrats for training of personnel in DRSTP.

Economic benefits

The JDA has commenced activities and a number of Blocks in the JDZ have been awarded to successful bidders. Consistent with the practice in the industry, companies were required to pay a signature deposit as a precondition for securing the Blocks. In 2004, Blocks 1–6 were placed on offer for bid by interested companies. Block 1 was offered to ExxonMobil and Environmental Remediation Holding Corporation (ERHC). Details are shown in Table 9.1.

A total of $406 million was raised from the signature deposit of companies contained in Table 9.1. In addition $1.51 million was raised through the sale of bid forms and seismic data information. Other Blocks would in the near future be placed on offer for bid and attract inflow of sustainable revenues into the treasuries of the collaborating states. Chevron has also announced plans to invest $20 billion in the African region. Part of the investment will be in the Gulf of Guinea. The revenues generated through signature deposit and other related activities will be applied in the execution of projects and programmes that will boost economic activities in the African subregion. The exploration and production programmes will also spin-off other revenue generating activities in the oil services subsector, fabrication, financing and the provision of other professional services. The aggregate expenditures in the Gulf of Guinea have the potential to stimulate a multiplier effect that could have profound positive economic impact in the region. The opportunity exists for other countries in the region with overlapping maritime boundaries to emulate the spirit of cooperation demonstrated by Nigeria and the Democratic Republic of São Tomé and Príncipe.

Table 9.1 JDZ oil Blocks and signature deposit

SN	*Block/Company*	*Signature deposit*
1	Block 1 – ExxonMobil and ERHC	$123 million
2	Block 2 – Devon/Pioneer/ERHC	$71 million
3	Block 3 – Anardako, Devon/ERHC	$40 million
4	Block 4 – Noble/EHRC/Conoil	$90 million
5	Block 5 – ICC/OEOC	$37 million
6	Block 6 – Filthim Huzad Oil and Gas	$45 million
	Total	$406 million

Source: JDA publication – *The Journey So Far*, March 2006.

References

1 Bruso, J., Getz, S.L and Wallace, R. 'Oil and Gas Journal'. 16 Feb. 2004.
2 Tukur, H.A. 2005 'A Historic Step of Trust, Economic Co-operation and Brotherhood in the Gulf of Guinea'. *JDZ News*, Sept. 2006, pp. 10–11, 32–34.
3 Dimka, S., 2005 'Milestone to History: JDZ Oil Exploration Activity Begins'. *JDZ News*, 2005, p. 5.
4 Ibid., p. 6.
5 'Austria–Czechoslovakia Agreement Concerning the Working of Common Deposit of Natural Gas and Petroleum signed at Prague on 23 January 1960'. *United Nations Treaty series*, vol. 495, 1964.
6 'Norway–United Kingdom Agreement Relating to the Exploitation of the Frigg Field Reservoir and the Transmission of Gas therefrom to the United Kingdom'. Executed in London 10 May 1976. Treaty series N. 113 1977, Cmnd 7043.
7 Barberis, J.A.. 'Los Recuros Minerals Compartidos entre Estados'. *El Derecho International. Derecho de la Integracion*, vol. 162, 1979.
8 'Kuwait–Saudi Arabia, Agreement Relating to Partition of the Neutral Zone', Signed on 7 July 1965. *United Nations Document ST/LEG. SER. B/15, 760.*
9 Prescott, V. 1989 'Maritime Boundary Agreements: Australia–Indonesia and Australia–Solomon Islands' *Marine Policy Reports*, vol. 1, pp. 37–45.
10 'Oxfam: Submission to the Joint Committee on the Treaties Inquiry into the East Timor Sea Treaty'. August 2002.
11 Obiorah, S.U. 2005 'Expanding Frontiers – A Case of Nigeria/DRSTP Joint Development Zone'. *JDZ News*, vol. 1, September 2005, pp. 15–16.
12 Treaty between the Federal Republic of Nigeria and the Democratic Republic of São Tomé and Príncipe 21 February 2001.
13 Ibid., p. 16.
14 Ibid., pp. 18–19.
15 Tumsah, K.M. Nigeria – DRSTP Joint Development Zone: A Legal Perspective. *JDZ News*, September 2005, vol. 1, p. 8.
16 'Nigeria–São Tomé and Príncipe JDA Petroleum Regulations 2003', p. 10.
17 Olisa, M.M. 1997 'Nigerian Petroleum Law and Practice' 2nd edn, Fountain Books, Ibadan, 1987, p. 23.
18 Nigeria–São Tomé and Príncipe, op.cit. p. 10.

10 Refineries and petrochemicals

Introduction

Major infrastructural developments in the downstream sector in practical terms started in 1965 with the construction of the pioneer refinery at Alesa-Eleme in Port Harcourt. Thereafter, similar projects were executed at Warri and Kaduna as subsidiaries of NNPC. The downstream sector of the Nigerian petroleum industry has expanded significantly and prominent among these developments is the award of 18 licences to indigenous companies for the construction of private refineries.[1] Subsequent sections of this chapter will discuss the establishment of the country's refineries and petrochemical companies. This will provide an opportunity to evaluate the characteristics of some of the major NNPC assets in the downstream sector which can be identified as follows:

- Port Harcourt Refining Company Limited;
- Warri Refining and Petrochemical Company Limited;
- Kaduna Refining and Petrochemical Company Limited;
- Eleme Petrochemical Company Limited.

Port Harcourt refinery

The Port Harcourt Refining Company Limited (PHRC) complex consists of the old refinery which has a capacity of 65,000 b/d and the new refinery which has a capacity of 150,000 b/d. The old refinery was constructed in 1965 and relies on the Crude Distillation Unit (CDU) and the Vacuum Distillation Unit (VDU) for the processing of crude oil. The zone covered by the old refinery is often referred to as Area 1. On the other hand, the new Port Harcourt refinery, which has a capacity of 150,000 b/d, was initiated in 1984 within proximate location of the old Port Harcourt refinery. This refinery was constructed by a consortium of companies, namely Spie Batignolles SA (France), Spie Batignolles Nigeria Limited, JGC and the Marubeni Corporation of Japan. The activities of the new Port Harcourt refinery commissioned in 1989 are located in Areas 2–5 as follows:[2]

Area 2

Area 2 as classified in the refining system consists of the Naphtha Hydrotreating Unit (NHU), Catalytic Reforming Unit (CRU) and Kero Hydrotreating Unit (KHU). The naphtha hydrotreating unit hydro-desulphurises the naphtha while the CRU is responsible for upgrading the naphtha to higher octane value reformate. In the KHU the kero product is upgraded to make it suitable for aviation fuel. The refinery system relies considerably on catalysts. Sometimes their value or catalytic power weakens, warranting a rejuvenation process. This process is carried out in the catalyst rejuvenation unit. The caustic treatment unit, hydrogen purification, water treatment and fuel gas vaporiser are all located in Area 2.

Area 3

Area 3 comprises the Fluid Catalytic Cracker Unit (FCCU) which cracks vacuum gas oil and heavy gas oil into premium motor spirit (PMS) and LPG. The catalytic cracking process breaks complex hydrocarbon into simpler molecules. This is aimed at producing high quality, lighter and more desirable products. The FCCU process rearranges the molecular structure of hydrocarbon compounds to convert heavy hydrocarbon feed stock into lighter fractions such as kerosene, gasoline, heating oil, LPG and other petrochemical feed stock. Essentially catalytic cracking is similar to thermal cracking. However, in the case of FCC, catalysts expedite the conversion of the heavier molecules into lighter products. The use of catalysts in the cracking process increases the yield of improved quality products under far more relaxed operating conditions than in thermal cracking. The catalysts used in the refining process are mainly solid materials such as zeolite, aluminum, hydrosilicate, treated bentonite clay, Fuller's earth, bauxite and silica-aluminia.[3] Typical temperatures in the FCC process range are between 850 and 950°F at 10–20 psi.

In the FCC a preheated hydrocarbon charge is mixed with a regenerated catalyst at its entry point into the riser leading into the reactor. The charge is mixed with recycled steam within the riser, vaporised and raised to a reactor temperature of 900–1,000°F with the aid of the catalyst. As the mixture moves up in the riser, the hydrocarbon charge is cracked at 10–30 psi.

Area 4

This area has within it dimersol, butamer isomerisation and the alkylation units. These units play a major part in producing high octane gasoline.

Isomerisation

The isomerisation process converts n-butane, n-pentane and n-hexane into their respective isoparaffins with substantially higher octane numbers. The

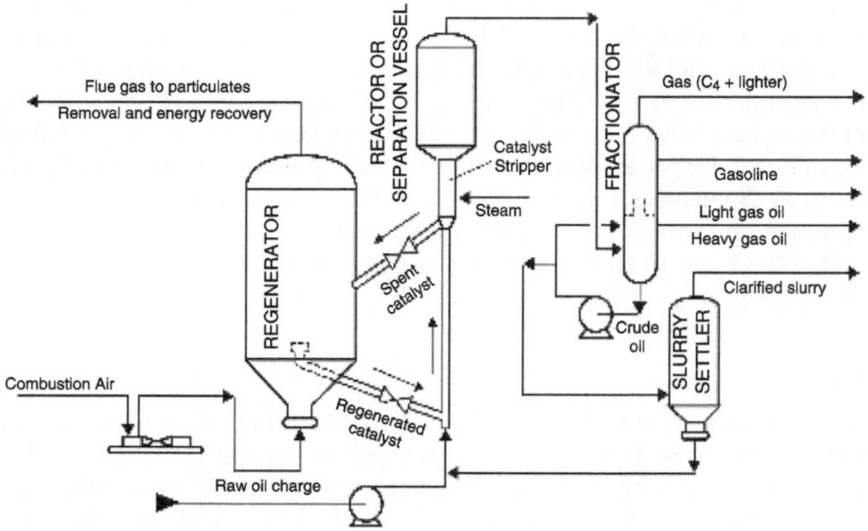

Figure 10.1 Schematic of Fluid Catalytic Cracker (FCC) unit.
Source: Set Laboratories 2007

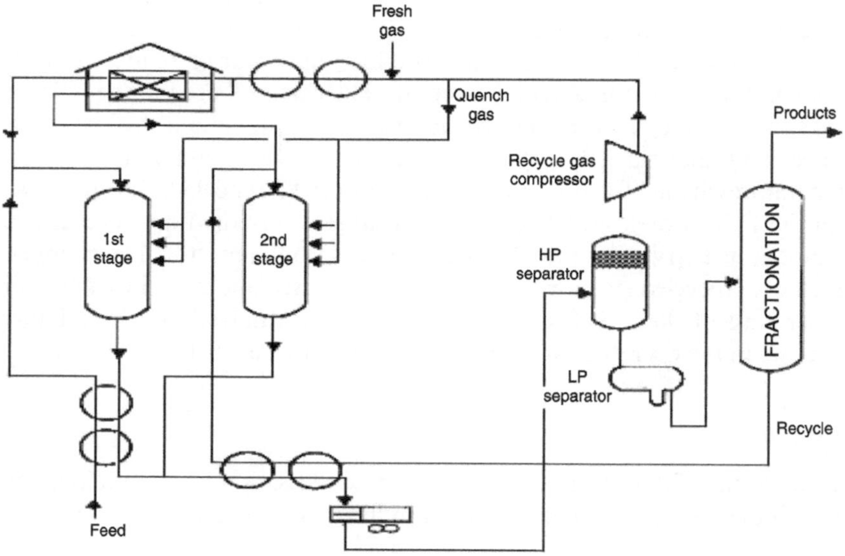

Figure 10.2 Schematic of Two-Stage Hydrocracking Unit.
Source: Set Laboratories 2007

straight-chain paraffins are usually transformed into branch-chained counterparts. These equivalents of the paraffins contain similar atoms but are arranged in different geometric structures. The isomerisation process is essential for the conversion of n-butane into iso-butane, which serves as feed stock for the alkylation unit and the conversion of normal pentane and hexanes into higher branched isomers suitable for gasoline blending. Isomerisation is identical to catalytic reforming in the sense that hydrocarbon molecules are rearranged. Two distinct isomerisation processes which involve butane (C_4) and pentane/hexane (C_5/C_6) are common in the refining process. Butane isomerisation produces feed stock for alkylation. The process which takes place at low temperatures utilises aluminum chloride catalyst in addition to hydrogen chloride for the conversion. Pentane/hexane isomerisation increases the octane number of the light gasoline components. In a standard C_5/C_6 isomerisation procedure, dried and desulphurised feed stock is mixed with a small portion of organic chloride and recycled hydrogen. This mixture is then heated to a temperature of about 230–340°F and at a pressure of about 200–300 psi. In further sequence the mixture is then passed over a metal catalyst in the first reactor where benzene and olefins are hydrogenated. The feed proceeds into the isomerisation reactor in which the paraffins are catalytically isomerised into isoparaffins.[4]

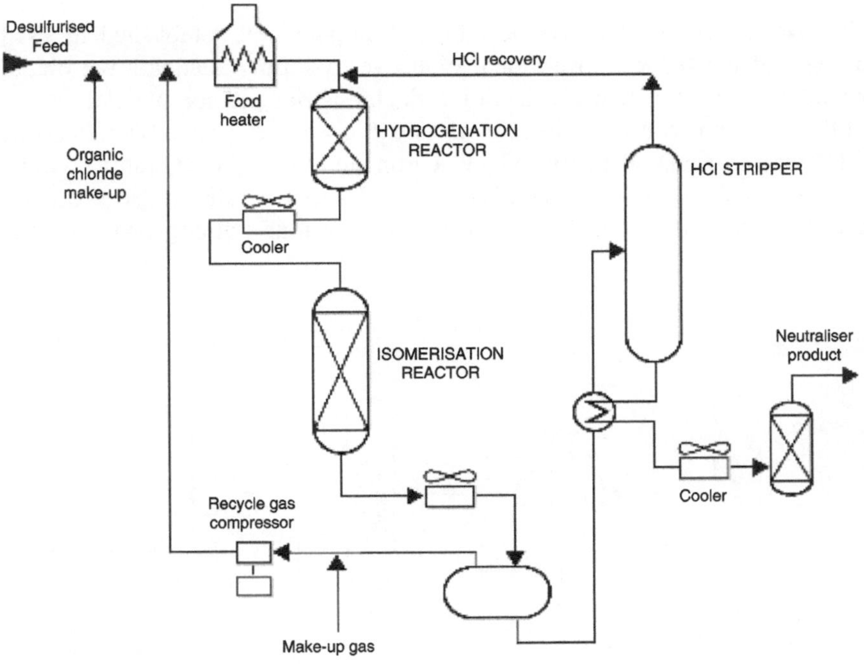

Figure 10.3 Schematic of C_5 and C_6 isomerisation.

Source: Set Laboratories 2007

Alkylation

Alkylation combines low molecular weight olefins (mainly propylene and butylene) with isobutene in the presence of sulphuric acid or hydrofluoric acid which acts as a catalyst. The resultant is an alkylate product which is composed of a mixture of high-octane and branched-chain paraffin hydrocarbons. Alkylate is a suitable premium blending stock in view of its robust anti-knock properties.

In the above process the feed stock (mainly propylene, butylene, amylene and fresh isobutane) is introduced into the reactor and comes into contact with sulphuric acid (H_2SO_4), a catalyst in concentrations of 85 to 95 per cent. The reactor is separated into zones such that olefins are introduced into each zone through distributors, and within the same period the sulphuric acid and isobutanes flow over the reactor content from zone to zone. The reactor effluent is separated into hydrocarbon and acid phases in a settler. The old and new Port Harcourt refineries have a combined processing capacity of 210,000 b/d. The facilities, design capacity and production slate of the Port Harcourt Refinery are outlined in Tables 10.1 and 10.2 respectively.[5]

Warri refinery

The Warri Refining and Petrochemical Company was established in 1978 as part of the federal government objective to supply adequate petroleum products to meet demand in the domestic, industrial and commercial sectors of the Nigerian economy. The refinery essentially processes two types of crude oil: Escravos crude oil, supplied by Chevron; and the Ughelli Quality Control Centre (UQCC) crude oil, supplied by Shell. Occasionally it also processes the Forcados blend crude oil. Through the various refining processes the

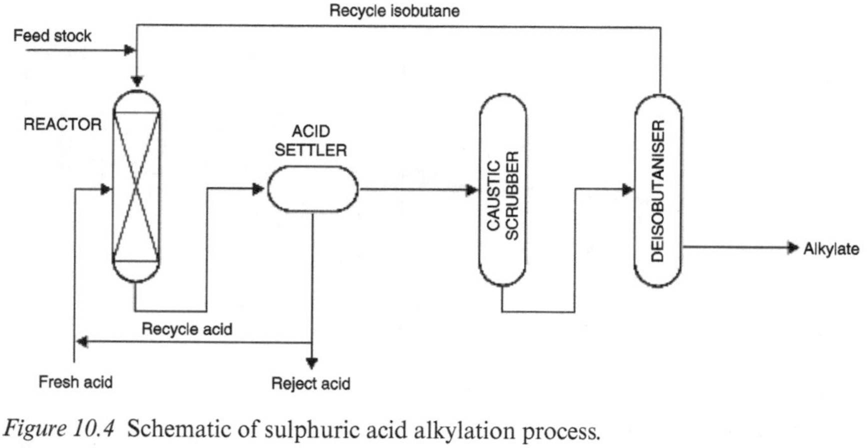

Figure 10.4 Schematic of sulphuric acid alkylation process.
Source: Set Laboratories 2007

Table 10.1 Port Harcourt refinery production slate

Feed processed (Crude)	*MTPA*	*%WT*
Crude products	9,335,000	100
C_3	22,000	0.2
C_4	34,200	0.4
LPG	53,500	0.8
PMS	3,080,350	33.0
DPK	1,416,550	15.9
AGO	2,458,960	26.3
LPFO	974,500	13.4
HPFO	801,300	0.6
Fuel gas	373,400	4.0
Flare loss	25,000	0.3
Net liquid loss	90,300	1.0
TOTAL	9,335,000	100.0

Source: Port Harcout Refinery Bullet in 2005.

Table 10.2 Port Harcourt Refining Company facilities and capacity outline

Areas	*Units*	*Capacity*	*Licensor*	*Year*
Area 1	Crude distillation	150,000 b/d		1989
	Vacuum distillation	53,560		1989
Area 2	Naphtha hydrotreating unit	33,000 b/d	UOP	1989
	Catalytic reforming unit	33,000 b/d	UOP	1989
	Kero hydrotreating unit	14,500 b/d	IFP	1989
Area 3	Fluid catalytic cracker unit	40,000 b/d	UOP	1989
	Gas concentration unit	40,000 b/d	UOP	1989
	Gas treating unit	LPG: 11,478 b/d Fuel gas: 16,458 nm^3/hr		
Area 4	Dimersol unit	4,850 b/d	IFP	1989
	HF alkylation unit	7,200 b/d	UOP	1989
	Butamer unit	3,610 b/d	UOP	1989
Area 5 (OPHR)	Crude distillation unit	60,000 b/d	UOP	1965
	Platforming unit	6,000 b/d	UOP	1965
	LPG plant	60 Tons/day	UOP	1965
Power plant and utilities	Steam boilers (4NOS)	120 Tons/hr each	BABCOCK	1989
	Turbo generators (4NOS)	56MW (14 mw each)	ALSTOM	1989
	Skid mounted nitrogen plant	500 nm^3/hr (gaseous)	ATLAS COPCO	1999
	Nitrogen plant	600 nm/hr (gaseous/liquid)	LINDE	1989

Source: Port Harcourt Refinery Bulletin 2005.

plant produces LPG, PMS, Dual Purpose Kerosene (DPK) and Automotive Gas Oil (AGO – diesel), Low Pour Fuel Oil (LPFO) and High Pour Fuel Oil (HPFO). The Warri refinery had an initial capacity of 100,000 b/d and was subsequently de-bottlenecked to operate at 125,000 b/d.

Figure 10.5 Section of refining and petrochemical company.
Source: Port Harcourt Refining Company Limited, 2004

Processing units

The plant consists of the primary and secondary processing units. The atmospheric distillation unit and the gas plant constitute the primary processing units. On the other hand, the secondary processing unit comprises the naphtha hydrotreating unit, the naphtha reforming unit, kerosene hydrotreating unit, FCC, the Merox units, and the Hydrofluoric (HF) and the alkylation unit.

CDU

The process at this stage involves the sequential pumping of crude oil from the storage tanks through heat exchangers into a heating furnace. During the heating process salts and other impurities are removed electrolytically after which the hot feed stock is passed into the distillation tower. In the tower the vapours are separated from the liquids. The vapours are extracted from the top of the tower, cooled in a condenser and transferred to the gas plant. During the same process, the liquid fractions are extracted at various levels in the tower and also at the bottom. The liquid fractions are naphtha or straight-run gasoline, heavy naphtha, kerosene, AGO and heavy atmospheric gas oil.

- Naphtha – this product comes out in direct useable form and requires no further processing. It is usually channelled to storage.
- Heavy naphtha – this is sent to a naphtha hydrotreating unit where it is further processed.
- Kerosene – this product is hydrotreated in the kero hydrotreating unit to obtain high grade Dual Purpose Kerosene (DPK) which is used as aviation fuel.
- AGO – this comes out in final useable state and is subsequently transferred to storage tanks.
- Heavy atmospheric gas oil – this fraction of the distillation process requires further processing in the FCC.
- Atmospheric residue – this is the final set of liquids which is normally channelled to the VDU for additional processing.

The gas plant

Gas is one of the products of the refining process. The gas from the CDU is processed to separate butane from propane, both of which are in liquid state and high pressure. Butane and propane are often referred to as LPG. The separation process takes place under high pressure but lower temperature than the CDU where furnaces are involved.

Vacuum distillation unit

The function of the VDU is to process the atmospheric residue under high temperature and low pressure to produce vacuum gas oil which is moved to the FCC for further cracking into appropriate hydrocarbons. The vacuum residue is utilised for blending into fuel oil.

Naphtha Hydrotreating Unit (NHU)

The NHU subjects the heavy naphtha to hydrogen treatment in a chemical reactor filled with a catalyst. The primary aim is to desulphurise the naphtha and prepare it for further processing into a high grade gasoline pool. The hydrotreated naphtha as a prerequisite is heated to a high temperature before being fed into the chemical reactor where the desulphurisation takes place. The process also requires high pressure to allow the catalyst to be effective.

Naphtha Reforming Unit (NRU)

Heavy naphtha has a Research Octane Number (RON) of less than 60 compared to regular gasoline which has a RON of 92.5. In order to improve the heavy naphtha quality as a gasoline component it is treated with hydrogen in a catalytic reformer. The NRU allows the desulphurised naphtha to be further mixed with hydrogen. This mixture is then sent through four furnaces and four catalytic reactors under moderately high pressure in an alternating fashion. The outcome of this process is the reformate which serves as a high grade gasoline blending stock. The NRU produces a small quantity of gas which is channelled to the gas plant or the refinery fuel gas stream. Excess hydrogen is also produced and this is utilised in naphtha and kerosene hydrotreating units respectively.

Kerosene Hydrotreating Unit (KHU)

The KHU provides a suitable medium for the mixing of the raw kerosene from the CDU with hydrogen under catalytic conditions. This process produces Dual Purpose Kerosene (DPK) which is of a higher grade and suitable for use as aviation fuel. The DPK can be used as domestic fuel except that it is more expensive and therefore not economical.

The Fluid Catalytic Cracker (FCC)

The FCC is central to the refining process because its functions contribute significantly to improve the overall yield in terms of products volume. The heavy gas oil and the vacuum residue are mixed and raised to high temperatures with the aid of heat exchangers in a furnace and fed into the FCC. The FCC, which relies on catalysts, cracks the heavy residues into lighter hydrocarbons. The hydrocarbon vapours are channelled into the main fraction tower where they are separated into gas, gasoline, decanted oil and slurry oil. The gas produced in the process is further processed to produce two grades of LPG, namely propane and butane. The decanted oil serves as a feed stock for the carbon black plant but can also be blended into fuel oil if the plant is non-operational. The sulphur in the LPG and gasoline from the FCC are catalytically removed in the Merox unit.

HF (Hydrofluoric) alkylation unit

The HF alkylation unit is usually fed with butane streams from the gas plant and the FCC. The unit converts the isobutane and the olefin content of the butane in a catalytic process into alkylate. Alkylate is a high grade gasoline blend stock used in the production of lead free gasoline.

The petrochemical plant

The petrochemical plant of Warri refinery was commissioned in 1988 and consists of the Polypropylene (PP) and the carbon black plants. PP is one of the primary products of the plant. It is a thermoplastic polymer formed through the linking together of many molecules of PP. It was developed in 1954 by G. Natta of Italy and was produced in a factory setting for the first time in 1957 by Montecatini Corporation of Italy (www.icett.or). It was at

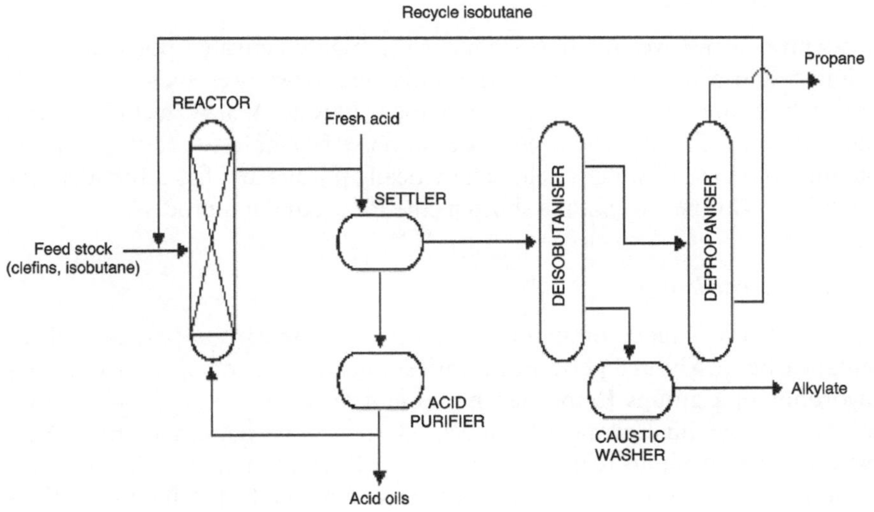

Figure 10.6 Schematic of hydrogen fluoride alkylation.

inception developed for sale as a fibre. The fibres obtained through melt spinning were light and strong. However, they were not amenable to dyeing; therefore their use in the garment industry was minimal. At present, PP is extensively used in moulded products, namely:

- industrial materials – pipes, containers, mechanical and electrical appliances;
- automotive wares – battery cases, bumpers, dash boards;
- household articles – furniture, containers, refrigerators and crates;
- fibre-flat yarn, mono-filament, staple fibres, blankets, mats etc.

Innovation in technology has moved the manufacturing process to a higher pedestal requiring greater concentration on catalyst improvement. The PP manufacturing process has gone through first and second generations. Consequently, it has arrived at a third generation which is closer to a final process.

FIRST GENERATION PROCESS

This process relies on TiC13 organic aluminum compound catalyst. The catalyst is usually dispersed in a hydrocarbon solvent and PP is polymerised at about 10 kgf/cm^2 at a temperature between 60°C and 80°C.[6] Once the catalyst is broken down with the aid of a decomposition solution material such as methanol, polymer is washed and PP is separated from the solvent. The material is converted into a finished product by drying and making it into pellets.

SECOND GENERATION PROCESS

This process involves the development of a highly activated polymerisation catalyst and a process for extracting polarised substances such as H_2O, O_2 etc. which are known to deactivate the catalyst. A new technique was developed to eliminate the need to decompose the catalysts and relies on the gamma type TiC13 which is characteristically porous and has a broader surface area. This new technique shortens the PP production process.

Carbon black plant

Carbon black is made of about 90–97 per cent fine carbon particles (elemental carbon). About 3 per cent comprises phenolic, lactonic, carboxylic and hydroquinonic groups. Embedded in the amorphous mass is an intra-structure of micro-crystalline arrays of condensed rings. The pattern of the arrays within the amorphous mass is characteristically random and the carbon percentages as well as the particle size vary depending on the production method engaged. Carbon black is a product of incomplete combustion and it is often described as the dark component of smoke. Essentially all carbon black processes commence with the production of smoke, as decanted oil and vaporised propane are used as the feed stock for the plant (NNPC publication in

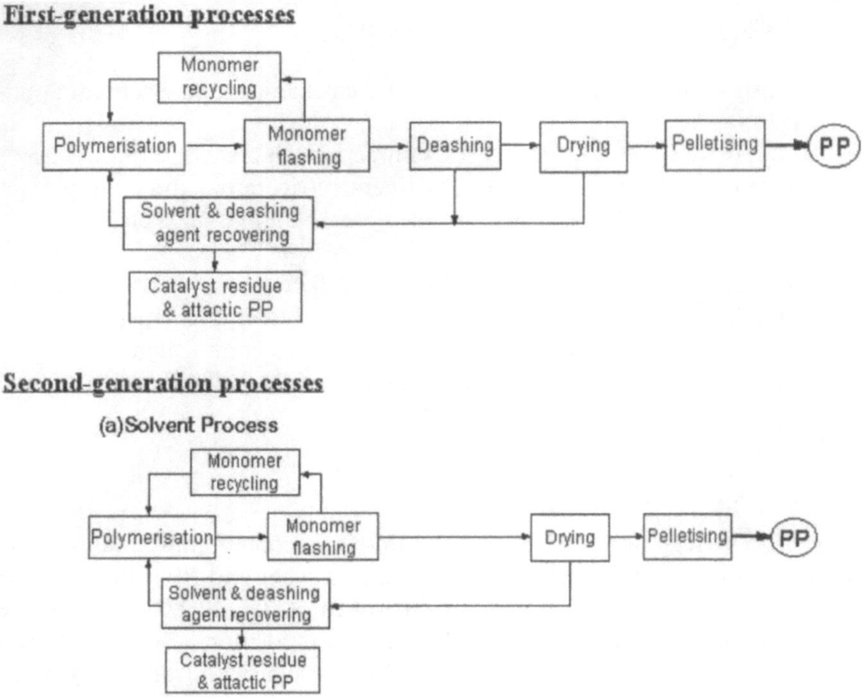

Figure 10.7 First and second generation PP manufacturing processes.

WRPC 2000). Carbon black production relies on a reactor in which the conversion oil is decomposed to carbon black. In practice, the preheated decanted oil is introduced into the reactor through spraying nozzles. In this medium the fuel gas and preheated atmospheric air are burnt to produce a hot 'blast' which mixes with the decanted oil to produce thick smoke. The process leads to the formation of carbon black.

Kaduna refinery

The Kaduna Refining and Petrochemical Company was commissioned in 1980. It had an initial capacity of 100,000 b/d and was subsequently de-bottlenecked to operate at a capacity of 110,000 b/d. The refinery utilises Venezuelan, Kuwaiti and Arabian light crude oil to produce various grades of petroleum products. It has four processing units, namely:

- the fuel plants with an installed capacity of 60,000 b/d;
- tin and drum unit;
- the petrochemical unit;
- the lubes plant with an installed capacity of 50,000 b/d.

The primary products of the plant are Linear Alkyl Benzene (LAB), kerosolvents, base oil, waxes, reformated benzene, toluene and toluene concentrates. The plant, built by Chiyoda of Japan, was designed to produce 30,000 MT/yr of LAB, 2,700 MT/yr of heavy alkylates, and 38,000 MT/yr of deparaffinated kerosolvents. All these products are consumed by the domestic market. Heavy alkylates serve as basic input for the production of oils and grease. LAB is a basic ingredient for detergents while the solvents are used in the manufacture of paints. The feed stocks for the petrochemical plant which comprise kerosene and naphtha are derived from the Kaduna refinery. Petroleum products from the plant are evacuated by PPMC while the petrochemical products are marketed directly to companies that use them as feed stock for their plants. Tins and drums which are also manufactured at the plant are marketed directly to end-users. The petrochemical products are produced using contemporary techniques some of which have been discussed earlier.[7]

Operational constraints

Port Harcourt, Warri and Kaduna refineries have similar operational constraints. For this reason, such constraints will be discussed collectively. The problems of the refineries accumulated over a period of time but became quite serious in the 1990s and reached crisis level by 1999. It is a known fact that the refineries were not maintained on schedule. For this reason the key processing units and other equipment deteriorated. This led to low performance and frequent breakdowns. Apart from the main TAM, availability of spare parts for isolated repairs of pumps or other critical equipment was a problem and ailing equipment could not be regularly repaired; therefore the plants experienced down-time. Issues of maintenance and spare parts have over the years been extensively examined, and it was concluded that lack of government understanding of the operational system of the refining plants accounted for the delays experienced in the release of funds needed to carry out major TAM. It is also contended that some of the delays in spare parts procurement emanated from bureaucratic bottlenecks in NNPC. Chief executives of the refineries argue that the idea of forwarding all major requests for procurement of critical spare parts to the corporate headquarters caused delays that are detrimental to the smooth operation of the plants.

In view of this, the CEOs of the refineries have delegated both administrative and financial authority so as to allow timely intervention to effect critical repairs in the refineries. This point is well appreciated by management but no definite action has been taken to empower refineries to procure key materials single-handedly for the operation of the plants. It is important to indicate, however, that the refinery procurement system was not centralised from inception. Refineries enjoyed some measure of autonomy with the result that TAM was carried out smoothly. However, in subsequent years the rules of accountability in refinery operations were broken. Consequently, issues of misappropriation of funds, procurement of substandard spares as well as outright inflation of contract prices became a serious problem in local refinery administration. These issues, to say the least, eroded the confidence of top management in the administrative integrity of the refinery leadership. This is not to say that all refinery CEOs were guilty but the dent created by their predecessors was deep enough to erode confidence in the minds of corporate management.

These problems collectively led to poor equipment performance which then caused a drastic drop in the overall output of the refineries, in some cases to about 40 per cent. The situation was rather precarious and the government decided to privatise Port Harcourt and Kaduna refineries. This decision was hinged on the fact that the huge investments in TAM were not producing satisfactory results. Port Harcourt refinery was earmarked for privatisation by the Bureau for Public Enterprises (BPE) and bids were on two occasions submitted by interested buyers. Unfortunately, the offers in both cases fell far short of the reserve price. In 2005, NNPC embarked on project PACE which was designed to infuse a new orientation in the work ethics, corporate governance and adherence to systems and processes. To execute project PACE the consultants Accenture and Shell Manufacturing Systems (SMS) carried out a diagnostic study of the corporation. Major operational gaps and flaws were detected. Based on the diagnosis, remedial measures were introduced which have stopped some of the excesses experienced both at the corporate level and at the SBUs. In consideration of these positive developments, NNPC appealed to the federal government to rescind the decision to privatise Port

Harcourt refinery. The request was accepted and as a result the refinery was momentarily taken off the privatisation list. In June 2007 it was rescheduled and sold to Transcorp. The sale was opposed by Trade Unions in the oil industry (NUPENG and PENGASON) through strike actions. Other labour unions also embarked on a nationwide strike demanding that the federal government reverse the sale of the refinery. In addition the unions demanded that an increase in the price of premium motor spirit from ₦65 to ₦75 and VAT from 5 per cent to 10 per cent be reversed. The strike actions caused serious economic and social disruptions. In an effort to mitigate the negative impact of the strike on the citizenry the federal government granted some concessions and the price of PMS was reduced to ₦70 while the increase in VAT was suspended. In the case of the Port Harcourt refinery, BPE offered 10 per cent equity interest each to the workers and the host community respectively. This gesture on the part of BPE notwithstanding, the demand for the reversal of the sale of the refinery from members of the public continued unabated. However, Kaduna refinery is still scheduled for privatisation. An in-depth assessment of the situation revealed that the refineries underperformed due to external factors such as pipeline vandalism. Warri and Kaduna refineries have been victims of such vandalism perpetrated by unknown persons in the Niger Delta and these acts prevented the transportation of crude oil to the plants. Both refineries experienced interrupted crude oil supply for an extended period of over one year. These operational disruptions adversely affected the revenue flow of the plants thereby making them suitable candidates for privatisation.

Eleme Petrochemical Company

The Eleme Petrochemical Company Limited (EPCL) was commissioned in 1996. It was initiated by the government as a priority project to be developed in three stages. Stage 1 included olefins plant, polyethylene plant, PP plant, butene-1 plant, utilities and offsite facilities, plastics development centre etc.

The complex was, among other things, designed to provide:

- basic petrochemical raw materials for local industries;
- utilise the abundant natural gas as a feed stock;
- provide jobs within the complex and spin-off job creation;
- act as a catalyst for industrial expansion; and
- save foreign exchange.

The estimated cost of the plant was ₦754.15 million plus $1 billion. A consortium of banks, namely Japanese (JFC and Japan Export Import Bank – JEXIM), Italian (Credito Italiano) and French (Banque Française du Commerce Exterieure–BFCE) financed 85 per cent of the offshore loan which amounted to $857.38 million. The project was configured to repay the loans using export sales revenues and the mechanism set up for this involved the establishment of a trust account with the Bank of Tokyo-Mitsubishi Trust Company (BTMT), New York as the front trustee to collect export sales revenues for distribution to lenders. This arrangement notwithstanding, several unforeseen events caused project delays. Consequently, the project was unable to generate funds to effect loan repayment. In this regard, a loan deferral agreement was signed in 1994 and 1996 respectively between NNPC and the financiers to allow the outstanding loan of $970 million to be rescheduled for payment in subsequent periods.[8] The agreement also provided as follows:

- All sales proceeds of Itochu corporation (a member of Japan Finance Corporation–JFC) 20,000 b/d crude oil liftings were to be paid into an existing FGN account with Bank of Tokyo-Mitsubishi, Tokyo.
- A minimum balance of $50 million would be maintained by the central Bank of Nigeria.
- Other sources of funding including domestic channels were to be explored by NNPC, CBN and the federal government for loan repayment.

Feed stock supply

NGL, which served as feed stock for the complex was supplied by NNPC (60 per cent), NAOC (20 per cent) and Phillips Oil (20 per cent). The NGL plant at Obiafor Obrikom was built exclusively to supply the feed stock to EPCL and the pricing of the feed stock was determined such that the investment on the NGL plant plus a 15 per cent mark up could be recovered at the end of 15 years. The Agip and Phillips portions were to be paid in foreign currency while the NNPC portion would be paid in local currency. In

Table 10.3 Eleme petrochemical plant capacity and configuration

	Plant/Facility	*Capacity MT*	*Contractor*
1	Olefins plant	300,000 ethylene, 126,000 propylene	Chiyoda, Japan
2	Polyethylene plant	270,000 LDPE/HDPE	Kobe Steel Ltd, Japan
3	Polypropylene	80,000 PP	Tecnimont, Italy; JGC, Japan
4	Butene-1	22,000	Kobe Steel Ltd Japan
5	Utilities and offsites	N/A	Chiyoda
6	Infrastructure	N/A	Spibat France; Spibat Nigeria

Source: EPCL Operational Update Document 2008.

comparative terms experts are of the view that EPCL NGL is expensive in relation to similar feed stock in Saudi Arabia and other OPEC countries. In addition to NGL the plant relies on PP Rich Feed Stock (PRF) from the Port Harcourt Refinery FCC to meet 55 per cent of PP requirements.

Petrochemical products

Marketing, plant maintenance and other exigencies of the project marketing agreements were signed with Nova Chemicals of Canada and Tecnimont of Italy. These companies were to market polyethylene and PP products respectively. The agreements which were for 10 years duration had an inbuilt 7.5 per cent sales commission which was to be deducted at source. As at 1988, 87,500 MT of Grade 'A' polyethylene (PE) and PP, and 33,900 MT of Grade 'B' PE products had been sold in the export markets. These sales generated $42.4 million and $12.22 million respectively from the Grade A and Grade B products. EPCL products were exported to the UK, the Netherlands, Turkey, Egypt, India, Côte d'Ivoire, Algeria, the US, Spain, Belgium and South Africa. At the domestic level a total of 84,000 MT of PE and PP resins had been sold at the end of November 1988. This generated a total of ₦6.74 billion. Although the demand for the products was strong and sustainable, several factors militated against the smooth operations of the plant. These problems will be discussed below.

Operational constraints

The operations of EPCL were beset by a number of problems which included the following:

- working capital;
- spare parts;
- loan burden;
- feed stock supply;
- TAM.

Working capital

EPCL was initiated by the government as a priority project in consideration of the capacity of the plant to stimulate industrial development. The plant was expected to promote import substitution, gas utilisation, technology transfer and employment generation, and also to create a multiplier effect through the activities of small scale industries. At project inception in 1990 it was envisaged that the plant would run smoothly and generate enough revenue from sales to fund its operations. The initial financial projections turned out to be unrealistic due to a decline in prices, and as a result the projected revenues could not be realised. Meanwhile, the operations of the plant were funded with proceeds from domestic sales and Grade 'B' export sales. The working capital requirements by far exceeded the revenues generated through domestic sales and Grade B export sales. In 1998 projected revenues were estimated to be ₦10.65 billion and $7.56 million respectively, while actual receipts amounted to only ₦3.8 billion and $4.61 million respectively. During this period the plant experienced a revenue shortfall of $35 million. This situation prevented EPCL from procuring chemicals, catalysts etc. to keep the plant in continuous operation. Subsequently, the circumstances led to capacity underutilisation and high down-time of plants. The NGL feed stock from Agip experienced unscheduled interruptions due to community disturbances, and operational exigencies also prevented Port Harcourt Refinery (PHRC) delivering PRF to the plant. The aggregate of the problems served as serious obstacles against the smooth operation of the plant.

Spare parts

It is customary to stock two years' spare parts prior to commissioning of the plant, but this practice was not adhered to in the case of EPCL. Lack of spare parts militated against scheduled maintenance of equipment thereby jeopardising the integrity of the plant.

Loan burden

EPCL had an outstanding loan of about $970 million which had to be amortised. As pointed out earlier, the sales revenues were grossly inadequate to pay for portions of the loans that were due for amortisation. In this regard the company was unable to raise funds both locally and internationally to sustain the operations of the plant.

Feed stock supply

Constant availability of feed stock is critical for the smooth operation of the plant. Interruptions in NGL supply warranted shut down which lasted 10 to 15 days. Interruptions in NGL supply are often caused by local communities which make demands on NAOC and ConocoPhillips. Similar disruptions were experienced in the PRF supply from the Port Harcourt refinery. Delay of feed stock supply from there was mainly equipment related.

TAM

The mandatory TAM was due in 1999 at an estimated cost of $30 million. However, this amount was not provided, which prevented maintenance from taking place. This delay affected segments of the plant, causing shut downs and deterioration of critical equipment.

EPCL privatisation

It is important to state that the problems of EPCL escalated such that the operations could not generate sufficient revenues to sustain the activities of the plant, pay salaries, as well as defray the overhead expenditures of the company. The dismal performance of the plant coupled with the huge loan outstanding caused serious concerns for the federal government. In view of this, BPE was directed to schedule the plant for privatisation. Accordingly, the company was privatised in 2005 and Indorama became the new owners. Formal operations under Indorama commenced in October 2006 and although the company was successfully handed over to the core investors, the circumstances surrounding the process have been the subject of enduring debate among trade unions, law makers and members of the general public. The privatisation process has benefited only a few companies which have succeeded in acquiring major public assets, namely hotels, the Nigerian Telecommunication Company (NITEL) etc. The federal government hotel chain and NITEL are huge national assets. EPCL is situated on vast hectares of land (hitherto used for farming) donated by communities in good faith to fulfil national objectives. The communities argue that the vast parcels of land in question were not sold to the government but were simply volunteered for a worthy cause.

References

1 Department of Petroleum Resources (DPR) Lagos, Nigeria.
2 Port Harcourt Refinery publication, 1996, pp. 1–4.
3 Set laboratories, www.setlaboratories.com
4 Op. cit.
5 Port Harcourt Refinery Bulletin, 2001, p. 8.
6 www.icett.or.jp pp.1–3.
7 Kaduna Refinery Bulletin, 1996, pp. 1–13.
8 Eleme Petrochemical Company Limited, EPCL, Operational Update Working Documents, 1998, pp. 1–17.

11 Products marketing companies

Introduction

Petroleum product marketing in Nigeria was introduced in 1908 by Socony Vacuum Oil Company which later became Mobil Nigeria Plc. In the early years the 'Sunflower' brand of kerosene was marketed in Lagos, Kaduna, Enugu and Ibadan. In subsequent years other MNCs – Shell, Texaco, Esso and British Petroleum – marketed petrol, kerosene, diesel and Aviation Turbine Kerosene (ATK). In 1965 the old Port Harcourt refinery with an initial capacity of 35,000 b/d was jointly established by Shell and British Petroleum to refine crude oil locally to produce petrol, kerosene, diesel and other related products. The capacity of this pioneer refinery was later expanded to 65,000 b/d in response to increasing domestic demand. Between 1974 and 1976 incidents of product scarcity were experienced and a study conducted by the government established that the scarcity derived from low supply as well as inadequate distribution channels. More importantly, it was confirmed that the demand for petroleum products had increased significantly, so in 1978 Warri refinery was constructed and the southern pipeline network was completed. In 1980 Kaduna refinery was commissioned and within the same period the northern products pipelines and depots were established to transport, store and distribute petroleum products from the refinery. In the area of products marketing the subsector was dominated by National Oil, African Petroleum, Unipetrol, Texaco, Elf, Agip and Mobil. Some of these companies have changed ownership: Conoil and Oando have acquired National Oil and Unipetrol respectively. Elf has also metamorphosed into Total.

In addition to these developments, the government has liberalised and partially deregulated the activities in the downstream sector, which has created avenues for both major and independent marketers to import products. They have also been licensed to construct storage depots to support their businesses. Although prices of petroleum products are yet to be determined by market forces, marketers have had an opportunity to interact with the Petroleum Products Prices Regulatory Agency (PPPRA) to negotiate the price of products at the pump. There is a general feeling among marketers that the 'regulated price regime' cannot create a conducive business environment for the importation of products. It is argued that the regulated prices are not sufficient to cover the cost of importation and haulage of the products to the distribution points. For this reason some marketers have indulged in manipulation of their distribution systems to underdeliver products to customers. NNPC and DPR are against these sharp practices. Defaulting marketers have been subjected to severe penalties which include closure of their petrol stations. This setback notwithstanding, one can contend that the downstream sector has experienced phenomenal growth. Apart from the major marketers, many independent markets have registered their presence in the subsector.

At present it is estimated that over 1,490 independent marketers operate in the downstream sector. Some of the stations are small and feature two pumps while others have standard infrastructure and operate eight pumps or more. The proliferation of independent marketers and stations derived from the products scarcity era, which promoted black market activities, and some of the stations were constructed with a view to securing approval for higher products allocations by PPMC. Some marketers secured approval for an increase in the volume of products allocated to their companies through the expansion of the marketing outlets, but others who built additional stations in anticipation of allocation of higher volumes were not so successful, which jeopardised their speculative investments. The products scarcity has been brought under control and therefore some of the stations remain redundant because black market activities can no longer be sustained. Such stations are on offer for sale in the market but the standards of construction and indeed the locations fail to attract potential buyers. The 21 depots across the country are linked by over 4,950 km of pipelines,[1] and products are pumped from one location to the other with the aid of pumping stations. Products marketing is a strategic activity of the downstream sector; therefore subsequent sections of this chapter will focus on the activities of the PPMC.

Creation of PPMC

The PPMC was established in 1988 as a strategic subsidiary of NNPC. Its establishment was a direct government intervention aimed at providing suitable infrastructure for the transportation and distribution of petroleum products nationwide. As demand for petroleum products has increased from a modest level of 9 million litres/day in the 1970s to about 30 million litres/day in 2006, PPMC has become the focal point for all petroleum products-related activities, allocating products to and coordinating the activities of the major marketers – Conoil, Oando, AP, Total, Mobil and Texaco. It also interfaces with 1,490 independent marketers. The major marketers control about 60 per cent of the domestic market while the independent marketers control the remainder. On the whole, both major and independent marketers operate over 7,000 retail outlets across the country but local refineries, operating at less than full capacity, account for aggregate production of only about

18 million litres/day. In a situation where the total demand is 30 million litres/day, the balance of 12 million litres daily requirement is imported. This is required to create equilibrium in supply and demand.

The scarcity situation in 1999 created an opportunity for elaborate black market activities. Products allocated to specific stations within Nigeria were diverted, smuggled to neighbouring countries and sold at outrageously high prices. These artificial profits instigated the construction of petrol stations in order for the speculators and profiteers to apply to PPMC for higher products allocation. The aggressive pursuit of these objectives led to what might be considered an overindulgence in the establishment of petrol stations. Some of the stations were constructed without proper location analysis and survey. Consequently, they failed to attract adequate patronage from motorists. Sadly, these stations when supplied with products, manipulate their dispensing pumps and underdeliver products to customers. These malpractices are against the operational ethics of PPMC and DPR and offenders have been subjected to severe penalties, including closure of their stations. PPMC has a large staff strength, a wide pipeline network covering 4,950 km, pumping stations and 21 products depots dispersed throughout the country.[2] The task of crude oil transportation to the refineries, products distribution through the network of pipelines, and sale of products at the depots presents a high challenge. In carrying out this assignment PPMC has to maintain the integrity of the pipelines and depots, coordinate marketers and assure safety of the operations.

Pipelines and depots network

The primary mandate of PPMC is to transport and distribute petroleum products to various locations in the country. This assignment is carried out through a network of integrated pipelines and depots. Some of the refineries and depots are linked by pipelines and the movement of products is facilitated by the use of pumping stations and booster stations located at Abudu and Abaji. The depots are located at Port Harcourt, Aba, Enugu, Markurdi, Yola, Warri, Benin, Ore, Ibadan, Mosimi, Lagos, Satelite (Ejigbo), Atlas Cove, Suleja, Minna Kaduna, Gombe, Kano, Gusau, Jos, Maiduguri and Calabar. Products from these depots are despatched in 33,000 litre tankers operated by both major and independent marketers.

The pipeline system

The pipeline network which serves as the conduit for petroleum products destined for storage at the depots was constructed in stages. The 21 depots have a total capacity of about 651,169 m^3 of PMS, 257,000 m^3 of DPK, 467,900 m^3 of AGO (diesel) and 6,350 m^3 of ATK. The four refineries have enough capacity to store 379,749 m^3 PMS, 244,700 m^3 Household Kerosene (HHK), 33,530 m^3 AGO and 63,500 m^3 of ATK. Although the tank farms

Figure 11.1 Products loading bay at a depot.
Source: NNPC –Pipelines and Products Marketing Company Ltd, 2004

have the capacity to store the above volumes, the aggregated output of all the refineries cannot measure up to this level. The underlying reason for low domestic products build-up is the low capacity utilisation of the refineries, thereby warranting importing products to supplement output from the local refineries. For operational purposes, the network is divided into five categories as shown in Table 11.1.

Marine transportation and storage

Port Harcourt and Warri refineries are located on the coast and products from both refineries are despatched by trucks and coastal vessels. They are transported from Port Harcourt and Warri to Lagos and discharged at Atlas Cove. Calabar is also fed from Port Harcourt through coastal vessels. Excess products from the refineries are routinely transferred to the M.T. *Tuma* anchored off the shores of Bonny which serves as a floating storage tanker. M.T. *Oloibiri*, a vessel acquired by NNPC, is engaged as a floating storage facility for crude oil and is stationed about 24 km off Pennington in Warri. The vessel is currently on regular charter, one of which was a five year charter to the Texaco Petroleum Company Ltd. In carrying out its functions the PPMC interfaces with various organisations in order to ensure that quality products are available for public consumption. It liaises with the DPR to ensure that marketers obtain petroleum products storage and distribution licences, prior to securing products allocation from PPMC, and then dispense accurate quantities to consumers. It interacts with the Customs Department in order to prevent smuggling of petroleum products across the borders into neighbouring countries, and the services of the Department of Weights and Measures are regularly engaged to ensure that calibration of meters at depots,

Table 11.1 NNPC pipeline network

s/n	*Pipeline system*	*Locations covered*
1	2A	Warri–Benin–Ore–Mosimi
2	2AX	Auchi–Benin
3	2B	Atlas Cove–Mosimi–Ibadan–Ilorin Mosimi–Satelite (Ejigbo in Lagos) Mosimi–Ikeja
4	2C	Escravos–Warri–Kaduna (crude oil line)
5	2D	Kaduna–Zaria–Kano Zaria–Gusau Kaduna–Jos Jos–Gombe–Maiduguri
6	2E	PH–Aba–Enugu–Makurdi
7	2EX	PH–Aba–Enugu–Makurdi–Yola
8	2CX	Enugu–Auchi Auchi–Suleja–Kaduna Suleja–Minna
9	2DX	Jos–Gombe

refineries and retail outlets is accurate. PPMC also works closely with the Nigerian Ports Authority to ensure that pilotage services are rendered to incoming and outgoing vessels, boats and crafts. It is also significant to indicate that the company closely interacts with the Power Holding Company of Nigeria (PHCN) which supplies power to the depots and other facilities across the country.

The downstream sector is dominated by eight major marketing companies, namely:

- Total Nigeria Plc;
- Texaco Nigeria Plc;
- African Petroleum Plc;
- Mobil Oil Nigeria Plc;
- Oando Plc (formerly Unipetrol);
- Conoil (formerly National Oil);
- Agip Nigeria Plc (initially acquired by Unipetrol and now Oando);
- ElF Nigeria Ltd (now Total).

Prior to 2000 the equity holding in the above companies was between foreign companies, private Nigerians and NNPC and was as follows:

- Total Nigeria plc (Total 60 per cent and Nigerians 40 per cent);
- Texaco Nigeria Plc (Texas Petroleum Co. 60 per cent and Nigerians 40 per cent);
- African Petroleum (NNPC 40 per cent and Nigerians 60 per cent);
- Mobil Oil Nigeria Plc (Mobil Oil Corporation 60 per cent and Nigerians 40 per cent);
- Agip Nigeria Plc (Agip 60 per cent and Nigerians 40 per cent);
- National Oil and Chemical Marketing Co. Plc (NNPC 40 per cent, Shell 40 per cent and Nigerians 20 per cent);
- Unipetrol Nigeria Plc (NNPC 40 per cent and Nigerians 60 per cent);
- Elf Oil Limited (Total 60 per cent and Nigerians 40 per cent).

In view of the privatisation process, the government divested interests hitherto managed by NNPC. Consequently, major transformation and restructuring took place in the original companies. Some of the companies, namely National Oil, Oando and Elf, changed names as a result of the acquisition by new owners. Conoil acquired National Oil, Ocean and Oil acquired Unipetrol, which had earlier acquired Agip Nigeria Plc; and Sadiq Petroleum acquired African Petroleum. The downstream sector, especially in the area of products marketing, was dominated by the major marketing companies. In 2003 the major marketers controlled about 52 per cent of the market and individual companies had market shares which ranged between 7 and 12 per cent. The main products marketed by the companies were petrol, diesel, kerosene, LPG and ATK.

The dominance of the major marketing companies continued until the mid-1980s and the activities of these companies were observed to revolve around the major cities. This arrangement did not conform with the government objective of transforming the rural areas into major production centres, so in order to guarantee product availability in rural areas, the DPR granted operating licences to independent marketers for the first time in 1980. It is currently estimated that about 1,500 independent marketers operate in the downstream sector and as at 2001 these marketers controlled about 43 per cent of the products market. The rapid expansion of their ranks can in part be attributed to the prolonged period of scarcity of products experienced between 1996 and 1999. Scarcity of petroleum products during the period under reference created a parallel products market often referred to as the 'black market'. The 'black market' orchestrated high products prices and instigated malpractices among the marketers. All allocations to markets were expected to be sold at designated petrol stations within the country. However, it was observed that some of the products allocated for local petrol stations were diverted to the neighbouring countries for a premium price. This practice of diversion further aggravated the supply situation and inevitably raised the price of products. The situation created serious consequences for the domestic economy as many factories failed to operate at full capacity due to lack of materials to sustain their production systems. The scarcity era created artificial profit in the downstream sector especially among the independent marketers. These artificial profits provided funds for the building of new petrol stations, on the basis of which higher product allocations were made by PPMC. In this regard, the activities of the independent marketers caused a proliferation of petrol stations, most of which were used as cover to obtain higher product allocations which were smuggled across the borders for huge profits.

The products supply situation has stabilised since around 2001 and most of the stations which were constructed to take advantage of the protracted scarcity currently remain grossly underutilised. Some of the stations were hastily constructed, thereby ignoring location economics and traffic flow which are key determinants of successful products marketing. In view of the partial deregulation and liberalisation of the petroleum industry both major and independent markets can now import petroleum products. However, available 2006 products importation records of NNPC indicate that both major and independent marketers are yet to take full advantage of the new policy. They are of the view that the prevailing prices at the pump are not adequate to cover the aggregate cost components incurred in the importation and distribution of the products. This indicates that product availability in the rural areas will for some time to come be constrained by an inappropriate product price structure, as stated by the Independent Marketers Association.

NNPC products retail business

The construction of the first Port Harcourt refinery in 1965 signalled active marketing of petroleum products in Nigeria. The years preceding 1965 witnessed importation of petroleum products by affiliates of the major oil companies such as Shell, Mobil, Texaco etc. Although NNPC in fulfilment of its mandate constructed three additional refineries at Port Harcourt (New PH Refinery), Warri and Kaduna, it was not directly involved in petroleum products marketing in petrol stations. However, its involvement in the area of products marketing was indirect in the sense that it acquired equity interest in National Oil, Unipetrol and African Petroleum. The equity interest in these companies existed until 2000 when they were, at the request of the government, divested to the general public. In taking this action, NNPC activities in the oil and gas industry were narrowed down to exploration and production as well as refining. It is important to note, however, that the primary aspiration of NNPC is to become an integrated world-class oil and gas company. To acquire this status it must of necessity expand its operations to include petroleum products marketing as this is a standard practice among major oil and gas companies. The need to expand the scope of operations, especially as it relates to petroleum products marketing, is buttressed by the fact that all NOCs of the OPEC family operate petroleum products retail outlets. Specifically, Aramco of Saudi Arabia, Kuwait NOC, Pertamina of Indonesia, Petronas of Malaysia etc. all operate petrol stations as part of their overall operational strategy. Aramco, Kuwait NOC and the Libyan NOC, in addition to operating petrol stations at the national level, have acquired equity interest in international products marketing companies in various countries across the globe. The fundamental purpose of the extension of the operations of NOCs to other markets offshore is to diversify and grow the revenue base of the various companies.

NNPC for many years operated as a typical government company relying to a great extent on the government subsidy in some operations and management commission or fees for services rendered to the federal government. It is important to note that a number of reforms have been introduced at federal government level with a view to engendering transparency and prudence in the allocation and utilisation of available resources. Key aspects of the reforms are the monetisation and privatisation programmes. The privatisation programme is aimed at divesting government interest in public sector enterprises so as to allow private ownership. It is the thinking of the government that private participation would infuse the required discipline and managerial experience which predicate the success of private enterprises. In this regard all existing subsidies would be withdrawn and enterprises would therefore operate purely on a competitive basis. It implies therefore that only those which record entrepreneurial success can survive.

In the case of NNPC, the federal government has stated in very clear terms that its operations must be run in a cost efficient manner in order to guarantee its continued existence. As a subtle demonstration of the new government orientation, subsidy on the 445,000 b/d crude allocation to the domestic refineries has been withdrawn. This implies that crude oil for the refineries would be obtained at prevailing international market rates. In the context of NNPC the federal government reform programme simply emphasised the need for the entrenchment of commercial practices in all operational areas of the corporation. In pursuing commercial objectives NNPC seeks to expand and consolidate its revenue base in order to strengthen the corporation and transform into a truly integrated world-class oil and gas company.

Compelling factors

The establishment of the NNPC retail business was primarily in response to the challenge of operational diversification and revenue generation. A retail division has been established in the investment directorate to administer and monitor the activities of the products retail business, with the ultimate objective to construct 36 Mega stations – one of which will be constructed in each State of the Federation. A project team for the retail business was constituted in 2002 to develop a comprehensive plan for NNPC participation in the products retail business. The required strategy was developed and in the same year NNPC commenced products marketing at the Mega station, Kings Way Road, Lagos. Abuja Mega station was also commissioned in 2002. Data

Figure 11.2 NNPC petroleum products Mega station.
Source: NNPC – Retail Marketing Operations, 2006

made available in 2008 indicated that 30 Mega stations had been constructed while six additional stations were under construction.

The ultimate objective is to construct 36 Mega stations, one of which will be located in each State of the Federation. As part of the marketing strategy, plans have been drawn up to acquire existing stations and transform them into NNPC branded standard stations or to build new ones. Each Mega station has a capacity of about 400,000 litres and features about 20 pumps, while a standard station would operate an average of 12 pumps. It is important to note that the Mega stations are city based and fed with contractor operated tankers. NNPC in consideration of the transportation needs of the riverine (coastal) communities has embarked on the construction of 12 floating stations, each with a capacity of 300,000 litres. As at 28 April 2008, 30 Mega stations and 9 floating stations had been completed in response to the increasing demand for petroleum products. In the riverine areas, five floating stations were deployed to various locations in readiness for operations, of which the first was officially commissioned on Friday, 6 October 2006. When fully mobilised, the floating stations will operate in six States of the Niger Delta.

The riverine areas of the Niger Delta face serious logistic problems in marine transportation considering the fact that remotely located towns and villages depend on Eket, Port Harcourt, Warri and Calabar for petroleum products. In view of these transportation problems, the unit cost of the commodity is relatively high and products obtained through informal and uncoordinated sources, apart from being expensive, risk adulteration. In this regard therefore, one would posit that the introduction of floating petrol

Figure 11.3 NNPC floating Mega station.
Source: NNPC – Retail Marketing Operations, 2006

stations in the Niger Delta region will gradually eliminate the products scarcity problem and facilitate the smooth flow of logistics and other marine activities. NNPC Retail has from its inception positioned itself as a high volume turnover business determined to operate the land based stations on a 24 hour operational scheme. This is designed to make products available to the customer at any preferred time. The introduction of the floating Mega stations in the Niger Delta region will market the products at the official price and promote commerce. This will also assuage the problems of the perennial black market prevalent in the region.

Socio-economic value of Mega stations

Nigeria is globally recognised as a major oil and gas producer. This notwithstanding, it has for over a decade been experiencing serious setbacks arising from petroleum products scarcity. Prior to the divestment of the government equity in the products marketing companies, NNPC enjoyed the cooperation of major and independent products marketing companies. These companies carried out the directives of NNPC and non-compliant behaviour attracted sanctions. However, with the privatisation of the major marketing companies (listed in the preceding section) the powers of NNPC diminished. Consequently, enforcement of the government approved price of petroleum products became a problem and both major and independent marketing companies joined labour unions to embark on nationwide strike action. During such industrial action all privately owned stations remained closed, thereby depriving members of the public and vital commercial organisations of the opportunity to buy petroleum products. Under such circumstances the NNPC Mega stations served as intervention outlets for the sale of petroleum products to consumers. Another significant factor is the fact that the pump price at these stations is slightly lower than all other stations nationwide. In this regard one can posit that the government relies on the Mega stations as mitigation mechanisms during periods of scarcity. The stations were also utilised to indirectly administer a petroleum products subsidy. Although the action of the government was well intentioned, it was observed that the main beneficiaries of the subsidy at the pump were the middle and upper class income earners who own automobiles. In this context, experts contend that the subsidy is not enjoyed by the low income earners; therefore its removal would allow NNPC to recover full cost and a marginal profit which can be used to provide critical social services. In view of this, attempts have been made by NNPC and the Petroleum Products Price Regulatory Agency (PPPRA) to increase product prices in order to reduce the huge deficit shouldered by the NOC (NNPC) on behalf of the government. Although increase in the pump price could generate higher revenues which can be channelled to provide social services, it has always met stiff opposition from organised labour. The unions contend that price increase is often implemented without there being a proportionate increase in the wages of workers. Such increase promotes escalation of transport fares both for workers and farmers who move agricultural products from rural areas to urban centres. The net effect of price increase, it is further argued, is an amplification of the hardship index among the low income earners. In response to this, the federal government offered to provide 'social palliatives' which include provision of convenient and affordable public transportation, transport allowances and car loans to qualified workers etc. Although the government has expressed good faith in its determination to provide such reliefs, the labour force has often ignored such offers on the grounds that similar promises made in the past remained unfulfilled.

Future of Mega stations

Mega stations are widely acknowledged by the general public as suitable intervention mechanisms for the provision of petroleum products. However, the privatisation programme in some ways poses a threat to the future survival of the stations. Under the present arrangement, the supply of products to the Mega stations is guaranteed as long as the refineries and indeed PPMC are directly controlled by NNPC. However, under the privatisation programme the refineries and the 22 products depots would be sold to companies which operate petrol stations. Once the refineries are privatised, the investors controlling 51 per cent equity interest would constitute the new management and as such dictate the quantities of products allocated to other marketing companies, including the NNPC-operated Mega stations. It is important to indicate at this juncture that other marketing companies have always regarded the Mega stations as price distorters. This is because the stations have consistently maintained a lower pump price which is widely appreciated by the consumers. For this reason, Mega stations are highly patronised by motorists who seek to save some money no matter how small. The stations are known to have constant supply of products by virtue of their special relationship with PPMC and the pumps dispense accurate volumes contrary to the practice in some major and independent marketing outlets. Sadly, however, Mega stations do not import products directly. This portends danger for the stations in the post-privatised refinery era. The major and independent marketers are fully-fledged commercial entities which have long standing cordial relationships with banks. Obtaining loans or initiation of letters of credit for products importation for these companies is an easy affair. They have the necessary collateral to support such loans. On the contrary, Mega stations operate under a department in the NNPC corporate structure. Consequently, they lack the flexibility, authority as well as the collateral to secure robust loans from banks for petroleum products importation. It could be posited therefore that products importation for the smooth operation of the stations could be a serious challenge for the Mega stations in the post-refinery privatisation era. Each station has a stock capacity of 300,000 litres which is often depleted within two days. Sourcing large volumes of products from refineries controlled by competitors will be a major challenge for NNPC.

Availability of products (petrol, kerosene, and diesel) at the Mega stations is of primary concern to the government since it serves as a suitable outlet for the administration of subsidy to the public. The continued existence of the stations is important; therefore it would be necessary for NNPC to put in place robust contingency plans aimed at ensuring that the outlets continue to enjoy uninterrupted products supply from the refineries when they are operational. More importantly, Mega stations should be structured to independently access loans and associated letters of credit for the importation of the commodity. Any action in this direction would be aimed at preventing these vital facilities and the socio-economic benefits they provide from being subjugated to the manipulations of 'competitor' core investors.

References

1 PPMC Profile, 2006, p. 5.
2 'Investment and Business Guide Publication', 2005, pp. 1–25.

12 Gas monetisation

GLOBAL OUTLOOK

Starting from the mid-1980s, gas has progressively become the fuel of choice in major energy markets. Aggregate global demand increased from 1,656 billion cubic metres per year in 1985 to approximately 2,093 bcm/yr in 1995. This presents an average growth rate of 2.4 per cent per annum. This growth rate has long been exceeded due to increasing use of gas in industrial and domestic facilities. In regional context the 2.4 per cent figure is quite low considering the fact that Asia Pacific recorded a growth rate of over 7 per cent per annum (i.e. from 108 bcm/yr to 215 bcm/yr). Between the mid-1980s and 2004 gas demand grew much faster than primary energy. Aggregate growth in primary energy during the period under reference was about 1.6 per cent per year compared to well over 2.4 per cent per year for gas. In the Asia Pacific region growth in primary energy was estimated to be 4.8 per annum compared to an annual growth rate of 7 per cent per annum for gas. It was observed that in the developed economies such as Europe only a modest growth rate of 0.4 per annum was achieved for primary energy compared to 2.5 per cent year annum for gas demand.[1]

Several factors have been responsible for the rapid growth of demand for gas for power generation and other activities. Some of the factors are derived from the environmentally benign characteristics of gas while others are triggered by events in the broader economy. Most importantly among the factors which served as catalysts for increased utilisation of gas is the need to mitigate the rapid ozone layer depletion. In recent years, especially since the Kyoto Protocol Initiative, environmental protection has become a major political consideration. In this regard, the clean nature of gas has endeared its use to consumers who desire to mitigate the impact of their fuel consumption on the environment. In relative terms gas is a clean burning fuel compared to coal and oil because it emits low levels of sulphur dioxide and nitrous oxide and virtually zero particulates.[2] Gas is also preferred as fuel because it requires minimal handling and storage before and after consumption. It leaves no sludge or ash which is characteristic of oil and coal combustion respectively. It is also contended that gas has flexible use especially in the power sector where change in technology has paved the way for rapid gas penetration. The search for alternative sources of fuel to oil continues to intensify and this creates an opportunity for the increase in the use of gas as the fuel of choice. Intensified environmental lobbying has created obstacles for the construction of nuclear and hydroelectric power plants. The setbacks of nuclear and hydroelectric power plants have created avenues for increased gas utilisation in power generation. The preceding factors coupled with competitively priced gas have led to significant increase in the global gas market share.

Future scenario

Experts contend that the profile of natural gas as the energy of the future will continue to increase and as a result capture a sizeable proportion of the global energy market. Rushby,[1] in a study, estimated that global gas demand would expand from 2,100 bcm/yr in 1995 to 3,000 bcm/yr in 2010 (i.e. annual growth rate of 2.7 per cent). It was further estimated that gas demand in the Asia Pacific region would increase from 200 bcm/yr in 1995 to 470 bcm/yr in 2010 (i.e. about 8 per cent annual growth rate). In addition to the factors discussed earlier, it is envisaged that gas market penetration will be aided by factors such as market deregulation, changes in economic structure and technological innovation.

Gas market deregulation

Globally countries with well-developed energy markets are introducing deregulation policies in order to allow market forces to determine gas price. Such measures involve the abolition of State imposed monopolies which mandate State owned companies to be the sole purchasers/marketers of gas.[3] The deregulation process has also dismantled existing subsidies which have historically distorted inter-fuel competitions, making it difficult for an efficient global energy market to develop. Evolutionary trends of energy markets in the USA, the UK, Australia and Argentina indicate quite clearly that deregulation in the gas market has triggered significant increase in the demand for the commodity. With the rapid economic expansion in China, India and other South Asian countries which depend on cheaper and cleaner sources of fuel the demand for gas will escalate.

Changes in economic structure

Changes in economic structures which lead to development stimulate energy consumption. Such changes also lead to increased urbanisation which is linked with increased energy demand in residential and commercial sectors. As the population experiences some affluence arising from increased income in urban centres, consumer preference shifts from coal and other non-commercial fuels (wood and biomass) to cleaner sources of fuel such as gas. Heavy industries traditionally rely on coal and fuel oil for both power generation and other industrial activities. Changes in economic structure create opportunities for the emergence of lighter industrial activities. Light industries and manufacturing traditionally depend on cheaper forms of fuel such as gas. In this regard the expansion of light industries will increase the consumption of the commodity.

Technological innovation

Technological innovation has expanded the horizon for gas utilisation in sectors which traditionally use gas and other new areas of application of the commodity. It is observed that research into fuel cell technology may enhance gas fired power generation. Innovation in gas fired air-conditioning is also likely to pave the way for gas penetration of the sector thereby increasing use of the resource in this kind of production. It is further expected that current research will bring about gas conversion (GTL) technology which will unlock stranded gas in remote locations for commercial use at competitive prices.

GAS IN NIGERIA

Nigeria is widely known for crude oil production having joined OPEC in 1971 and is often referred to as the seventh largest exporter of crude oil. It is active in OPEC and has held key positions in the organisation. Exploration and production activities in the country are primarily targeted at crude oil, with associated gas produced along with the crude oil. Until recently the gas had little use, thereby warranting the flaring of as much as 75 per cent (i.e. over 3.0 bcf/d or equivalent of 400,000 b/d). Gas utilisation in Nigeria was until recently limited to power generation, fertilizer manufacturing and the firing of furnaces in cement companies and, these activities notwithstanding, the level of gas utilisation was extremely low. As shown in Figure 12.1, aggregate demand was less than 200 mm scf/d in 2005. The establishment of the NLNG at Bonny in Rivers State opened a new chapter in gas utilisation. The gas supply forecast in Figure 12.2 indicates that supply will move in an upward direction to correspond with both local and international demand.

The global emphasis on environmental protection initiatives derived from the Kyoto Protocol has played a major role in limiting the flare level in Nigeria. Currently it is estimated that about 43 per cent of gas (i.e. about 230,000 b/d equivalent) produced is flared with a gross monetary loss to the economy amounting to about $2.5 billion annually and a total loss of about $50 billion in 20 years. Gas flaring continues in the Niger Delta (Figure 12.3), but it requires total eradication in view of the adverse environmental consequences. Carbon dioxide, nitrous oxide and sulphur dioxide, which are

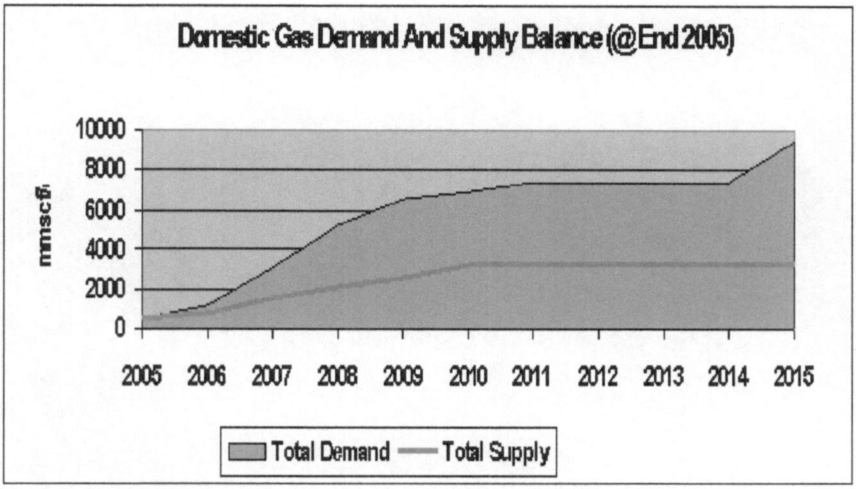

Figure 12.1 Domestic gas demand and supply balance.

Source: NNPC Gas Master Plan 2006

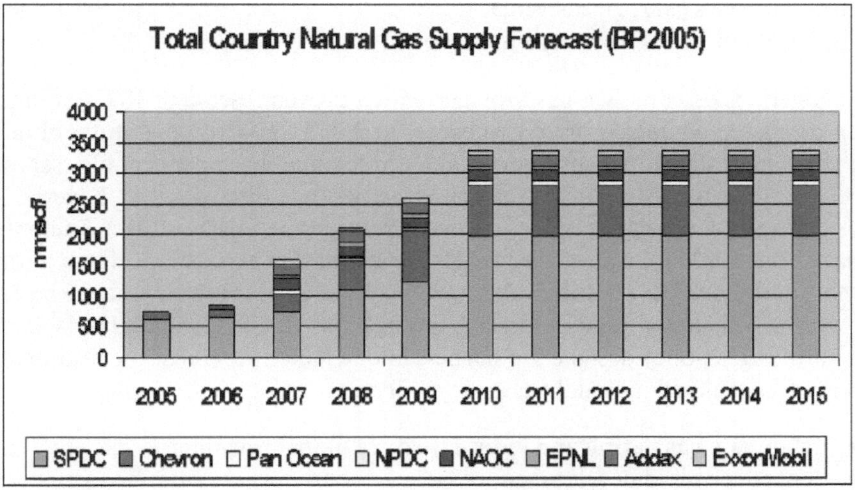

Figure 12.2 Future gas supply forecast by IOCs.

Source: NNPC Gas Master Plan 2006

derivatives of gas flaring, have serious environmental consequences and global studies on the environment indicate that gas flaring in Nigeria results in the discharge of about 70 million MT of carbon dioxide into the atmosphere annually (Table 12.1). This singular action negatively impacts on the local and regional environment through emission of sizeable proportions of greenhouse gases (GHG).[4]

Figure 12.3 Gas flare in the Niger Delta 2006.
Source: NNPCN – Group Public Affairs Division Photo Archive

Nigeria is essentially a gas province with a proven reserve of 187 TCF and an estimated aggregate reserve in excess of 250 TCF. This constitutes about 5 per cent of the global gas reserves and approximately 50 per cent of reserves in the African continent. As indicated earlier, gas discovered in Nigeria is as a result of oil exploration and production activities. No direct attempts have been made to explore exclusively for gas and experts are of the opinion that undiscovered potential reserves may be much higher than known reserves. The use of gas in Nigeria is low and this can be attributed to lack of a natural gas policy designed to harness and enhance the use of the resources. A well formulated national gas policy will bring about the following:

- eliminate gas flaring and reduce treats of global warning;
- engender rapid development of IPPs;
- make gas available to both industries and homes;
- provide gas storage facilities to mitigate the effect of low demand;
- create avenues for distribution of gas, LPG and CNG to remote regions of the country, West African sub-region and the international market;
- guarantee the availability of supply to meet current and future contracts.
- facilitate the production of oil and NGL recovery;
- create value addition through local processing of gas through LNG, GTL, fertilizer and other gas based projects.

Failure to harness the gas resources in Nigeria promptly has negatively

Table 12.1 Major oil producing countries gas flare rates

Country	*Flared gas (TCF)*	*Share of world total (%)*
Algeria	6.8	6
Angola	4.3	4
China	3.2	3
Egypt	0.9	1
Indonesia	4.5	4
Iran	10.5	10
Nigeria	17.2	16
Mexico	5.6	5
North Sea	2.7	3
Russia	11.5	11
Venezuela	4.5	4
United States	2.8	3
Other Countries	33	30
WORLD	107.5	100

impacted on the power sector where unreliable gas supply has contributed to low power generation and low capacity utilisation in the organised private sector of the Nigerian economy. Lack of a reliable power source has created a culture of reliance on diesel and petrol powered generators in factories and individual households. The quantity of diesel produced from the local refineries in most cases falls short of domestic demand and this situation has often warranted importation of the commodity by some major and independent marketers.

Natural gas is relatively cheap; therefore it has the potential of stimulating rapid growth in the non-oil sector of the Nigerian economy, but the national economy itself requires an integrated strategy for the development of gas and power in order to guarantee economic stability and the creation of employment opportunities. The establishment of the necessary gas infrastructure will provide urban and rural areas access to environmentally friendly cheap fuel for the manufacturing sector, power generation and domestic use. Gas, if properly harnessed and utilised, will reduce the pressure on fast depleting resources such as oil, wood and biomass.

Replacement of wood as a fuel source would halt desertification and further enrich the environment through oxygenation emanating from rejuvenated forests. The federal government has set 2009 as the year for zero flare in the Nigerian oil and gas industry and this has been clearly communicated to all major oil producing companies. In addition, they have all been mandated to draw up gas utilisation plans as integral components of their production activities. Funding remains a major constraint in the implementation of the gas flare-out programme and experts are of the opinion that this requires careful re-evaluation by policy makers and engineers drawn from various sectors of the industry. The initial target date of 2008 has arrived and flaring continues unabated. While government has in recent years pressed for compliance with gas utilisation directives, a UNDP/World Bank study identified potential obstacles to full gas policy actualisation as follows:[5]

- lack of a clearly defined long-term vision for the gas sector;
- lack of an appropriate gas sector development strategy and implementation plan;
- lack of robust fiscal, legal, contractual and regulatory framework and institutions to interface with foreign investors;
- lack of capacity to evaluate, correlate and provide accurate opinion on the bankability of proposals received from potential investors;
- lack of infrastructure for the transportation and storage of gas for the local and regional markets.

The study showed that the success of the gas sector development dependen on:

- provision of appropriate legislation, funding and backing for projects in power sector refurbishment, IPP development and promotion of LNG projects;
- introduction of fiscal concessions as incentives for potential investors (this should also be linked with appropriate penalties against erring producers who fail to embrace the policy guidelines of the government);
- develop coordinated fiscal incentives and legislative frameworks that promote new projects that can enhance gas utilisation;
- provide guidelines for minimum plant size so as to optimise the utilisation of gas on a short-term or long-term basis.

A review of unsatisfied demand by sectors provides an insight into priority areas where gas could be engaged in large volumes to justify the cost of gathering and transmission and the distribution pipeline networks needed to satisfy demand. In this regard the study classified large volume gas utilisation sectors as follows:

- gas to LNG plants;
- gas to Power;
- gas to Pipelines for export;
- gas to fertilizer plants, factories etc.

In consideration of the above, subsequent sections will focus on gas utilisation projects in Nigeria. The primary objective of the discussion will be to assess gas utilisation efforts in the country and the potential for revenue generation.

Gas utilisation projects

The global environmental remediation efforts of the United Nations and other agencies were fully endorsed by the federal government. A clear demonstration of government support was the signing of the Kyoto Protocol which was introduced to eradicate gas flaring. The outcome of such a policy is a safer and less polluted environment that would guarantee global climate stabilisation. Apart from protecting the environment it became apparent that gas flaring was a colossal economic waste. In consideration of both factors and others not discussed in this section the federal government through NNPC introduced gas monetisation projects. The primary objective of the projects was the conversion of gas into viable economic activities. Alternatively gas would be utilised to produce energy that can be engaged in various production and household activities. Over the years some gas based projects have been introduced and vigorously pursued to full completion. Others have reached advanced stages of completion. These projects can be identified as follows:

- Nigeria LNG Project;
- NAFCON Fertilizer Project;
- Kwale IPP;
- Geregu 1 and 2 IPP;
- Papalanto 1 and 2 IPP;
- Omotoso 1 and 2 IPP;
- Omoku Rivers State IPP;
- Alaoji Abia IPP;
- West African Gas Project;
- Brass LNG Project;
- OK-LNG Project;
- Chevron Texaco GTL Project;
- Mobil LNG Project.

In years ahead LNG, power and fertilizer will be the main gas consuming sectors in Nigeria. The percentile contributions of the various sectors and contributions of power generation to gas monetisation are portrayed in Figures 12.4 and 12.5.

EMERGING GLOBAL LNG BUSINESS

Gas continues to increase in profile as the preferred energy of the future and global trade in LNG is primarily driven by increasing demand in consuming countries. In 1997, nine LNG exporting countries shipped about 4 TCF (i.e. 83 million MT) to LNG importing countries, while in 2002 the volume of shipment from 12 exporting countries increased to 5.4 TCF (i.e. 113 million MT). Current trends of industrial expansion in China and India, and the

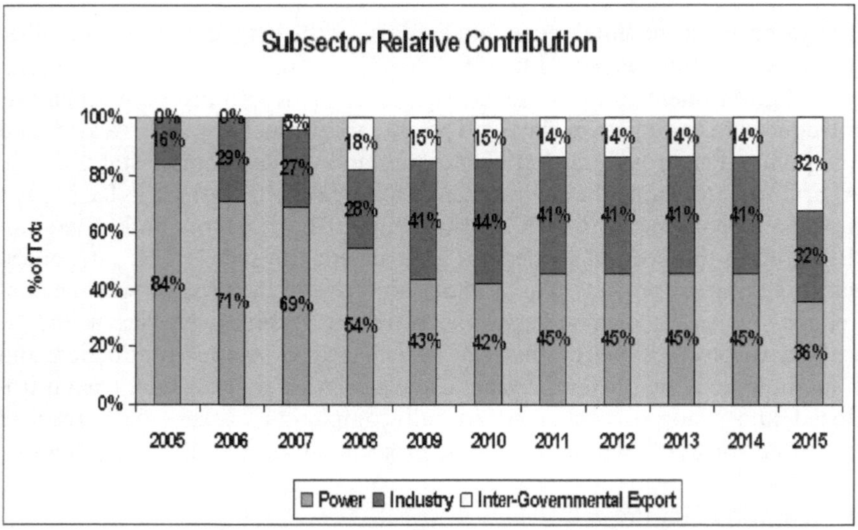

Figure 12.4 Sectoral contributions to gas monetisation.

Source: NNPC Gas Master Plan 2006

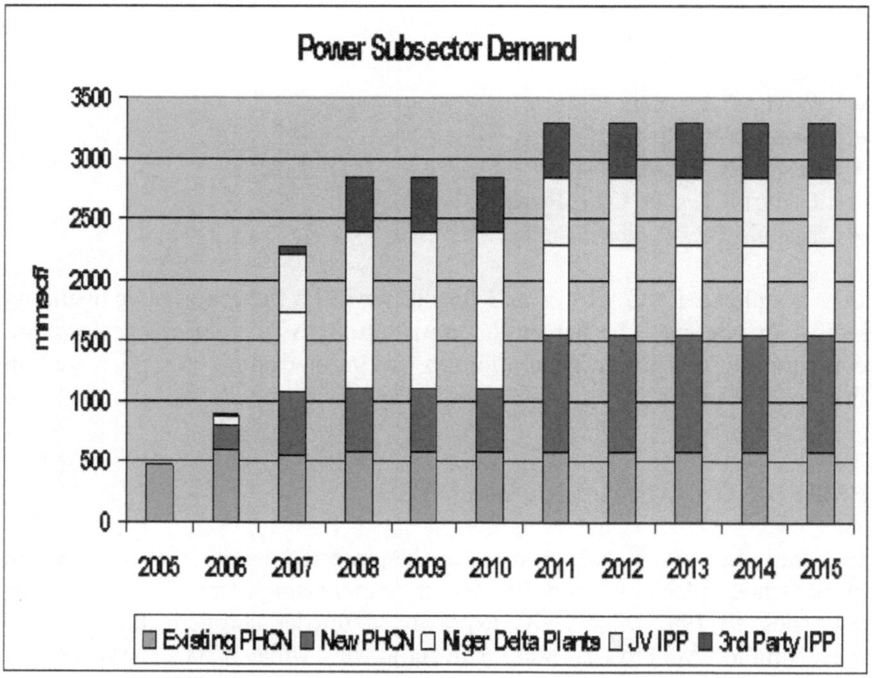

Figure 12.5 Power subsector projects estimated gas demand.

Source: NNPC Gas Master Plan 2006

energy switch to gas in the production processes, indicate that the LNG market will continue to grow. Domestic gas resources in developed countries are presumed to be declining, thereby warranting the importation of LNG. The trade is also driven by the desire of producing countries to commercialise and monetise the commodity: world LNG trade accounts for about 7 per cent of the global natural gas consumption. However, the proportion of LNG in the energy mix of consuming countries varies. In the US it accounts for 2 per cent while in Japan LNG constitutes 100 per cent of the gas supply.[6]

Global LNG exporting centres

The increase in the demand for gas has in recent years stimulated investments in gas monetisation projects in LNG exporting countries. The main sources of LNG are the Pacific Basin, Atlantic Basin and the Middle East which account for about 49 per cent, 29 per cent and 23 per cent respectively of the global supply of the commodity (see Table 12.2). In 2003 two new LNG Trains came on stream in Trinidad and Tobago and Malaysia which elevated liquefaction capacity in the world to 6.6 TCF (or 135 million MT). New facilities are also being constructed in Norway, Egypt, Australia and Russia. Other expansion programmes are expected to boost liquefaction capacity by as much as 2.8 TCF (57.7 million MT) annually, effective from 2007. Oil and gas producing countries, namely Angola, Venezuela, Yemen, Bolivia, and Peru monetise their gas resources significantly.

The Pacific Basin

The Pacific Basin comprises exporters such as Australia, Malaysia, Indonesia, Brunei, the US and Russia. In this category Indonesia, with a liquefaction capacity of 1.4 TCF (30 million MT) at Bontang and Arun, was by far the largest LNG exporter in 2002 exporting 1.1 TCF (22.5 million MT) which amounted to 21 per cent of the global LNG exports. Japan is the largest consumer of Indonesian LNG with South Korea and Taiwan consuming smaller volumes from the same source.

Australia

In 2002 Australia exported 364 bcf (7.6 million MT) of LNG from the southwest shelf project to consumers in Japan. Based on increased demand, investors have embarked on the construction of additional facilities which will add 205 bcf (4.2 million MT) annually to the existing capacity. Conoco-Phillips plans to increase liquefaction capacity by 175 bcf (3.5 million MT/yr) at Darwin. This project is designed to monetise large volumes of gas in the Timor Sea which falls under the Australia and East Timor Joint Development Zone. In pursuance of its gas monetisation programme Conoco-Phillips embarked on the construction of the 258 bcf (5.25 million MT/yr) capacity Green Sunrise project in collaboration with Osaka Gas, Shell and Woodside Petroleum. It is important to state that the LNG programme in Australia is also promoted by ChevronTexaco in collaboration with Shell and ExxonMobil through the construction of the two-Train 487 BCF (10 million TM) Gordon project.

Brunei

Brunei operates a two-Train liquefaction terminal at Lumut. This plant has a capacity of 351 BCF (7.3 million MT) with 90 per cent and 10 per cent of the output supplied to consumers in Japan and South Korea respectively.

Russia

Russia has abundant gas reserves, the majority of which are stranded, but some of the gas in the region is marketed through extensive pipelines. In an effort to monetise its gas Russia commenced work on its first two-Train LNG plant on Sakhalin Island situated off its east coast. The facility is designed to hold a capacity of 465 bcf (9.5 million MT), with approximately half of this volume being exported from the first Train (effective in 2007). Sale and purchase contracts for the export volumes have been firmed up with Japanese utilities companies covering a period of 20 years. Ongoing discussions indicate the existence of potential outlets for gas exports from Russia to the US and other Asia Pacific countries.[7] Indonesia and Malaysia are major LNG exporters in the region.

Atlantic Basin exporters

In 2002 Atlantic Basin oil and gas producing countries exported 1.5 TCF (32 million MT) which accounted for 29 per cent of global LNG production. The annual capacity increased to 2.1 TCF (42.5 million MT). It was believed that the facility expansion programmes in Algeria, Nigeria, Norway and Trinidad and Tobago would by 2008 increase LNG liquefaction capacity in the Atlantic Basin beyond the 3.3 TCF (73 million MT) mark.

Algeria

Algeria is considered a large gas province and was classified as the second largest LNG exporter in 2002. In the same year it shipped 935 bcf (19.58 million MT) to Europe and the United States. It is important to note that Algeria marketed about 1.0 TCF of its gas through pipelines to Europe as far back as 1964 and gradually expanded its operations to include four liquefaction plants. The country embarked on extensive renovation of its plants in 1999 thereby raising the production capacity to 1.2 TCF (23 million MT) per annum.

Trinidad and Tobago (TT)

TT exported 189 BCF (4.0 million MT) in 2002. The liquefaction facility comprises three Trains with an annual feed gas requirement of 486 bcf (9.79 million MT). In order to further monetise available gas the government approved the construction of the fourth Train with a gas requirement of 253 bcf (5.1 million MT) per annum. The LNG produced in the island state is exported to the US, Spain, the Dominican Republic and Puerto Rico.

Libya

Libya is one of the LNG exporting countries in the Atlantic Basin. In 2002 it exported 21 bcf (0.4 million MT) from a 131 bcf (2.7 million MT) facility located at Mersa Brega.

Egypt

Egypt has embraced the gas monetisation movement and embarked on the construction of a one-Train liquefaction plant with a capacity of 243 bcf (5.0 million MT) at Damietta which started production in 2004. It also executed a two-Train plant with an installed capacity of 175 bcf (3.6 million MT) at Idku. The LNG produced from the Idku plant is fully utilised by Gaz de France.

Norway

Norway has also joined the league of LNG producers in the Atlantic Basin. It embarked on a programme to export 200 bcf (4.1 million MT) from its liquefaction plant located at Melkoye as from 2006. Gas produced in Norway is targeted at markets in France, the United States and Spain.

Venezuela

The idea of initiating an LNG project in Venezuela has been discussed since the early 1970s. Shell and Mitsubishi reached a preliminary accord to establish a 228 bcf (4.08 million MT) per annum Mirisal Sucre plant to develop offshore reserves. As part of the development strategy the two companies explored with TT the possibility of transferring Venezuelan gas for processing at their Atlantic LNG plant. The finalisation of these discussions would depend on the resolution of basic economic, logistic and to a lesser extent political issues.

Angola

Angola has in recent years become an active theatre for exploration and production activities. Oil and gas have been discovered in commercial quantities in the territory and the NOC, Sonangol, has engaged world-class consultants to guide its investment strategies. Shortly after it commenced production it introduced shipping and insurance into its investment portfolios. Discussions between ChevronTexaco, ExxonMobil, Total, BP and Sonangol for the establishment of an LNG plant have reached advanced stages. The proposed single-Train plant with an initial capacity of 195 bcf/yr (4.0 million MT) would process offshore associated gas which will be linked to the plant with a network of pipelines. Available gas reserves can sustain additional Trains if the partners decide to expand the scope of the initial investment.

Equatorial Guinea

Equatorial Guinea has a major gas field at Alba. Steps were taken to monetise the gas reserves in the region in 2003, 17 years after a Sale and Purchase Agreement (SPA) was completed between Marathon Oil and British Gas to deliver 165 bcf (3.39 million MT) annually to the Lake Charles regasification plant in the US. The sponsors of the project have executed engineering feasibility studies to ascertain the viability of the project. The final investment decision was taken some time in 2004.

Middle East

The Middle East is highly endowed with large quantities of hydrocarbon resources and gas constitutes a significant proportion of these reserves. In 2002 LNG exporters in the Middle East produced 1.2 TCF (24.8 million MT) LNG which accounted for 23 per cent of world production. Up until 2003 the main Middle East LNG exporters, namely Qatar, Oman and the United Arab Emirates, collectively had 1.4 TCF (29 million MT) annual capacity. The planned expansion of facilities in Qatar and Oman will boost LNG capacity with additional 619 bcf (13 million MT) per year. This new facility will increase aggregate capacity in the Middle East to 2.0 TCF (42 million MT) annually by 2008.

Qatar

Qatar is the fourth largest LNG exporter in the world and has an annual capacity of 726 BCF (14.9 million MT) operated from two liquefaction plants owned and operated by the Qatar Gas and Ras Laffan LNG (Ras Gas) partnership. Qatar Gas undertook de-bottlenecking to expand capacity while two additional Trains are programmed to be added to the Ras Gas plant. Both exercises were expected to add about 458 bcf (9.4 million MT) to existing capacity annually by 2006 and maintain the growth path to accommodate demand. Qatar has a strong customer base in Japan and South Korea. However, short-term cargoes are also channelled to consumers in Europe and

Table 12.2 Global LNG projects

Project	*Participant*	*Size MT/yr (existing)*	*Probable market*
Algeria Gassi Touil	SONATRACH + 12	5(+22)	Europe
Angola	ChevTex, Sonangol, Total, Bp, Esso	8	Europe, US
Australia	Northwest Shelf JV	8	NA
Equatorial Guinea	Marathon	3	US
Egyptian LNG	BG, BP	7 + more	Gdf, BG-Italy and US
Egypt (SEGAS LNG)	Union Fenosa (+ENI), Egypt	5	Spain
Nigeria LNG*	NNPC, Shell, Agip, Total	30	Europe, US
Nigeria Brass	NNPC, Total, Agip, ConocoPhillips	10	US, Europe
OK-LNG	NNPC, Chevron, Shell, BG	22	US, Europe, Asia
Indonesia	PT Badak NGL	25	NA
Malaysia	Malaysia LNG	24	NA
Nigeria (WND)	ExMob, ChevTex, ConocoPhillips	10	US, Europe
Norway	Statoil, Total, Gdf	(4)	Europe, US
Trinidad	BP, BG, Repsol, Tracteble (Tr 1), NGC (Tr 1)	5 +(10)	US, Spain
Russia	Gazprom, Norsk Hydro	Big	US,
Venezuela	Shell, Mitsubishi, PDVSA	5	US, Mexico

Source: Compiled from various sources.

the US. Qatar is noted for low oil and gas production cost; therefore it is sufficiently leveraged to monetise its huge gas resources through the production of LNG. The country is poised to expand its LNG export capacity beyond 2.9 TCF (60 million MT) by 2015.

Oman

Oman started the LNG business in 2002 with two Trains and a dedicated export terminal. The two-Train facility has an annual capacity of 356 bcf (7.28 million MT) per annum. Kogas of Korea is the major off-taker while small volumes of the commodity are shipped to customers in the United States, Japan and Europe. Plans were concluded to introduce the third Train with an installed annual capacity of 161 bcf (3.3 million MT) by 2006. Gas reserves in Oman are limited; therefore its expansion programmes in LNG exports could be hampered by declining feed gas.[8]

United Arab Emirates (UAE)

UAE are globally ranked the fifth largest natural gas province and rank ninth in LNG exports. They have a 278 bcf (5.7 million MT) LNG processing plant which markets 90 per cent of the gas in Japan. In 2004 UAE had no immediate plans to expand the capacity. However, the steady increase of gas in the global energy mix could warrant additional investments in the LNG project.

World LNG shipping capacity

Available statistics indicate that the global LNG market had 151 tankers in operation in 2003. By way of categorisation 16 vessels out of the total had capacity of less than 50,000 m^3, 15 had 50,000 to 120,000 m^3 range and 120 vessels were 120,000 m^3 and above. About 100 ships are under construction and out of these at least 46 are designed to carry a minimum of 138,000 m^3 of LNG (i.e. 2.9 bcf of natural gas). It is anticipated that vessels capable of carrying 250,000 m^3 (equivalent to 5.2 bcf of natural gas) could be introduced into the market barring constraints imposed by existing loading and regasification terminals. With the addition of new LNG vessels, global fleet capacity will increase by 44 per cent from an estimated 17.4 million m^3 of liquid (i.e. equivalent to 336 bcf of natural gas) in 2003 to 25.1 million m^3 of liquids (equivalent to 527 bcf of natural gas) in 2006. In comparative terms shipping, depending on distance, accounts for between 10 and 30 per cent of the delivered LNG volume as against 10 per cent for oil shipment. The gap between the cost of LNG transportation and oil derives primarily from the high cost of LNG vessels. At present, 138,000 m^3 vessels cost between $150 million and $160 million compared to VLCCs which cost about $100–120 million but carry four to five times the energy equivalent carried by the LNG vessels.

Although the 138,000 m^3 LNG vessel peaked at $280 million in 1995 the cost has declined to about $150 million in recent times. LNG vessels are tied to dedicated long-term routes and contracts whereas oil tanker acquisition is based on speculation. However, in recent times international companies engaged in import and export of LNG, such as Shell, BP and Tokyo Gas, have ordered vessels which are not dedicated to projects. Experts are of the opinion that availability of uncommitted LNG tankers is a key element in catalysing the development of the LNG market in the short term. The construction of LNG vessels is highly specialised; therefore, globally, eight shipyards with locations in Japan (3 shipyards), Europe (2 shipyards) and South Korea (3 shipyards) are dedicated for the construction of LNG vessels.

Poland, China and India have perfected plans to acquire the capacity to include LNG ship construction in their shipyards.

Nigeria LNG Company

Gas has in the past been used for power generation, fertilizer manufacturing and the firing of furnaces in cement companies. Although these activities have been going for a number of years the level of utilisation from these sources was extremely low. The establishment of the NLNG plant at Bonny Rivers State ushered in an era of major gas utilisation. The NLNG Company is a JV company between NNPC (49 per cent), Shell (25.6 per cent), Total (15 per cent) and ENI (10.4 per cent). Shell Gas BV (SGBV) is a subsidiary company of the Shell Group registered in the Netherlands. It has broad business interests in LNG projects in Malaysia, Brunei, Oman and Australia. On average, Shell Gas sells over 80 billion m^3 of gas annually. Total is an international integrated oil and gas company and sustains a complex operational network in about 25 countries, whereas ENI is an Italian NOC with a subsidiary company – Nigerian AGIP (NAOC) – operating in Nigeria. NLNG was registered in 1989 and the Final Investment Decision (FID) signed by the shareholders in November 1995. In December of the same year a consortium of engineering firms – Technip, Snamprogetti, M.V. Kellogg and Japan Gas Corporation (TSKJ) was awarded a Turnkey contract involving Engineering, Procurement and Construction (EPC) for the construction of the plant, residential quarters and the gas transmission system. The NLNG project was executed in phases. The first phase of the project which commenced in February 1996 involved the construction of Trains 1 and 2 (Base Project). Train 2 was completed in August 1999 and in September 1999 it commenced production of LNG. Train 1 came on stream in February 2000. The Base Project has the following features:[9]

- land coverage – 2.72 km^2 of reclaimed land;
- LNG jetty;
- materials off-loading jetty;
- dedicated gas pipelines traversing 110 communities;
- two condensate storage tanks (capacity – 36,000 m^3 each);
- two 65,000 m^3 LPG refrigerated tanks for propane and butane;
- two LNG storage tanks (capacity – 84,200 m^3 each);
- residential area covering 2.08 km^2.

Trains 1–3 each had a capacity of approximately 4.9 million MT per annum, providing an aggregate production capacity of about 14.7 million MT per annum. Beyond Trains 1–3 the project proceeded to the NLNG Plus phase which comprised of Trains 4 and 5. Trains 4 and 5 each had a production capacity of about 4.0 million MT per annum in addition to 0.7 million MT of LPG and condensate per year. The construction of Train 3 progressed without major setbacks and was completed in November 2002. It is important to indicate that all major milestones were met on schedule thereby making it possible for Trains 4 and 5 to be completed in 2005 and 2006 respectively. Trains 4 and 5 include the following:[10]

- extended LNG and LPG jetty facilities;
- 2X GE – 7 gas turbine;
- air cooled C3/MR design;
- parallel PCI cryogenic heat exchangers;
- enhanced liquefaction pressure;
- acid gas incinerator;
- expanded LPG chilling unit.

Bonny Gas Transport (BGT)

Bonny Gas Transport Limited (BGT) is wholly owned by NLNG which is designed to operate about seven Trains with an aggregate capacity of about 30 million MT/yr. The company has SPAs with Enel of Italy, Gas Natural Spain, Botas of Turkey, Gaz de France and Transgas, Portugal amongst others. BGT was established to complete the normal gas liquefaction and transportation chain. Beyond this, the establishment of a transport arm guarantees uninterrupted supply of the commodity to customers, and creates an opportunity to capture available margins in LNG transportation. BGT acquired vessels in response to the transportation needs of the NLNG and the company currently has eighteen vessels in its fleet. Thirteen of the vessels are owned by BGT while five are chartered from different companies. Gas is considered clean and is referred to as the energy of the future, while its environmentally friendly characteristics endears it to a broad range of consumers. An increase in the demand for gas in industrial activities in developed countries leads to a corresponding increase in the demand for LNG transportation.

At the global level in 2004, total active LNG carriers were estimated to be 175 with a capacity of 21 million MT/year. In view of the expanding proportion of gas in the global energy mix the number of new orders for LNG carriers increased from 56 in 2004 to about 107 vessels in 2005. Based on current consumption patterns, experts are of the view that LNG demand will expand to approximately 200 million MT/year in 2010. The evolving demand pattern will lead to the introduction of an additional 161 new carriers into the global LNG tanker market between 2005 and 2010. This will expand the LNG tanker fleet to 336 at the end of the period under reference. The envisaged increase in the demand for LNG tankers simply signifies the increase in the demand for gas. In practical terms sustained increase in the demand for gas is a good development for the BGT since this will lead to high vessel engagement rate. The increase in the demand for gas and the long distance covered require the introduction of larger vessels (increase from

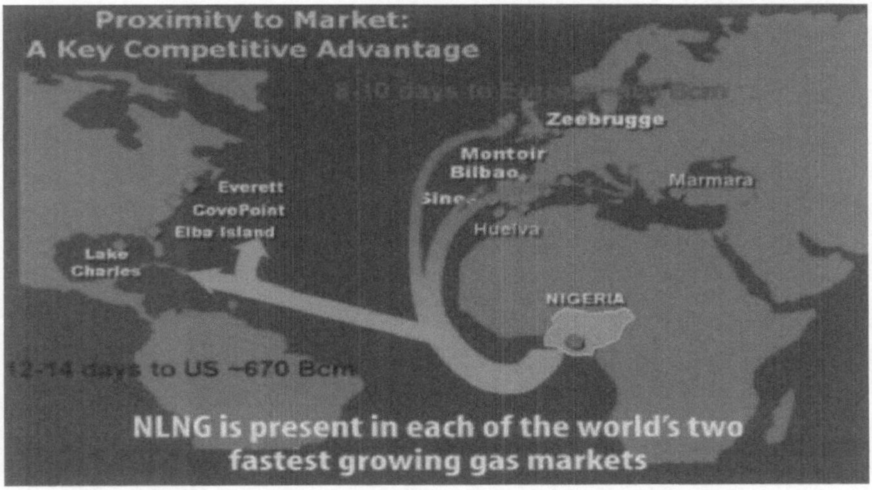

Figure 12.6 Bonny Gas Transport (BGT) market outlets (2005).
Source: Facts and Figures on Nigeria Liquefied Natural Gas (NLNG) 2005, p. 14

145,000 m^3 to 250,000 m^3) which will take advantage of economies of scale and improve the profitability of the company. BGT has acquired operational capability and it was allocated about 127 cargoes out of 157 available NLNG cargoes carried in 2005.

Shipping finance

NLNG adopted a practical approach to vessel acquisition bearing in mind the lead time required to take delivery of a vessel. The first set of vessel acquisition occurred in 1990 when a loan of $132 million was secured through Citibank for the purchase of 4 carriers for the NLNG Base Project. BGT acquired the charter party pending the commencement of business at the Bonny plant. At the third Train expansion stage the company also secured an additional loan of $160 million in 1999 through Credit Suisse First Boston for the purchase of two vessels. In view of the increase in demand for vessels to export LNG liquids, BGT acquired two additional vessels from the parent company at a cost of $210 million. Finally, in an effort to meet the transport needs of Train 3 adequately the company, in September 2001, raised $100 million through Credit Suisse First Boston to acquire LNG *Bayelsa* which was delivered in February 2003 under the NLNG Plus Project.

The construction of Trains 4 and 5 also increased the demand for shipping services. To satisfy this need BGT in 2003 raised the sum of $460 million through ABN AMRO Bank, Fortis ING Bank, Credit Lyonnaise, HVB Verein, West Bank and West LB to part-finance the acquisition of eight vessels estimated to cost $742 million. The balance of $282 million was financed through internally generated revenue and shareholders funds. Some

Figure 12.7 Bonny Gas Transport LNG vessel.
Source: Facts and Figures on Nigeria Liquefield Natural Gas (NLNG) 2005, p. 18

Nigerian banks, namely First Bank, FSB International Bank (now Fidelity Bank), Guaranty Trust, Union Bank and United Bank for Africa, financed about $100 million in the loan package. Vessels for Train six are to be chartered from Bergesen of Norway and NYK shipping of Japan.

Fleet management

The management of the BGT fleet was contracted to STASCO (Shell International Trading and Shipping Company Limited) and the Anglo-Eastern Group. STASCO is a major fleet management company which has operated in the marine industry for over 100 years. It has a team of marine experts who provide professional advice on various types of marine related projects. The portfolio of vessels currently managed by STASCO include about 24 oil tankers (VLCC, Aframax, Panamax and Handymax, product carriers etc.) and 27 LNG vessels. The tonnages range from 10,000 to 300,000 DWT. The Anglo-Eastern Group was established in 1974 and operated initially as a unit of a ship brokering operation. It later metamorphosed into a complete ship management company. It has grown over the years and is currently ranked as one of the largest vessel management companies globally. In 2001 the company merged with Denholm Ship Management Company which resulted in further consolidation of the group as a famous ship management company. It has requisite depth of expertise and experience and its services encompass ship crew management, technical support, superintendence, insurance, agency services, accounting and procurement. The vessels maintain high safety and maintenance standards which account for the high average availability rates of about 86.7 per cent and 88.9 per cent recorded in 2005.

Nigerianisation in NLNG

One of the primary objectives of the federal government is to create opportunity for active indigenous participation in the Nigerian maritime industry. In furtherance of this objective NLNG shareholders in September 1991 endorsed a Nigerianisation scheme. This issue was re-evaluated and updated in 2004 for subsequent implementation. Under the Nigerianisation scheme it is envisaged that adequately trained Nigerians will hold key management positions in the company with effect from 2013. In order to accomplish this objective OND/HND graduates were recruited and trained as technicians and operators. University graduates were recruited and trained to perform specific functions in the company. The recruitment of young men and women was designed to develop a workforce that will imbibe the work ethics and philosophy of NLNG. Other cadres of technical staff were also provided requisite training with a view to ensuring the acquisition of skills needed for supervision, operations and management positions in the organisation.

Beyond developing the blueprints for Nigerianisation, NLNG has also embraced the NCD policy of the government. In furtherance of this, the company has put in place the NCD plan. Under this scheme the company intends to source local inputs (i.e. skills, materials, services etc.) in executing major and minor jobs in the organisation. Active engagement of local input would invigorate industry-related production activities, transfer of technology and catalyse a multiplier effect in the earning potentials of various levels of workers and other participants in the oil and gas industry. To achieve this collaborative objective effectively, NLNG intends to execute a local content survey periodically in Nigerian cities with a view to identifying indigenous companies capable of providing goods and services that will serve as input in the various activities of NLNG programmes.

Brass LNG

Brass LNG Ltd was incorporated in Nigeria in 2003 as a JV between NNPC (49 per cent), Total (17 per cent), ConocoPhillips (17 per cent) and ENI (17 per cent). The facility consists of two Trains and has a total gas requirement of 12.6 TCF. The company was established in 2003 primarily to buy gas, liquefy it, then ship and sell the LNG to end users in Europe, the US and the Far East. The project consists of two Trains each with a capacity of 5 million MT per annum, thereby creating a total annual plant capacity of 10 million MT and a prospect of further expansion. Total and Agip will supply the gas and the plant is estimated to come on stream in about 2010.

The main features of the plant are as follows:[11]

- two Trains each with a design capacity of 5 million MT of LNG per annum;

- condensate extraction and storage for the feed gas composition;
- two Trains of LPG separation and processing;
- berthing facilities for LNG and LPG tankers;
- 130,000 m^3 LPG storage;
- 250,000 m^3 LNG storage.

OK-LNG

OK-LNG is a Green Field LNG project located in the Olokola Free Trade Zone. It is a JV project between NNPC (49.5 per cent), Chevron (18.5 per cent), Shell (18.5 per cent), and British Gas (13.5 per cent). The project consists of four Trains each having a capacity of 5.2 million MT/year. It will require 25.2 TCF of gas for a 20 year period and the first LNG cargo is expected to be shipped in about 2010.[12] The gas monetisation project such as the Mobil LNG project and others proposed by producing companies are being anticipated. On the whole the preceding gas monetisation projects will utilise approximately 78.5 TCF over a period of 20 years. These projects, when finalised, will allow NNPC to achieve zero flare in 2009 and prevent further emission of about 70 million MT of carbon dioxide annually into the Nigerian environment. The monetisation process has yielded impressive financial dividend. Between 2000 and 2001 NLNG recorded a net profit of about $780 million and accumulated reserves of $905 million. In view of planned expansion programmes earnings from NLNG will expand substantially.

ChevronTexaco (ChevTex) LNG Project

As part of its gas utilisation programme ChevronTexaco anticipates the construction of an LNG plant in the West Niger Delta area. The project, when executed, could consume 1.5 bcf/d of gas as will be supplied by ChevronTexaco, ExxonMobil and ConocoPhillips.

ExxonMobil (MPN) LNG

ExxonMobil also has plans to establish a Deep Water 6.5 million MT/year floating LNG plant in Erha axis northwest of Escravos. The project will consume about 1.0 bcf/d of gas in addition to other volumes to be dedicated to the proposed IPP project.[13]

Statoil LNG

Statoil anticipates the construction of a 5 million MT/year floating LNG plant to be located at the Nnwa/Doro field. The daily consumption is estimated to be 0.8 bcf/d.

Other gas utilisation projects

ExxonMobil along with Shell and Sasol has credible GTL technologies whereas companies such as Syntroleum and Synergy Technologies are promoting new technologies for the conversion of stranded gas into diesel, naphtha and other distillates. Whereas ExxonMobil considers 50,000 b/d the break-even capacity, Syntroleum claims that a 20,000 b/d plant with netbacks of $1.0/MBTU is a possibility. Synergy Technologies relies on the Cold Plasma Technology premised on the Glid Arc Plasma physics principle. Synergy Technologies GTL technology is designed to produce 30,000 b/d of high grade diesel and naphtha. Both Syntroleum and Synergy Technologies approached NNPC for term gas supply contracts for the execution of the projects. The prospects for the expansion of the projects aimed at promoting gas utilisation are encouraging. Gas monetisation in Nigeria will attain a higher horizon once concrete agreements have been reached in the area of gas supply for GTL projects.

Escravos Gas Projects 1–3

ChevronTexaco has embarked on elaborate gas utilisation programmes fondly referred to as EGP1–3. EGP1, designed to gather associated gas, was implemented in 1997. It uses an onshore plant which processes gas dedicated for piping to NGC for distribution to local industries. EGP1 also has an ancillary project which produces LPG. The first shipload of 30,000 MT (i.e. 334,000 barrels of oil equivalent) was exported on 30 September 1997.

West African Gas Pipeline (WAGP) – EGP2

The WAGP is a subregional project initiated by Nigeria, Ghana, Benin and Togo. It covers about 630 km stretching from Lagos to Takoradi Ghana. The above countries are represented by their National Energy Companies (NECs) – namely NNPC (25 per cent), Ghana (16.3 per cent), Société Beninoise du Gaz (2 per cent) and Société Togolese du Gaz (2 per cent) – on the one hand, and International Oil Companies (IOCs) – ChevronTexaco (36.7 per cent) and Shell (SPDC, 18 per cent) on the other. The project is designed to supply gas primarily to Ghana and at a secondary level to the neighbouring countries, Benin Republic, Togo and Côte D'Ivoire. The initial capacity of the project which stands at 200 mm scf/d is expected to expand to about 500 mm scf/d in years ahead. Ghana is the largest regional market which depends on fuel oil and hydrodynamics technology for power generation. The supply of gas from Nigeria will release locally produced fuel oil for export or further cracking for enhanced product yield. The project was initially estimated to cost about $540 million with the World Bank providing a grant to leverage the participation of Ghana, Benin and Togo. FID was concluded on 21 December 2004 and the first gas delivery occurred in about the second quarter of 2007. EGP3 would be dedicated to the Escravos gas-to-liquid plant.

ChevronTexaco gas-to-liquid plant (GTL) – EGP3

The 33,000 b/d GTL plant programmed to start in 2010 would be located adjacent to the EGP and will at its peak convert 300 million scf/d of gas into high grade petroleum products. The Escravos GTL project is a JV between NNPC/ChevronTexaco and Sasol of South Africa. Chevron and Sasol have entered an area of mutual interest agreement and established Sasol Chevron Holdings Global JV with a 50:50 equity interest. The primary aim of the JV is to actively monetise stranded gas using the Fischer-Tropsch and Chevron's Isocracking technologies. GTL technologies are emerging globally and have the potential of producing high grade products which conform to the new environmental standards requiring low carbon dioxide, nitrogen oxide and particulate emission (Table 12.3). These products are proven to be devoid of sulfur and carcinogenic aromatics. Diesel derived from the GTL has higher cetane values and manifests better performance than conventional diesel. The Sasol technology combines methane and oxygen to produce synthetic gas under catalytic conditions. The synthetic gas is further heated in a slurry phase reactor to a temperature of 450°F or 240°C and subjected to Fischer-Tropsch conversion. The process yields condensates and waxy syncrudes. The Isocracking process is a Chevron proprietary technology which upgrades the waxy syncrudes. The process produces light premium fuel and naphtha with near zero content of sulfur, nitrogen and carbon monoxide. The project would promote gas monetisation as well as mitigate the negative impact of carbon dioxide emission arising from gas flaring. Other benefits include job creation, skills development through FEED, expansion of shipping facilities and ports, communication and transportation systems.[14]

The components of the plant are as follows:

- Haldor Topcon's autothermal reforming technology to produce synthesis gas from natural gas;
- Sasol's Fischer-Tropsch slurry phase reactor technology to convert synthesis gas into waxy hydrocarbons;
- Chevron's hydroprocessing technology to convert waxy hydrocarbons to finished liquid products;
- cryogenic air separation and hydrogen production facilities;
- product storage, process cooling, effluent treatment and power generation facilities;
- main air compressor and booster compressor;
- air filter, air chiller and molecular sieve absorption unit;
- main heat exchanger;
- low pressure (LP) and high pressure (HP) columns with associated heat exchangers;

- liquid-oxygen pumps;
- steam turbine drivers for main and booster compressors;
- liquid oxygen storage and vaporisation facilities;
- control system;
- pre-assembled modular structures etc.

Trans-Sahara gas pipeline

The Trans-Sahara gas pipeline is a collaborative project between Nigeria and Algeria. The project requires the laying of 2,500 km pipeline from an existing gas pipeline at Ajaokuta in Nigeria through Niger to the In Salah Development area gas pipeline which is currently being constructed in Southern Algeria by Sonatrach and BP. The distance from Nigeria to Algeria is relatively shorter than the distance from the Siberian fields. It therefore creates an advantage in the competitive pricing of the gas in the European market. The laying of the 2,500 km pipeline to link the Algerian gas infrastructure will create the required integrated distribution backbone system for the satisfaction of domestic markets in Nigeria. Once appropriate tariffs are set the Trans-Sahara pipeline project and the domestic gas transmission system can operate simultaneously. It is estimated that the project will cost $3 billion for throughput of 1 bcf/d and escalate to $3.6 billion when throughput increases to 1.3 bcf/d.

Power generation

The Nigerian Electric Power Authority (NEPA), now Power Holding Company of Nigeria (PHCN), is a major consumer of gas in the Nigerian market. However, two factors have made it an unattractive customer to the Nigerian

Table 12.3 World-wide GTL activities

S/No	*Name of company*	*Plant capacity*	*Location*
1.	Shell	140,000 b/d	Qatar
2.	ConocoPhillips	164,000 b/d	Qatar
3.	Syntroleum & Yakutagazpron	13,000 b/d	Russia
4.	SasolChevron & Qatar Petrol	66,000 b/d	Qatar
5.	Sasol Chevron	130,000 b/d	Qatar
6.	Syntroleum & Marathon	90,000 b/d	Qatar
7.	Rentech	16,000 b/d	Indonesia
8.	Chevron & NNPC	34,000 b/d	Nigeria
9.	Rentech	10,000 b/d	Bolivia
10.	Sasol 1	8,000 b/d	Sasolbury
11.	Petro SA	47,000 b/d	NA
12.	Shell	12,500 b/d	Secunda
13.	Sasol II/III	160,000 b/d	Secunda
14.	Sasol Qatar Petroleum	34,000 b/d	Qatar

Source: David Nissen – Potent and Partners 2004.

Gas Company (NGC), the sole transmitter of commercial gas in the domestic market. NEPA was known to default on gas SPCs thereby leading to huge unsettled debts. Secondly, the capacity utilisation in the existing plants dropped drastically, resulting in undelivered volumes. The drop in capacity utilisation was partly attributed to the poor state of equipment in the power generation and transmission centres. In 1999 Nigeria had seven power generation plants (three hydro and four thermal plants). The total installed capacities of the existing plants (including upgrades) were estimated to be 6,200 MW.

Over the period the performance of the plants declined due to poor maintenance. In 1998 an ExxonMobil study put the functional capacity of the existing plants at 4,528 MW. The four thermal existing plants (Table 12.4) require 821 mm scf/d of gas. Capacity utilisation further declined to 2,257 MW in 2000 whereas aggregate demand exceeded 6,000 MW. The drastic shortfall in supply created a crisis thereby causing a significant drop in industrial output. The low levels of power output serve as a disincentive for potential investors whose operations depended on reliable power. In an effort to reverse the rapid decline of power generation, the federal government, through a policy initiative dependent on gas monetisation, embarked on the National Integrated Power Project (NIPP). The main thrust of the project was to put in place seven new single cycle gas turbines (with provision for conversion to combined cycle operation). To achieve this objective, the National Assembly enacted the Electric Power Sector Reform (EPSR) Act 2005 which provided the legal premise for the Power Holding Company of Nigeria (PHCN) to be incorporated and acquire the assets of the abrogated National Electric Power Authority (NEPA). Under PHCN, 18 companies would be formed to be solely responsible for generation, transmission and distribution.[15]

The seven new power plants (Table 12.5) will have a combined capacity of 3,458 MW and a peak operation gas requirement of 884 mm scf/d. Alongside

Table 12.4 Existing power plants

N	*Location*	*State*	*Type*	*Capacity (MW)*	*Cum. cap.* (MW)*	*Gas use (mm scf/d)*
1	Kainji	Kogi	Hydro	760	760	Nil
2	Jebba	Kwara	Hydro	580	1,340	Nil
3	Shiroro	Niger	Hydro	600	1,940	Nil
4	Sapele	Delta	Steam/gas	1,020	2,960	81
5	Egbin/AES 1	Lagos	Steam	1,320	4,280	350
6	Egbin/AES 2	Lagos	Stream	270	4,550	90
7	Afam	Delta	Gas	276	4,826	90
8	Ughelli	Delta	Gas	840	5,666	210
			Total	5,666		821

Source: PHCN 2006.
*Cumulative Capacity

Table 12.5 National integrated power plants

S/N	*Location*	*Type*	*Capacity (MW)*	*Cum. capacity (MW)*	*Gas use (mm scfld)*
1	Calaber	Gas	500	500	140
2	Egbema	Gas	350	850	100
3	Gbaran	Gas	250	1,100	70
4	I-Abasi CS	Gas	300	1,400	80
5	I-Abasi 1	Gas	188	1,588	60
	I-Abasi 2	Gas	500	2,088	60
6	Ihovbor	Gas	500	2,588	140
7	Sapele	Gas	500	3,088	140
8	Omoku	Gas	120	3,208	24
	Omoku	Gas	250	3,458	70
		Total	3,458		884

Source: PHCN 2006.

the NIPP are the GOPA Power Plants located in Kogi, Ondo, Ogun and Abia States. These plants have an aggregate capacity of 2,998 MW and total gas requirement of 940 mm scf/d. The new policy initiative has also coopted the IOCs to channel all flared gas to produce power and other hydrocarbon products. Accordingly Agip, Chevron, ExxonMobil and Shell embarked on the establishment of IPPs in various locations across the country. Available records indicate that the four companies will operate plants that will produce 3,286 MW. Similarly, the plants will require 655 mm scf/d of gas at peak operation.[16] A total assessment of the gas utilisation programmes in the area of power generation indicated that the existing plants, NIPP, GOPA and the IOC (IPP) plants will utilise 3,300 mm scf/d of gas and generate 15,742 MW. In this regard the three LNG projects (NLNG, Brass and OK-LNG) and all the power plants discussed in the preceding section would utilise 12.9 billion scf/d. It is essential to note also that in 2000, aggregate gas utilisation arising from the existing power plant was 0.82 billion scf/d. This compares to about 12.9 billion scf/d gas requirement from all the LNG plants, the NIPP, GOPA and IOC (IPP) plants. The huge quantum growth in gas utilisation and power generation in the Nigerian context derives from policy reforms in the oil and gas as well as the power sectors. The policy reforms in both sectors are complementary such that policies in the upstream sector which promote activities to unlock gas reserves create opportunities for higher power generation through gas utilisation. IPPs have been commissioned at Geregu, Omotosho and Kwale to add more power to the national grid.

The level of power generation in Nigeria declined drastically and this became a disincentive for potential investors whose operational activities depended on steady sources of electricity. Power supply is generally erratic, thereby warranting high dependence of private petrol and diesel powered

Figure 12.8 Kwale Independent Power Production plant (IPP).
Source: NNPC, 2005

generators. The proliferation of private generators has created distortions in the estimates of actual demand for power in the country. Per-capita power consumption is about 152 KWh compared to Algeria and Malaysia with per-capital consumption of about 500 KWh and 1,000 KWh respectively. With effect from 1999 the federal government introduced a new policy geared towards generation of adequate power in the country for both commercial and domestic needs. The major oil Companies, namely Shell, Agip, ExxonMobil, ChevronTexaco and Total, identify with the government objective of generating between 3,000 to 10,000 MW and have for that reason drawn up plans to establish IPPs. Federal government NIPPs at various locations across the country are as follows:

- Kwale/Okpai IPP (480 MW) was constructed under NNPC/Agip JV and commissioned in 2005;
- Afam (I–IV) IPP (660 MW);
- AFM V IPP 276 MW;
- Geregu 1 and 2 (828 MW);
- Papalanto 1 and 2 (835 MW);
- Omotosho 1and 2 (835 MW);
- Alaoji (500MW);

- NNPC/ChevronTexaco JV IPP is currently at feasibility stage;
- ExxonMobil LNG/IPP to be located at Bonny will utilise gas and produce 300 MW;
- Total IPP (NNPC/Total JV) anticipates producing approximately 400 MW of power for commercial use.

The establishment of IPPs is a positive signal for the increased monetisation of gas in the Nigerian economy. Considering the abundant gas resources available in the country and the attractive netback for power estimated to be $0.02–0.025/KWh reliance on gas for future power generation is advantageous. Gas fired turbine generators have a construction period of about 12 months as compared to hydroelectric plants which take many years to clear the environmental approval from relevant agencies. The emergence of IPPs, Build Operate Transfer (BOT), and Refurbish Operate Transfer (ROT) is providing the required vibrancy for the sector. To sustain development and gas monetisation in the power sector there is need for the government to introduce a sustainable regulatory framework that will guide activities in the sector. More importantly, to attain the level of an international framework and standard, a World Bank study of the Nigerian energy sector recommended that the power sector should essentially be split into generation, transmission, distribution and retail subsectors. This split, it was argued, will allow easy implementation systems and policies.

The Liquefied Petroleum Gas (LPG) sector

The environmental problems associated with gas flaring have been addressed a number of times in of this book. In an effort to minimise the impact of gas flaring on the environment, the federal government set in motion the following agenda:

- attain zero gas flare by the year 2009;
- address environmental issues;
- develop the domestic gas market;
- create 'value addition' in the gas sub-sector;
- generate significant revenues through gas monetisation programmmes.

These objectives are being pursued through various gas monetisation programmes, namely LNG, GTL, IPPs, butanisation, fertilizer plants etc. The four existing refineries have an aggregate name plate capacity (manufacturers specified production capacity) of 4 million MT LPG production per annum. From inception the refineries were earmarked to export excess LPG to markets in the West African subregion. This objective was not achieved due to protracted underperformance of the refineries especially in the area of capacity utilisation.

For nearly a decade the downstream sector was characterised by scarcity of petroleum products, low integrity of products distribution infrastructure and low stock levels at deposits. This situation warranted importation of petroleum products including LPG. Domestic supply of LPG from the local refineries declined to 160,000 MT per year in the late 1980s and decreased further to 40,000 MT/yr in 2002. In 2003 the supply sources experienced total collapse due to long down-time of the refineries, thereby necessitating the importation of about 50,000 MT of LPG. The poor state of the domestic refineries created an unprecedented scarcity of the product with the attendant escalation in the price of the commodity. Successive administrations have made efforts to engender sustainable supply of cooking gas through a comprehensive rehabilitation of the refineries and the execution of the butanisation projects.

The LPG supply situation at the national level was chaotic and dependence on the commodity among some urban dwellers shifted to wood and coal. In consideration of the grave consequences of massive dependence on wood on the environment and more particularly desertification, the government commissioned a number of studies and organised stakeholders' workshops to proffer strategies to harness the country's abundant gas resources. The primary objective was to satisfy local LPG demand and broaden the consumption pool among the general public. The collaboration between the government, foreign interest groups, input of sector studies and interactive workshops aimed at generating alternatives strategies for a secure and dependable gas LPG supply system spanned three years. During the period under reference per capita LPG consumption among a population of about 140 million was a dismal 0.2 kg/year (Table 12.6). This was about the lowest in West Africa and especially in relation to Senegal and Gabon which had per capita LPG consumption of 10.3 kg/year and 12.8 kg/year respectively.[17] The stakeholders' workshop held in 2003 as a first step recommended the setting of a target for the recovery of the LPG subsector to allow per capita consumption to increase to the West African subregional average of 3.7 kg/ year. The target so set was an ambitious 100 per cent increase which if achieved could stimulate a market potential of ₦47 billion annually. The stakeholders' workshop also recommended the formation of a steering committee to proffer alternatives on the rejuvenation of the sector. Specifically, the committee was mandated to put in place the following:

- LPG policy and regulation;
- institutional framework;
- safety and standards;
- modalities for ensuring availability and distribution;
- establishing investment needs and opportunities;
- recommending sustainable prices of the commodity.

The committee worked assiduously culminating in the development of frameworks and strategies as input for the development of a national gas policy. In the light of the gas sectoral anomalies, increased LPG utilisation will play a substantial role in implementing the NGL as well as LPG gas flares elimination policy. Local refineries are expected to supply LPG in the domestic market, but due to chronic operating problems they have proven to be an unreliable source of supply. Indeed, although Nigeria set out to be a net exporter of LPG, the circumstances have warranted NNPC to resort to importing the product on a regular basis to guarantee adequate domestic supplies.

Butanisation project

Along with other petroleum products, LPG (cooking gas) is produced at the three refineries located at Kaduna, Warri and Port Harcourt. In view of the establishment of LNG plants it is anticipated that over 3 million MT of LPG will be obtained from these sources and other private refineries which have been approved for construction. LPG is well established in Nigeria as a popular cooking and industrial fuel. It is environmentally friendly, portable and does not produce sludge or particulates. However, the typical yield of LPG at the refineries is only 1–3 per cent of the volume of crude oil which is processed. Due to inadequate distribution infrastructure sectoral market demands frequently exceed supply especially in Lagos and in other population centres remote from the refineries. As pointed out earlier this situation warranted NNPC to import the commodity to satisfy domestic demand. A strategic step in the butanisation programme was the installation of a Merox unit to enhance LPG supply from the Kaduna refinery.

The project is a component of federal government priority programmes designed to stimulate the marketing and use of LPG as a domestic fuel in

Table 12.6 West African gas consumption

	Total LPG consumption (MT)	*Residential Commercial (%)*	*Res/comm LPG (MT)*	*Population (million)*	*R/C LPG consumption per capita (kg/year)*
Cameroon	28	95	27	14.9	1.8
Côte D'Ivoire	50	85	43	15.7	2.7
Ghana	40	85	43	19.2	1.8
Senegal	100	98	98	9.5	10.3
Angola	50	90	45	11.7	3.8
Congo, Dem. Rep.	1	90	1	49.6	0
Congo, Rep.	4	90	4	2.7	1.3
Gabon	17	90	15	1.2	12.8
Nigeria	**58**	**40**	**23**	**125.1**	**0.2**
Other countries	13	90	12	65.8	0.2

Source: Nigeria Strategic Gas Plan (ESMP 279) World Bank Study. Report 279/04, Feb. 2004.

Nigeria. A feasibility study commissioned by NNPC recommended the establishment of a network of primary distribution depots for LPG along the zones of the existing petroleum products depots. However, the relatively small volumes of LPG to be transported between the refineries and the depots ruled out a pipeline connection in the short term. Eight of the nine locations recommended in the feasibility study and accepted by NNPC already have petroleum products. The construction of the depots was expected to improve significantly the supply and distribution of LPG to satisfy national demand up to the year 2020. The depots and the new LPG facilities were built in close proximity to each other. At the ninth location – Lagos – the LPG depot was built on a stand-alone basis. The locations for the primary distribution depots and their LPG storage capacities are shown below:

- Lagos – Lagos States, 8,000 m^3 – 4,000 MT;
- Ibadan – Oyo State, 2,000 m^3 – 1,000 MT;
- Ilorin – Kwara State, 2,000 m^3 – 1,000 MT;
- Gusau – Sokoto, 2,000 m^3 – 1,000 MT;
- Kano – Kano State, 2,000 m^3 – 1,000 MT;
- Enugu – Ebonyi State 2,000 m^3 – 1,000 MT;
- Calabar – Cross River State, 2,000 m^3 – 1,000 MT;
- Gombe – Gombe State 2,000 m^3 – 1,000 MT;
- Makurdi – Benue State 2,000 m^3 – 1,000 MT.

The objectives of the project are:

- to ensure availability of LPG on a regular basis in areas well beyond the refineries;
- to improve the infrastructure for primary distribution of LPG, bearing in mind a likely glut in the market;
- to reduce dependence on limited marketers storage facilities for primary distribution;
- to provide strategic reserve with which to sustain supplies during refineries shut-down;
- to minimise the distances marketers must travel to deliver the commodity to consumers;
- to encourage of use of LPG in substitution for wood fuel and kerosene; and
- to optimise the economic benefits of LPG production.

The butanisation project was classified as a priority project and funded directly by the federal government. Although the idea of making LPG readily available for homes across the country was a commendable idea, the refineries which were earmarked to supply LPG for the butanisation depots experienced near collapse and could therefore not stock the LPG depots. It is disappointing to indicate that all the nine depots were not officially commissioned for use. Sadly also the facilities were cannibalised by unpatriotic elements, which further compounded the problems of the programme. A facilities audit was conducted on the various depots and extensive looting of the key components of the facilities was reported. The problem of LPG supply continues to linger and current efforts of the government indicate that LPG will now be supplied through the NLNG plant in Bonny. In order to deliver quantities supplied from the NLNG plant effectively the LPG depots would require elaborate rehabilitation which could be associated with huge capital outlays. Considering the fact that most depots are remotely located from ports, logistics of tankers to stock the depots stands out as a major challenge to overcome in the overall strategy of surmounting LPG scarcity at the national level.

Fertilizer sector

Nigeria comprises an area of 90,770 km^2. About 34 per cent of this area is occupied by crops, 23 per cent by grassland and 16 per cent by forest. Approximately 13 per cent is taken up by rivers, lakes and reservoirs while the remaining 14 per cent is under other use. Agricultural land use involves three broad systems of production – traditional rotational fallow agriculture, semi-permanent or permanent agriculture based on intensive farming and mixed farming (i.e. crop/livestock agriculture). Food production in Nigeria has suffered setbacks ranging from lack of an entrepreneurial pool for agricultural investment, drought, infertile soil and inadequate agricultural inputs. Statistics provided by the Federal Ministry of Agriculture indicate that only 32 per cent of the households use fertilizers. Among those who do not use fertilizers 51 per cent consider the cost too high, 23 per cent do not know where to obtain fertilizers while 12 per cent felt they do not need fertilizers. Between 1992 and 1994 the supply of fertilizers increased from 100,000 MT to 500,000 MT. Aggregate national demand was estimated to be 6.8 million MT. The National Fertilizer Company of Nigeria (NAFCON) which was established in 1985 suspended operations in 1995 and was liquidated in 2005.

The aggregate demand for fertilizers (6.8 million MT) is quite reasonable and would require more than 55 mm scf/d feed stock to produce fertilizers to meet national and regional demand. The demand for gas could increase to 90 mm scf/d by 2008. Gas is abundantly available; therefore Nigeria is in a position to produce a large amount of high grade fertilizer brands and carve a niche market in Africa and in Europe. Establishment of private-sector driven fertilizer plants will create high demand for gas which is the primary feed stock for such plants and sustainable production of fertilizers will boost agricultural production. It will also stem importation of certain categories of food stuffs and other agricultural products in excess of about $2 billion/year which depletes scarce foreign reserves. More importantly, the use of gas for large scale fertilizer production will generate revenues which can be channelled to critical development projects in the country.

Benefits of gas monetisation

Gas monetisation is a key factor in achieving zero gas flare in the Nigerian oil and gas industry. Beyond environmental considerations gas monetisation objectives were pursued in order to create economic value for the abundant gas reserves in the country. In 1998 the objective of the federal government to establish an LNG plant was implemented through the incorporation of Nigeria LNG Limited at Bonny, Rivers State. The first two Trains came on stream in 2000 thereby signalling the commencement of LNG export from Nigeria. In 2001 Trains 1 and 2 jointly produced 6.03 million MT of LNG which amounted to 103 cargoes under a Supply and Purchase Agreement (SPA) with dedicated customers. Ten additional excess cargoes were disposed in the spot market. The expansion of the LNG plant has been incremental. In 2006 the sixth Train was put on stream. The expansion of the project to six Trains has also increased the market spread thereby bringing on board new customers mainly in Europe and North America. In 2004 three Trains were operational accounting for total production of 10.05 million MT. The total volume produced in 2004 accounted for 170 cargoes as well as 22 excess cargoes which were sold on either FOB or ex-ship basis.[18] In 2006 the NLNG plant produced 13.78 million MT from 5 Trains. About 227 cargoes were successfully loaded from the total production to meet contract obligations.[19] At the end of 2006 the NLNG had eleven long-term dedicated customers mainly in Europe and North America.

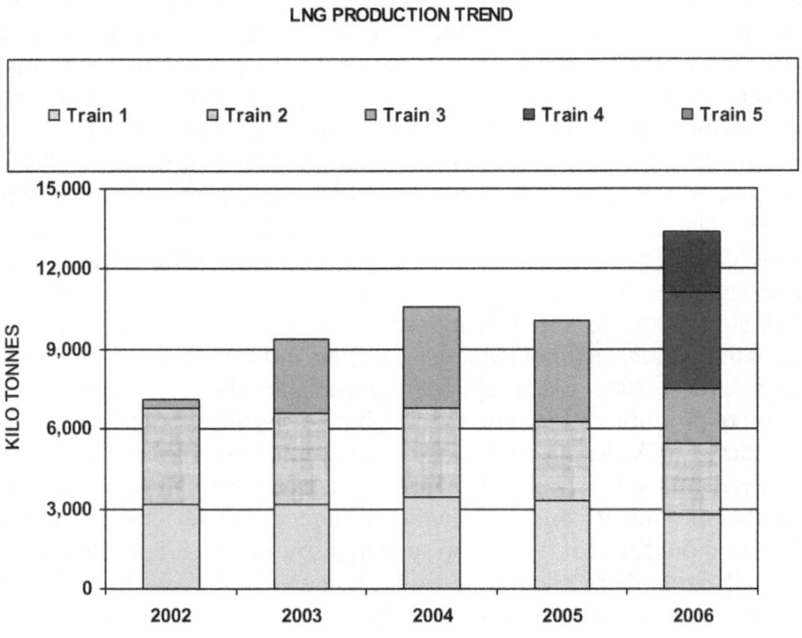

Figure 12.9 Nigeria LNG production trend.

Source: Nigeria LNG Annual Report 2006

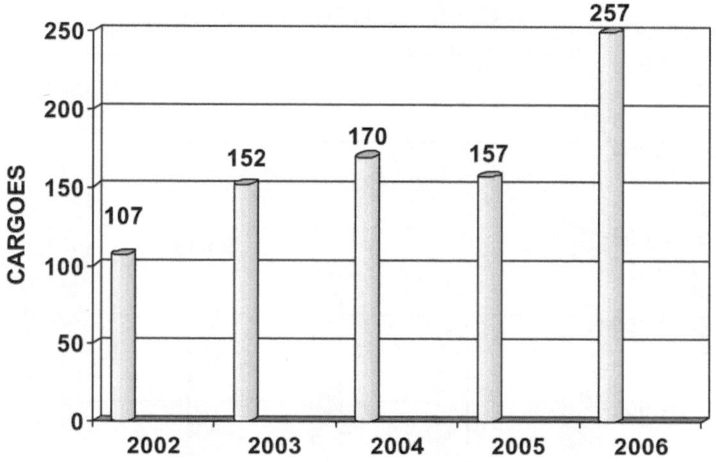

Figure 12.10 Nigeria LNG cargoes loaded.
Source: Nigeria LNG Annual Report 2006

Deriving from the increasing volumes of LNG produced at the plant is the expanding revenue base. In 2002 the company recorded a consolidated turnover of $1.097 billion and profit after tax of $478 million from the operations of two Trains of the base plant. In 2004 three Trains recorded a turnover of $2.28 billion and profit after tax of $1.33 billion.[20] The business further expanded and in 2006 a turnover of $4.60 billion as well as profit after tax of $2.81 billion were achieved respectively. It is interesting to note that the gas monetisation programme has recorded landmark achievements in terms of LNG production, export and the revenue levels achieved. Available records indicate that a 43 per cent increase (i.e. from $2.6 billion to $4.61 billion) was achieved in 2006 over the performance of 2005. Similarly a 44 per cent increase in net income after tax (i.e. from $1.56 billion in 2005 to $2.81 billion in 2006) was recorded. The performance indices remain impressive and the increased demand for gas as the preferred energy of the future indicates that LNG plants will continue to achieve higher results.[21]

In the area of job creation gas monetisation through the NLNG plant has achieved significant results. The project has provided training opportunities for marine cadet officers, plant operators and other staff in administrative positions. The company has also created about 500 jobs for Nigerians in various positions in the company, and other contractors have created job opportunities thereby contributing to the youth empowerment programme of the government. The earning capacities of the employees have increased resulting in an enhanced expenditure pattern in the local economy. The establishment of the NLNG has also created outlets for local procurement of materials and services which continuously impact the economy. In the

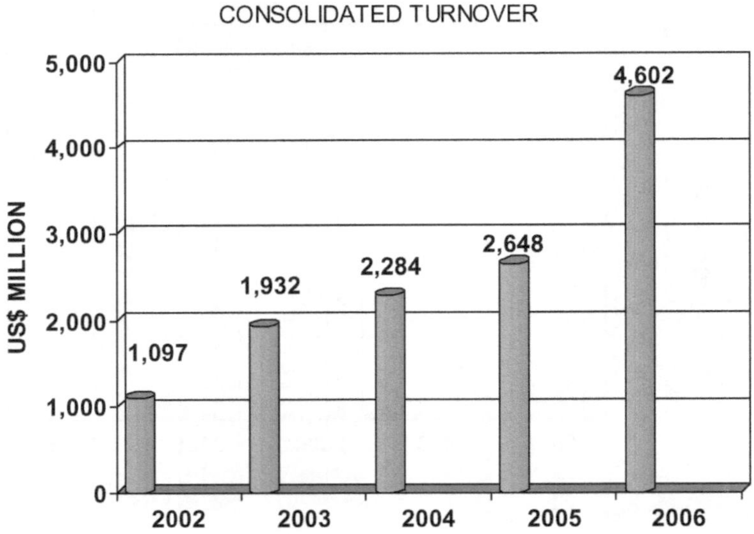

Figure 12.11 Consolidated turnover of Nigeria LNG – 2006.
Source: Nigeria LNG Annual Report 2006

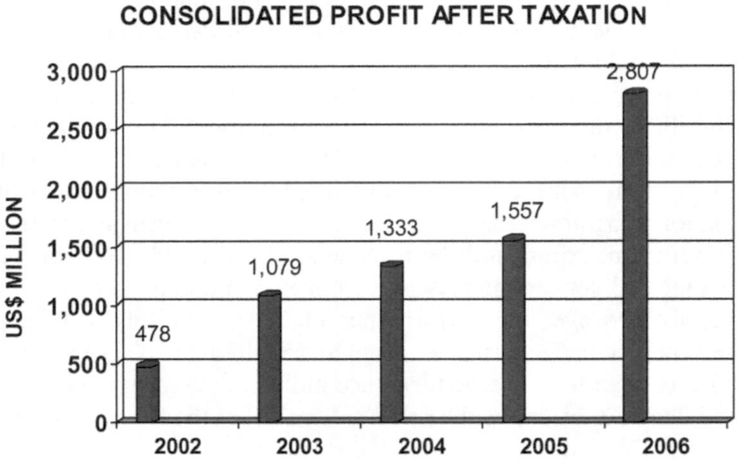

Figure 12.12 Consolidated profit after tax.
Source: Nigeria LNG Annual Report 2006

aggregate the government has, through the gas monetisation programme, achieved multiple objectives encompassing job creation, revenue generation and ozone layer depletion mitigation through zero flare enforcement in the oil and gas industry. The Nigerian government has a 49 per cent equity interest in the NLNG Company. The revenue profile of the company showed that a total of $7.25 billion net profit after tax was realised between 2001 and 2006.

The government would by virtue of its equity holding literally derive about $3.55 billion as a dividend. However, under normal circumstances part of the dividend accruing to the federal government would be lower due in part for the need for capital re-injection to consolidate the business. This notwithstanding, the performance indices of the company are impressive. It can be posited therefore that the decision by the federal government to establish two additional LNG plants – Brass LNG and OK-LNG – is born out of the encouraging positive performance of the pioneer Nigeria LNG plant. Other gas monetisation programmes are currently being explored by IOCs in order to maximise the earnings from gas. As mentioned before, gas is increasingly being referred to as the energy of the future. This implies that the demand for the commodity will increase significantly. It can be further posited that its monetisation would be a sustainable source of revenue for both investors and the producing countries. African oil and gas producing countries must rearticulate their policies and programmes in order to prudently utilise revenues from gas to provide infrastructure for the organised private sector. The governments of the affected countries face the challenge of rejuvenating their ailing economies with a view to improving the dismal per capita income prevalent in the region. Revenue derivation through the sale of gas is a laudable effort. However, NOCs should aspire to tap deep into the oil and gas production value chain and emphasis should be placed on value addition in the sale of gas. Purchasers should be persuaded to jointly establish gas based industries in the producing countries, as such ventures will achieve multiple objectives of revenue generation through gas sales, job creation, petrochemical resins export, GTL and other related products. The aggregate revenues from these activities will by far transcend the uni-directional earnings from LNG sales. More importantly, the establishment of gas based industries will create forward integration in the economy. Ultimately numerous small scale industries will emerge, utilising petrochemical products as feed stock in various manufacturing processes. The activities will create profound wealth and engender improved standards of living in the society through enhanced incomes among workers.

References

1 Rushby, I. 1997 'Natural Gas Challenges for the Next Decade', paper delivered at fifteenth World Petroleum Congress, Beijing, Topic 4, 1997, p. 1.
2 BP Statistical Review of World Energy, 1996.
3 Rushby, I., op. cit., p. 3.
4 'Nigeria Strategic Gas Plan'. *UNDP/World Bank Study*, EMS 279, February 2004, p. 1.
5 Ibid. p. 4.
6 Energy Information Administration (EIA). www.eia.doe.gov., 2006.
7 Ibid.
8 Ibid.
9 'Facts and Figures on NLNG 2004', pp. 9–18.
10 Ibid.

11 'Partnering for Sustainable Growth of Nigeria's Oil and Gas Industry'. NNPC World Petroleum Congress Publication, September 2005, pp. 71–74.
12 Ibid.
13 www.exxonmobilafrica.com
14 www.investorchevrontexaco.com.
15 'PHCN News'. Power Holding Company Nigeria, Lagos, January–April 2006, pp. 1–4.
16 'LNG Division, NNPC – IPP Status Report'. Sept. 2006.
17 Daukoru, E. 'Implementing the Blue Print for Sustainable Development of Liquefied Petroleum Gas in Nigeria'. Address delivered at the inauguration of Liquefied Petroleum Gas (LPG) Steering Committee. NNPC News, June 2004, p. 5.
18 Nigeria LNG Annual Report and Financial Statements, 2005, p. 18.
19 Ibid., p. 18
20 Nigeria LNG op. cit., 2006, p. 23.

13 Elements of petroleum law

Origins of Nigerian petroleum law

Exploration and exploitation of oil and gas activities in Nigeria have strong reference to 1908 when Shell D'Arcy and other pioneer companies flagged off their search for hydrocarbon in the territory. The activities of the pioneer oil companies were interrupted by World Wars I and II. At the end of World War II exploration activities resumed within the statutory purview of the Mineral Oils Ordinance 1914. The ordinance placed limitations on non-British citizens or non-citizens of British colonies. In this regard the ordinance stated in part:

> . . . no lease or licence shall be granted except to a British subject or a British company registered in Great Britain or in a British Colony, and having its principal place of business within Her Majesty's dominion. . . .[1]

The colonial ordinance was enacted to promote British interest and in particular to protect Shell-BP from external competition. The historical account of the Nigerian petroleum industry indicates that Shell-BP was granted a concession covering the whole territorial confines of Nigeria. Exploration and production rights were subsequently granted to companies in areas relinquished by Shell-BP. In the early period of petroleum exploration and production there existed a sharp dichotomy between rights of property and jurisdiction. In the United Kingdom, especially in the 1920s and 1930s, the issue of property rights over hydrocarbon fields was seen as an impediment in the exploration and production activities. The work programmes of oil and gas companies were disrupted by disputes over property rights by citizens who claimed ownership of the territories in which the activities were being executed. For this and other reasons, in 1946 the British Crown through its agents in the colonial administration in Nigeria passed the Minerals Ordinance contained in CAP 121 of 1955 Laws of Nigeria. The underlying intent of the ordinance was to vest in the British Crown the right of ownership of all hydrocarbon resources in the territory of Nigeria. In pursuance of this singular objective the ordinance portrayed the intent of the British Crown in very vivid terms. The ordinance stated that the entire property and control of all mineral oils in and upon the territory of Nigeria including rivers, streams and waterways shall be under the exclusive command and control of the British Crown.[2] The law was harsh in the sense that it was preoccupied by the control of property rights and made no provision for compensation for citizens whose property rights might be classified as encompassing an oil and gas field. The Minerals Ordinance of 1946 created a negative precedent which was to be inherited and perpetrated by subsequent governments in post-independence Nigeria. Currently, all mineral and hydrocarbon property rights are vested in the federal government of Nigeria. Citizens' rights are encumbered by the enforcement of the law and they cannot therefore seek compensation in a situation where oil is discovered within the confines of their natural abode. In 1955 Mobil Exploration (Nig.) Limited, a branch of American Socony-Mobil Oil Company, was granted a licence to explore for oil in Shell relinquished areas in parts of Sokoto, Kaduna and the Benue Plateau region. In subsequent years, Nigerian Agip, SAFRAP (now Total), Phillips Petroleum Company etc., commenced oil business in the southern region of Nigeria.[3] Oil production in 1956 was about 5,100 b/d but had escalated to 2.4 mmbd in 2006. Current oil and gas reserves are estimated to be 36 billion barrels and 187 TCF respectively. The repeal of the Mineral Oils Ordinance 1914 in 1958 and replacement by the Mineral Oils Ordinance Amendment No.5 1958 paved the way for the new non-British companies to search for oil in Nigeria.

Petroleum Profits Tax Ordinance

The Petroleum Profits Tax Ordinance 1958 in the colonial era came into existence in 1959 and was introduced in order to compel oil companies to pay 50 per cent of all profits derived from petroleum activities as a tax. Although this last piece of legislation in the colonial era sought to enhance the fortunes of the British Crown, the ordinance was retained in post-independence Nigeria. It created an avenue for the federal government to indirectly participate in the upstream sector by deriving revenues through petroleum tax. The reliance on tax revenues by NOCs was strongly advised against by OPEC which felt that direct participation in the upstream sector was a surer avenue to economic emancipation. Nigeria advanced from observer status at OPEC in 1964 to full membership in 1971 when the NNOC was established. The Mineral Oils Ordinance 1914 served the interest of the British Crown and was for that reason found unsuitable for the governance of the oil and gas sector in post-independence Nigeria. In consideration of this therefore the 1914 ordinance was repealed and replaced with the Petroleum Act 1969. It is important to note that although the Mineral Oils Ordinance 1914 was repealed, the Petroleum Act 1969 recognised and preserved the licences and leases granted under the ordinance.

Government participation

OPEC, in Resolution XVI of 1968, encouraged all NOCs in the OPEC family to actively participate in both the upstream and downstream sectors of their respective oil and gas industries. The main purpose was to bring about true sovereign control of the industries. In the Nigerian context and perhaps elsewhere, MNCs, through grants obtained in the 1960s and 1970s, enjoyed virtual ownership of the hydrocarbon resources (through exploration, production and marketing) while the federal government was limited to nominal ownership of the resources evidenced by derived taxes, royalties and lease rental fees. Control and supervision of activities in the industry were carried out by an ill-equipped unit of the Petroleum Resources Department (1970) under the Federal Ministry of Mines and Power, and a section of the Federal Ministry of Finance enforced compliance with fiscal obligations by the MNCs.[4] The rapid increase in global demand for oil and the increase in revenues from oil prompted government action in positioning itself for active participation in the industry and in furtherance of this, on 1 April 1971 the NNOC (established under NNOC Decree 1971) was merged with the Ministry of Petroleum. This paved the way for the formation of the NNPC, and this main company in the industry was established through Act CAP 320 of 1977. The establishment of NNPC allowed for the acquisition of an equity interest in the major MNCs, namely Shell Chevron, Mobil, Agip, EIF and Texaco. The federal government has equity interest ranging from 55 per cent to 60 per cent in the above companies, details of which are adequately discussed in subsequent chapters. The oil and gas industry has remained dynamic, and in response to global trend some structural changes have taken place warranting mergers among the companies. In 2005 Chevron and Texaco merged to form ChevronTexaco and EIF metamorphosed into Total. These changes occurred in order to pool resources and know-how and create appropriate synergy to tackle the increasing challenges of the petroleum industry.

The Nigeria A blazed the trail and offered 33 per cent equity interest to the federal government in an agreement signed on 19 October 1964. This offer was not formalised until 17 September 1971. Although Agip pioneered an offer of equity interest to the government, actual government participation in the activities of MNCs started with EIF (now Total) on 12 April 1971.[5] In the same year the government also acquired 33 per cent equity interest in the 50 per cent interest held by Phillips in OPL 34 and OMLS 60, 61, 62 and 63. The equity interest of the government has in subsequent years reached 60 per cent in the Agip/ConocoPhillips JV. As indicated earlier NNPC equity interest in the JV agreements is 60 per cent except in the NNPC/Shell/Agip and Total JV in which equity interest is 55 per cent. The 1958 ordinance stated that any person who is in lawful occupation of land shall be entitled to compensation from the grantee of an oil licence for disturbance of surface rights only. However, the ordinance carefully omitted the compensatory rights of the citizen in the event of oil and gas being found under the land they occupied. This approach was taken on the grounds that although oil could be found under the land being occupied by the citizen, he cannot lay claim to loss of resource over which he has no ownership rights. This posture has been maintained by successive governments thereby establishing a standard practice in Nigerian oil and gas law. It is important to state that the detrimental Mineral Ordinance Act of 1914 by extension laid the foundation of all the conflicts experienced in the Niger Delta.

Oil exploration and production

Oil exploration and exploitation in the country derived its roots from the German owned Nigerian Butimen Company when, in 1908, it pioneered such activitiy. Available information shows that the search for oil started from the hinterland and moved southwards to the Niger Delta region. This perhaps accounts for the large time gap between 1908 and 1956 when oil was first discovered at Oloibiri. The second major entrant in the search for 'black gold' in the territory was the Shell Petroleum Company and British Petroleum Company. These collaborating companies were granted a licence by the British Crown to explore for oil in an area estimated to be 357,000 sq. miles in size covering the entire mainland. Shell D'Arcy dominated the activities from 1938 to 1955. Activities in the industry have increased, which led to the consummation of JV agreements between the government and MNCs and the government has taken further steps to introduce PSCs and SCs into the industry. There are currently 25 PSCs and one SC which are governed by appropriate agreements. In practice the acquisition of an interest in each company is required to be formalised through the sequential execution of Heads of Agreement (HOAs), Participation Agreements (PAs) and then the Joint Operating Agreement (JOAs). PAs set out the terms and conditions of NNPC acquisition of a certain percentage interest (on behalf of the government) in any of the MNCs. In this regard one can refer to a PA as a legal document which legitimises the acquisition of a mutually agreed fractional interest in the petroleum concession(s) as well as the assets and funds engaged in the development of leasehold. The PA creates in law a relationship of concurrent ownership, coownership and cotenancy. In the majority of JV relationships the PA allows the company to assign to NNPC about 60 per cent of the interest in the business. However, the equity interest in the NLNG Company is 49 per cent. It is usually in exchange for a consideration which at the discretion of NNPC could be paid in oil. The value of the consideration is determined based on the fiscal net book value of the assets of the company as captured in the tax returns to the government.[6] In 1973 the government acquired a 35 per cent equity in Shell, Mobil and Gulf Oil (now ChevronTexaco). In 1974 the companies, in response to the acquisition, raised major points of concern and questioned the basis of the value of the consideration paid by the government. The disagreement was over updated net book value versus fiscal net book value.

The formalisation of equity acquisition suffered long delays occasioned by minor bureaucratic bottlenecks and principally by disagreements between the government and the oil companies over the valuation of interest. Consequently, the PAs and JOAs remained unconsummated from 1973 until 1985 when the PA was formally signed by all parties concerned. Essentially the PA was to provide a foundation and framework for the crafting of the JOA. In addition, it addressed fundamental issues such as the participatory interest of parties in the Oil Mining Lease (OML), the assets and the consideration paid to the oil companies as a remediation of disadvantages suffered as a result of the acquisition. Prior to the signing of the PA the operations of the JV companies were conducted on the premise of informal understanding between the parties concerned.

The Role of NAPIMS

Shortly after its incorporation NNPC established about 11 subsidiaries to undertake the various functions it was mandated to perform. One of the subsidiaries is the National Petroleum Investment Management Services (NAPIMS). It is a strategic entity in the corporate structure of NNPC. NAPIMS supervises the federal government interest in the JVs, PSCs and the SCs. Among numerous other duties it interacts with major oil companies to review their work programmes, capital and operating budgets. All major contracts in the industry undertaken by the oil majors are channelled through NAPIMS for NNPC approval. It also enforces the provisions of the JV, PSC and SC agreements entered into by NNPC and the MNCs. In performing its supervisory functions it earns a margin from the government to enable it to meet its obligations. In addition to the preceding, NAPIMS performs the following roles:[7]

- maximises PPT and guarantees a higher rate of return through efficient monitoring and cost reduction mechanisms;
- encourages gas utilisation and monetisation in the industry;
- promotes, in collaboration with the NCD, an acceleration of local inputs in the upstream and downstream sectors;
- ensures that a reserve base is maintained and reserve addition targets are attained: these are 36 billion barrels reserves in 2006, and 40 billion reserves and producibility of 4.5 mmbd respectively in 2010;
- promotes Nigerianisation content and expedites technology assimilation and transfer;
- efficiently manages Federation hydrocarbon resources;
- stimulates interest in the frontier areas among foreign and indigenous companies;
- administers the activities of the Nigerian electronic petroleum market;

- ensures the compliance of all partners in JVs, PSCs and SCs with laid down rules and regulations as well as provisions of the various agreements.

Joint Operating Agreement (JOA)

The JOA in the context of the Nigerian petroleum industry and to a greater extent in all other hydrocarbon industries outlines the legal relationships among the concurrent owners of the licences, concessions and leases. It also delineates the procedures and rules needed for their joint development for the greater utility of the owners. Typically the JOA designates one of the partners as the operator. The operator would conduct, manage and control the activities of the JV in the best way possible and in accordance with rules of the JOA in order to achieve the collective objectives of the partners. JOAs vary a great deal as the environment in which it is operated could have considerable influence on the content. Admittedly, one can posit that JOA is a globally accepted principle but no single JOA in terms of content can be deemed to be universally applicable. Some JOAs are known to be restricted to exploration, development and production with ownership of the oil produced exchanged at the wellhead. In other instances the JOA activities could include transportation of the crude oil beyond the wellhead to tank farms, export terminals and other downstream plants. In practice, the nature of the JOA depends on a number of factors which include the legal personality and sovereign status of the parties involved as well as the location of the activities of the JOA. A JOA could take a particular orientation if one of the parties involved is a NOC such as NNPC. The terrain of operation such as onshore, swamp, Deep Offshore, mangrove forest, frigid zone, farm-out/farm-in arrangement, unitisation of carried interest and many other factors can influence the legal provisions of the JOA. These intervening factors or material issues notwithstanding, a typical JOA contains some basic features which can be delineated as follows:[8]

- one of the collaborating partners is appointed the operator to execute the operations in such a manner that would benefit all the parties;
- enunciation of the rights, obligations and duties of the operator;
- periodic contribution of funds by participating parties to finance the programmes and the apportionment of the production equivalent of the equity holdings to the parties;
- formation of a management or operating committee (in a situation where an NOC is involved, its duties, powers and responsibilities are usually stipulated);
- limitation of the operator's authority especially with regard to the level of approval in matters involving award of contracts;
- affirmation of the non-operators' rights to have access to the operation

log books, accounts and records of the operations and periodic circulation of reports on the progress of activities by the operator;
- stipulation of a preferred accounting procedure.

The participation of an NOC in a JOA creates a peculiar situation. In the Nigerian context, NNPC, by virtue of its government representative status, is a party in all the JOAs. As a matter of propriety, and convenience, a single JOA covers all the functional OMLs jointly owned by NNPC and the party or parties to the agreement. For instance, the development of the OMLs jointly owned by NNPC, Shell, Agip and Elf is governed by a single JOA. The same would be true of the relationship between NNPC, Chevron and Texaco. As a matter of practice all the JOAs are uniform in character except in special situations which warrant the modifications to accommodate peculiar circumstances.

The JOA embodies a phrase 'Joint Property' which turns out to be central to the successful execution of the JOA. The term 'Joint Property' in the Nigerian context is deemed to refer to all fixed and moveable assets of the operator including exploration, development and production facilities. Others include storage, transportation, export facilities, offices, welfare facilities and staff houses. In this regard all funds contributed by the coventurers are mobilised to pay staff salaries of the operator. Such funds are expended for office rent, pension, gratuities, employee insurance, host community development, scholarship schemes and other professional and social obligations which are considered critical for the smooth execution of the joint programmes of the parties to the JOA. A careful assessment of the 'Joint Property' concept indicates that virtually all the expenses of the operator locally and the services of the home office are underwritten through the JOA funding arrangement. Experts are of the view that the expenses of the operator as delineated in the preceding sections should of necessity be funded because the operator is not in any other business except for the purpose of carrying out or administrating the joint interest of all parties associated with the JOA.

The discussions in the preceding sections indicated that the PAs signalled the commencement of the JVs initiated in 1973 and consummated in 1985. The JOAs which derive from the agreements contain provisions for change of operatorship. It is, however, disappointing to note that this provision of the JOA has to date not been effected in the existing JOAs involving:

- Shell Petroleum Development Company (SPDC);
- Nigerian Agip (NAOC), and Total;
- NAOC and Phillips;
- Total (Elf);
- Chevron Nigeria Limited;
- Texaco Overseas (Nig.) Petroleum Company; and
- Pan Ocean Oil Corporation.

The issue of change of operatorship is of paramount importance because it is the only avenue through which Nigeria can acquire full control of the activities of the upstream sector. It is observed with some measure of concern that the major oil companies contract all the exploration, development and production activities to contractors. Up until 2006 the key personnel of the OSCs were expatriates. The Nigerian personnel are not directly involved in the field operations of the major producing companies: the bulk of their activities are supervisory, administrative and not direct hands-on field practice. Office based supervision and administration are important but they cannot guarantee true technology assimilation and indeed acquisition. In consideration of this, it would be necessary to mandate the major companies to execute directly a certain percentage of their exploration, development and production operations. Direct participation by the oil companies in the execution of the programmes hitherto wholly executed by contractors will provide the Nigerian staff the opportunity to acquire relevant experience for the challenging tasks of crude oil exploration, development and production. Finally it is essential to note that the JOA embodies definite clauses which will be discussed below.

JOA clauses

Appointment of operator (Article 2.1)

The JV agreements in Nigeria involve two or more partners who have individual percentile interest in the OML. Although all parties to the agreement may have the capacity and expertise to explore, develop and produce oil it is more practicable and efficient to nominate one of the participating parties, on behalf of the others, to develop and produce the OML. Article 2.1 of the JOA states in part:

> . . . a company is hereby designated, and agrees to act, as the operator of the concession and contract area under this agreement and hereby assumes the duties and obligations of operator and shall have the rights of the operator hereunder.

In all known cases of JV agreements the MNCs serve as operators. As pointed out earlier, the JOAs were initiated in 1973 but remained unsigned until 1985. Operatorship comes along with unique privileges and responsibilities. The operator, as contained in the provisions of the agreement, is responsible for the execution of all approved joint operations on condition that all such activities are carried out in accordance with the applicable laws and regulations. The operator is also expected to conduct its affairs in a professional, cost effective and 'workmanlike' fashion. The privilege of operatorship allows the operator to carry out certain functions and outsource or contract others to selected contractors(s). The operator, regardless of the mode or channel of execution of the operations, remains liable for the final outcome of the job performed directly or through third parties. In addition to the above functions other duties of the operator include:[9]

- carrying out all joint operations with utmost care and in good faith and professional disposition as well as conform in practice with relevant laws. The operator is also expected to adhere to all the provisions of the agreement, adopt the uniform accounting procedures and other schedules associated with the agreement;
- settle all accounts payable promptly and avoid practices or circumstances that would warrant placement of lien, encumbrances and charges on all joint property;
- acquire surface rights when required for the joint operation and in doing so ensure compliance with governing laws and regulations;
- execute appropriate insurance policies for all joint property and personnel and renew same in a timely fashion;
- establish and maintain in Nigeria accurate entries in books of all operations accounting of joint operations, expenditures and records. The operator is also obliged to store data, produce statements and furnish all non-operators;
- maintenance of oil mining leases in the portfolio of operating agreement;
- regularly interact with non-operator(s) and make full disclosures concerning the status and progress of operations as well as appraise non-operators with evolving matters of importance concerning the operations;
- provide personnel, equipment, supplies, materials and services required for the execution of operations and keep in safe custody all joint property;
- prepare starting and annual work programmes, Authorisation for Expenditure (AFEs) and budgets taking into consideration uniform accounting principles and procedures as laid down in the operating agreement.

Expectations of the operator from the non-operators are quite extensive for the simple reason of ensuring that pooled resources are prudently utilised. For this and other considerations the operator is required to regularly place at the disposal of the non-operators specific information regarding progress reports and data on all wells drilled, field well performance data, monthly accounts of oil and gas extracted, tanker loadings as well as crude oil stock balance. In addition the operator is duty bound to promptly despatch to the non-operators vital information such as copies of all geological and geophysical reports, logs and surveys of every well covering survey and final depth measurements, fluid sample analysis, analytical reports, copies of drill stem tests as well as drilling completion and work-over reports. The operator is also required by the JOA to furnish the stakeholder's comprehensive information on major contracts scheduled for execution in order to advance the development and production of the field. The operator is obliged to provide information on contracts that are required to be awarded for the maintenance of auxiliary facilities. It is essential to indicate that the provision of the preceding information, and others not stated, is mandatory and in the spirit of full and frank disclosure by the operator. In order to keep abreast with the progress of work in the operations all non-operators are entitled to undertake a field tour of the facilities at their own discretion and peril. Clause 2.2.2 of the JOA seeks to protect the operator in circumstances where all activities have been undertaken with due diligence. The clause stated in part that:

> . . . the operator or its affiliate shall not be liable for any loss or damage which results from Joint Operations unless such loss or damage results from Wilful Misconduct on the part of its directors or supervisory staff. Wilful Misconduct in this context is construed to imply intentional, conscious, reckless or wanton disregard of any material provision of the JOA or any substantial part of the programme. This not withstanding the operator shall not be held liable if the provisions of the agreement are disregarded in compliance with explicit instructions from a government authority or in pursuance in good faith of an earlier decision of the operating committee.[10]

Akinrele, in commenting on the subject, portrayed wilful misconduct as:

> . . . an intentional and conscious, reckless or wanton disregard of (a) any provision of this Agreement, or (b) any programme; but shall not include an intentional and conscious disregard of either (a) or (b) above if the same is in compliance with the instruction of any government authority or is in pursuance in good faith, of a decision of the operating committee, or relates to safeguarding of life, property of joint operations, or any error of judgment or mistake made by any Director, employee, agent, or contractor of the operator in the exercise, in good faith, of any function, authority or any discretion conferred upon the operator.[11]

Fiduciary obligation

Scot defined a Fiduciary as

> '. . . a person who undertakes to act in the interest of another person (therefore) it is immaterial whether the undertaking is in the form of contract or that the undertaking is gratuitous'.[12] Williams further observed that the most important characteristic of a JV is that of a fudiciary relationship of trust and confidence.[13]

The relationship between the operator and the non-operator(s) is primarily governed by the provisions of the JOA. However, such relationships are in the broader context subject to the laws of trust in which property is held in its sole name as a trustee. An example of property held in trust is the joint bank account, which is defined in the JOA as the local and foreign bank accounts established, maintained and operated on behalf of the coventurers by the operator in which the stakeholders remit all funds needed to execute all joint operations in accordance with the 'cash call' obligations of the parties. In this regard therefore a 'joint property' refers to property (cash) held or set aside for the joint account. The fudiciary functions in the oil and gas industry evolved concurrently with the participation and operating agreements. It is significant to note that fudiciary functions of operators have so far not been subject of litigation in the Nigerian judicial system. This observation in some respects connotes harmony in the relationships of operators and non-operators. It has indeed been observed to be so. The fudiciary duties of the operator are inherent in the JOA and they are primarily premised on the trust reposed on the operator in carrying out the joint operations on behalf of the parties to the JOA.

Contracting powers of an operator (Article 2.2)

As part of the operational procedures the JOA delegates certain authority to the operator to enter into contract or purchase materials on behalf of the coventurers. In the event of delegation of such authority the operator is in a proper stead to enter into contractual agreements which bind the parties to the JOA. The commitment of the coventurers to a particular contract or placement of purchase order is subject to limits or restrictions clearly delineated and enforced by the operating committee. In practice therefore, if the authority exists for the operator to carry out contractual transaction(s) with a third party, it is not compelling on the third party to ascertain if the operator has the authority to enter into a contract. Therefore such contracts initiated by the operator shall be binding on the parties to the operating agreement. However, if the third party is privy to the lack of authority of the operator and undertakes or accepts to execute a contract which to the knowledge of the third party is over and above the authority of the operator, he stands the risk of being disentitled from obtaining compensation from the coventurers. Consequently, therefore, any operator who unilaterally awards a contract(s) or executes a purchase order, the value of which is above its limits of authority, stands fully liable and would for that reason not recover any contribution from the coventurers.

It is essential to indicate also that the selection of contractor(s) is subject to the 'Uniform Project Implementation Procedure' of the JOA. NNPC is a non-operator but wields considerable influence in the operating committee by virtue of its majority equity interest in the JVs which serve as source of the JOAs. Clause 2.2.8 (v) states in part:

> . . . the operator shall give preference to a contractor that is a company organised under the laws of Nigeria to the maximum extent possible

provided there is no significant difference in price or quality between such contractor and other contractors.

NNPC by virtue of its representative role of the government is entitled to perform an oversight function and in the process ensure that locally established companies benefit from the award of contracts, especially if the differential of their contract price is not significantly different from the quoted prices of other companies. NNPC has performed this oversight function through its subsidiary NAPIMS to the best of its ability. However, it was observed that the MNCs manipulated the process and channelled such contracts as would benefit local entrepreneurs to their affiliates. This unwholesome practice of manipulation and advancement of self-interest accounts in part for the dismal 5 per cent local content development achievement in the industry over a period of 40 years. The federal government is desirous to localise a substantial proportion of input in project execution in the upstream and downstream sectors. It is against this background that the Nigerian Content Policy is being enforced in the industry with the aim of ensuring that the proportion of local inputs in the form of materials and skills is elevated to 40 per cent in 2008 and 60 per cent by 2010. NNPC has in pursuance of this objective established the NCD at the Corporate headquarters to articulate elements of Nigerian Content, compliance procedures and to monitor adherence of MNCs to the enabling policy framework as enunciated in an industry-wide circular.

Finally, it is important to note that the contracting powers of the operator are primarily exercised as that of an agent of the principal parties to the JOA. In this regard, all transactions consummated by the operator are undertaken in its capacity as an agent. For this reason the contractor may at its discretion file proceedings against the operator or the silent coventurers notwithstanding whether or not the contract in question states the capacity in which the operator consummated it.

The operating committee (Article 3)

The operating committee is a body constituted by the co-venturers in order to supervise, control and direct all matters associated with the JV operations. The nature and composition of the operating committee as provided in Articles 3 of the JOA varies in some countries. However, its application is the norm rather than the exception. The operating committee is a standard mechanism or practice in JOAs in Nigeria. However, in the early years in the United States and Canada the operating committee concept was not widely applied as operators were given exclusive liberty to control and manage the operation within stipulated guidelines and regulations assembled by the parties to the JV. In such situations the operator is empowered to manage, control and execute exploration, development and production of OMLs or other activities. The operator is also empowered to operate the accounts with the foremost understanding that the non-operating partners shall be furnished with information about proceedings in a timely, accurate and transparent manner. The circumvention of the operating committee and endowment of the operator with such powers as delineated in the preceding section can of necessity occur in situations where all the partners are at par in terms of technology, know-how and professional expertise such that no single party has a dominant position. In that case it is taken as a given that all the practices and indeed expenditure patterns are well acquainted to the parties to the JOA. The inherent expertise and knowledge on the part of the coventurers serves as a check on the activities of the operator especially if all functions are carried out with utmost caution and in good faith. This privilege or position notwithstanding, contemporary JOAs are crafted in a manner that allows the engagement of the operator for as long as the majority of the stakeholders want it to continue but may also be removed by the principal parties with or without there being any culpable offence.[14] In this regard the functions of the operating committee can be summarised as follows:

- receive, evaluate, revise, consider and approve all or part of proposed programmes;
- scheduling of the selection, timing, scope, location testing, completion, plugging and abandonment of all wells and facilities for joint operations. The committee also approves the designation or change in status, use or classification of affected wells and facilities;
- determine and ensure strict adherence by the operator with appropriate schedules of the agreement governing uniform nominations, scheduling and lifting procedures (as contained in schedule D of the agreement);
- abandon and salvage JV property;
- consider for purposes of implementation or classification general policy issues, commission studies, research, procedures and methods of joint operation;
- conversion to expendable income part or all assets of the JV through outright disposal, exchange or sale;
- settle claims and litigation, taking cognisance of JOA approved limits and to the extent that such claims are not covered by insurance;
- deliberate on matters warranting expansion or reduction of the operational area delineated in the JV;
- undertake or attend to any other matters of JV interest not within the purview of the operator and other such matters that may be referred to it from time to time by any of the stakeholders.

The operating committee conventionally includes at least one representative from each of the co-venturers. Existing operating committees in the Nigerian petroleum industry consist of ten members appointed by parties to the JOA. Manifest practice indicates that NNPC has six members and the operator(s) has four representatives on the committee. Articles 3.2.1 and 3.2.2 state that for purposes of conducting meetings five members shall constitute a quorum, three members from NNPC and two members from the operator. The non-operator appoints the chairman of the committee while the operator appoints the secretary to record the proceedings of meetings and maintain a record of all actions and decisions of the committee. Voting rights in the operating committees are exercised in proportion to the participating interest of the parties. This may, however, not be applicable if only two of the parties to the agreement are involved in the committee. The JOA provides that the decisions of the operating committee shall be made by unanimous vote of all parties concerned irrespective of the number of parties and their proportionate interest in the venture. It is often observed that this arrangement endows on the operator undue advantage and powers and as a result exerts more influence over the joint operations than 'its participating interest would justify'. Indigenous operatives in the upstream sector who interact with the MNCs are of the view that the operators wield undue influence in the affairs of the JVs. For this reason NNPC is often called upon to place more stringent controls on the activities of the operators. The call for more control on the part of NNPC is born out of the impression that the operating companies are too frivolous in spending and to a great extent manipulate contracts in favour of their affiliate companies offshore.

Cessation of operatorship

As pointed out in the preceding section, the execution of the JOA requires the appointment of an operator who would on behalf of the non-operators execute all approved joint operations in accordance with applicable laws and regulations. In accepting this responsibility the operator is expected to carry out the operations with utmost care, good faith and in workmanlike manner. The operator would under normal circumstances enjoy the privilege of representing the parties to the JOA as long as all statutory policies pertaining to the operatorship are complied with. In the event of a serious default on the part of the operator the operatorship can be revoked. Article 2.4.1 of the JOA states the conditions under which the operator shall cease to be the operator:

- The operator assigns or purports to assign, other than to an affiliate, pursuant to clause 19.3 of the agreement.
- The affiliate to whom the operator has assigned general powers and responsibilities of supervision and management as operator ceases to be an affiliate of the operator.
- The party acting as the operator (or any affiliate of the operator which is a party) assigns or otherwise disposes of, other than to an affiliate, all its pertaining interest.
- A petition is presented to and agreed to be heard by a court and an order is made, or an effective resolution is passed or legislation is enacted, for dissolution, liquidation or winding up of the operator.

- The operator makes an assignment for the benefit of creditors.
- A receiver is appointed or a mortgagee takes possession of the whole or material part of the assets or undertakings of the operator.
- The operator is proved pursuant to a decision of a court of law to have committed wilful misconduct.
- The operator becomes insolvent or ceases or threatens to cease to carry on its business.
- The operator allows any final judgment to be filed without recourse to a further appeal whereby the operator is required to hold, dispose of or to convey its entire participating interest, or portion thereof for the benefit of a third party.

In addition to the above the non-operators who hold an aggregate of 60 per cent participating interest can in writing communicate to the operator the cessation of the operatorship.

Resignation of operator

The JOA takes into consideration the exigencies in the business life of an operator. This is to say that with time the circumstances of the operatorship as well as challenges experienced by the operator may differ significantly. In this regard, the operator is at liberty to resign its appointment upon giving six months prior notice to the operating committee. Article 2.5 of the operating agreement further states that:

> . . . resignation of the operator shall be without prejudice to any operator's rights, obligations or liabilities which accrued during the period when the operator acted as such . . . the departing operator could debit the joint account for all costs approved by the operating committee which are associated with the process of change of operatorship. The cessation of operatorship would, however, not deprive the discontinuing operator his right of ownership of participating interest in the concession, assets, working capital, benefits etc.[15]

Appointment of successor operator

In the event that an operator resigns, the non-operators may adopt the option of appointing a third party as operator or may appoint one of the other non-operators as a successor operator to take over the operatorship on the effective date of the operator's resignation or an earlier date which may be mutually determined by the operator and the non-operator(s). The JOA also provides that:

> . . . in the event one of the non-operators is to become the successor operator, it shall be entitled to nominate one of its affiliates as operator

> . . . such shall be bound by the provisions of this agreement with respect to the duties of the operator. Such affiliate shall remain as operator for so long as it is an affiliate of the operator unless the parties otherwise agree. The parties by mutual agreement may also appoint a third party, not being an affiliate of either party . . .[16]

Change of operatorship

Pursuant to the OPEC Resolution XVI Article 90 of 1968 which encouraged each member country to participate in their respective oil and gas industries, Nigeria initiated JV participation in the MNCs. The primary aim of government participation was to have greater control of the petroleum industry which had hitherto been dominated by foreign companies. In initiating the JV agreements, the government expressed its intention to assume the role of operator in the concessions. For this reason, the operating agreement caused to be enshrined in Article 2.7 the NNPC's wish to play the role of an operator at an appropriate time. To play the role of an operator effectively NNPC staff must have requisite training, skills and experience to perform the duties of an operator. For this reason the article under reference stated:

> . . . in order to achieve this objective, parties will agree on a programme to progress build-up of NNPC's operational experience and secure its access to the necessary technology and technical assistance, as required for the exercise of such responsibilities in a cost efficient and professional manner . . .[17]

Change of operatorship is a very important factor which determines the extent to which Nigeria can achieve the indigenisation and domestication of inputs in the oil and gas industry. The importance of this clause notwithstanding, it is observed that the enabling article is thin in content. Whereas sections which concern foreign operatorship are robustly protected through use of incontrovertible language, the change of operatorship to NNPC is rather casually expressed. In this regard Article 2.7 enunciated 'NNPC wishes to become more actively involved with the responsibilities of the operatorship in respect of the joint operations'. Much firmer language such as 'NNPC shall at a mutually agreed future date assume operatorship of the operations' would be more appropriate for a company that aspires to experience technology transfer and become a world-class integrated oil and gas company. NNPC's process of becoming an operator as crafted in the JOA is rather open ended.

The article under reference further stated:

> . . . to achieve this objective, parties will agree on a programme to progress the build-up of NNPC's operational experience and secure its access to the necessary technology and technical assistance, as required

> for exercise of such responsibilities in a cost efficient and professional manner . . .[18]

Although Article 2.7 provides for build-up of NNPC's operational experience, this would be rather far-fetched, especially in a situation where all the major jobs of the JV partners are contracted to third parties. Field operations involving exploration and development are suitable channels of skill acquisition and technology assimilation. However, it is observed that these critical upstream activities are outsourced to OSCs who are not party to both the JV and the JOA; therefore the NNPC cannot second its staff to the contractors to acquire skills and experience and for that matter achieve technology transfer. In consideration of this, one would suggest that a more pragmatic strategy should be adopted for 'build-up' of field related skills and experience which can match the responsibilities of effective operatorship. The need for acceleration of skills acquisition and experience accumulation calls for a paradigm shift in the pattern of programme execution and operators should be mandated to execute a mutually agreed percentage of the programmes directly in a given concession. Direct execution of jobs will create an avenue for staff of the NNPC to be seconded or attached to ongoing projects on production rigs and similar work environments in which hands-on experiences can be acquired. Engineers and technicians with hands-on experience, requisite skills and exposure derived through direct participation in challenging programmes stand the chance of forming the team which will assume the role of operatorship on behalf of NNPC.

Funding joint venture operations

Cash calls

JVs are mechanisms or arrangements which bring together two or more parties to pursue common objectives which are capable of generating financial benefits. Establishment of a JV therefore entails apportionment of mutually agreed percentile equity interest to the coventurers. The objectives of a JV are achieved through the execution of programmes so, consequently, JVs must be funded by the parties in a manner that corresponds with their participating interest. The JOA states that each party would be required to fund an equivalent of its participating interest share of all costs and expenditures to be incurred for the joint account. The mode and procedure of payment of parties' obligations entail the following:[19]

> The operator shall not later than fifteen (15) working days prior to the first day of the cash call month submit to each party relevant information which shall include:
>
> - an itemised estimate of such cost and expenditures as well as an

itemised return of the actual expenditures for the month which is two months preceding the cash call month . . .;

- an itemisation of the cash available or cash deficit in the joint bank account . . . as well as credit expected to be received in the cash call month; and
- such party's cash call for that month which shall be its participating interest share of the estimated expenditures adjusted by the cash or deficits and credit and deficits in (b) above.

The JOA stipulates that each party remits into the joint account its cash call on or before the due date which is designated as the first day of the cash call month. Article 6.2.3 further provides that the non-operator may have grounds to disagree with a particular cash call month if it exceeds the cost and expenditure which would normally be incurred or associated with the period under consideration (i.e. cash call month). This objection may further be based on the approved programmes and budget and the aggregate of all expenditure to a reference date in the period under consideration. Furthermore Article 6.2.3 affirms that:

> . . . in the event that the non-operators so dispute any portion of a cash call the non-operators shall give to the operator a notice in writing specifying the amount in dispute and the reasons not later than eight (8) days from the date of receipt of cash call. The non-operators may not, however, dispute any portion of a cash call required for the protection of life and property or for the prevention of pollution. . . .[20]

The disputed account notwithstanding, the non-operator is required to remit the undisputed portion of the cash call into the joint bank account on or before the due date and at the same time utilise their best endeavours to resolve the matter concerning the disputed account in an expeditious manner. Upon the resolution of the disputed account all non-operators shall remit the disputed portion or amount mutually into the joint account within ten days from the date of resolution of the dispute.

Cash call crude oil

The funding mechanism of JV operations calls for remittance of funds into a nominated joint account in a designated currency. The JOA provides some flexibility in the mode of fulfilling a party's funding obligation to the JV programmes. Of particular interest is the acceptance of the party or parties to the JOA to grant NNPC the option of paying for cash calls in kind. The standard practice is to pay in cash. In order to pay for cash calls in kind, NNPC is required to notify the operator at least 60 calendar days prior to the cash call month its desire to pay its cash call by assigning to the operator crude oil volume equivalent in value to the due cash call amount. In response to the notice the operator shall promptly provide NNPC with an accurate estimate of the due cash call for the cash call month under consideration. The estimate is to allow NNPC to compute the equivalent crude oil volume required to meet the cash call obligation taking into account the prevailing market price of the crude oil stream. Article 6.4.1 stipulates that the value of the lifting shall be mutually agreed by the parties in compliance with NNPC standard valuation procedures or any other alternative method agreed by the parties provided it is not lower than the amount computed using the NNPC pricing mechanism. In practice the assignment of the cash call portion shall be initiated by the originating party and such communication shall expressly request the operator to lift the assigned crude oil on behalf of NNPC. The assigned crude volume or cargo might have been scheduled for lifting or would be scheduled and proceeds derived from the transaction shall be applied to defray the obligations of NNPC. Any excess funds arising from the settlement of the cash call obligation would be paid into any nominated bank account of NNPC. However, if cash deficiency occurs as a result of the transaction such deficit shall be added to the next cash call.

Cash call default

Articles 6.2.2 and 6.4 stipulate that NNPC shall have the option of paying its cash call in kind and in alternative any cash call due shall be paid into a joint bank account. This provision notwithstanding, any party may, due to business exigencies or unforeseen circumstances, experience default in cash call obligations. In this regard any party that fails to furnish the operator a cash call crude notice as provided in Article 6.4 and 6.4.2 shall be deemed to have defaulted and therefore become the defaulting party and the following shall prevail:

- The operator shall as soon as practicable notify the defaulting party in writing (and if deemed necessary, request emergency meeting of the operating committee) . . . matter of the default and the remedy thereof shall be discussed in full . . . .[21]
- If the defaulting party remains in default for two months after the due date the operator shall, with the prior approval of the operating committee, take necessary steps to provide or borrow funds to meet the amount in the default . . . .[22]

Once a party has been declared a defaulting party as contained in Article 6.5, any of the non-defaulting parties could be at liberty to contribute an equivalent of its participating interest in relation to the funds not contributed (i.e. contribute part of the defaulter's obligation). The defaulting party shall in this regard pay interest to the non-defaulting party at a bank rate pro rata to the contribution made by such a party. The JOA further affirms:

> . . . during the continuation of any default, the defaulting party's participating interest share of a Joint Account credits shall be used to offset the amount due from the defaulting party . . .[23]

The prolongation of the default can cause more serious problems for the joint operations especially if it drags on for a period of more than four months. In order to forestall any serious consequences the non-defaulting party(s) shall be at liberty to invoke its right to seek remedy to forestall the collapse of the business.

Insurance

Individual insurance

The oil and gas industry is highly capital intensive and as such the plants and facilities on aggregate cost huge sums of money in foreign currency. In order to safeguard unpredictable calamities, oil companies, as is customary in other industries or even domestic lives, obtain insurance. Such insurance is mandatory in JVs and it covers physical damage to property, well control and other areas critical to the JV and the activities of the joint operations. Each party is obliged upon request to provide to the other party evidence to support the existence and validity of such insurance. The insurance obtained by the parties shall be distinct and shall not discriminate or diminish the significance of the insurance obtained by the operator for the joint account. Oil companies strive to maintain valid insurance in view of the large asset value involved in the operations. In order to avoid exposure of the assets each party whose insurance cover is to be cancelled is obliged to give the other party at least 30 days notice in writing of such cancellation. In the event of such cancellation the party which has taken out such policy is obliged to confirm to the other party that an alternative insurance exists in respect of the matter covered by the insurance. The party taking the policy is also obliged to confirm to the other coventurer(s) that the premium for the insurance has been paid. In the event that the affected party fails to prove the existence of an alternative insurance as well as evidence of full payment of premiums the party receiving the notice of cancellation could make payment of the outstanding amount to renew the premium on behalf of the defaulting party. In doing this, the party receiving the cancellation notice who also paid for the outstanding premium shall be entitled to the cost of renewal plus interest at the prevailing bank rate.

Joint insurance

In the preceding section it was pointed out that each party to a JV obtains an insurance to cover physical damage to property and other areas critical to the JV. However, the operator is mandated to carry out the operations of the JV pursuant to the provisions of the JOA. To play this role, the operator is required to administer the operations in a professional, effective and workmanlike manner and is expected to safeguard all foreseeable eventualities. In this regard the JOA permits the operator to obtain insurance for the joint property and joint operations regardless of any individual insurance obtained by the parties to the JV. The expenses incurred in obtaining such insurance are chargeable to the joint account. The insurance shall specifically cover:

- motor vehicle liability insurance;
- aviation liability insurance of US$10,000,000.00 or local equivalent;
- charterer's legal general liability insurance to provide coverage arising out of the use of any chartered barges or vessels;
- marine insurance;
- any insurance entered into by the operator in furtherance of joint operations including contractor's all-risk insurance.

The operator is obliged under the regulations of the JOA to obtain and maintain insurance to cover all the aforementioned areas. In the event that the operator by act of choice or omission fails to obtain the insurance as required, it shall be solely responsible for any loss, demands, claims or damages emanating from such acts of omission or negligence. However, the operator shall not be held culpable if it made all reasonable effort to obtain and maintain such insurance but was unable to do so. It is also the responsibility of the operator to ensure that any contractor or subcontractor engaged to carry out jobs for the joint operation maintains appropriate insurance that covers the scope and values of their jobs executed for the joint operations. In a similar manner the operator would also be required to use all reasonable effort to ensure that such vital equipment (such as marine drilling rigs and workboats used in joint operations) are insured by the owners of such vessels. Such insurance cover shall among others include hull, tackle and machinery collision, towers liability etc. The operator is duty-bound to discuss with the non-operators the type of insurance it intends to obtain and make full disclosure of the premium rate and policy terms and conditions including but without limitation to deductibles and insured values:

> . . . any loss arising from the transaction therefrom shall be promptly communicated to the non-operators and the operator is required to file all claims and make reasonable efforts to record proceeds and remit such proceeds to the joint account. For purposes of proper operations and record keeping the operator shall furnish the non-operators within 30 days from the date of issuance of an insurance policy or renewal copies of the policies or renewals issued by the insurance company.
>
> (Nigerian oil and gas industry MOU 2000)

References

1 Akinrele, A. 2005 *Nigerian Oil and Gas Law*. Sections 1–11, Oil, Gas and Energy Law, Dundee.
2 Akinrele, op. cit., Articles 1–12.
3 Schatzl, L. H. 1969 *Petroleum in Nigeria*. The Nigerian Institute of Social and Economic Research, Oxford University Press, Ibadan, pp. 1–20.
4 Akinrele, op. cit., Sections 1–16.
5 Olisa, M. M. 1997 *Nigerian Petroleum Law and Practice*. Fountain Books, Lagos, p. 68.
6 Ibid. p. 76.
7 'Nigeria: The Oil and Gas Industry Investment Opportunities', being a publication by the Nigerian National Committee on the World Petroleum Congress, 2002, p. 53.
8 Olisa, op. cit., p. 75.
9 Akinrele, op. cit., Sections 5–22.
10 Olisa, op. cit., p. 81.
11 Akinrele, op. cit., Sections 5–28.
12 Scot, A. 'The Fiduciary Principle'. Proceedings of the Twenty-Second Annual Meeting of the State Bar of California, 104, 1949, cited by D. A. MacWilliams in 'Fiduciary Relationship in Oil and Gas JVs'. *Alberta Law Review*, Petroleum Law Supplement Vol. VIII, 1970, p. 234.
13 Williams, H. R. 'Fiduciary Principle in Law of Oil and Gas' 1962. Thirteenth Annual Institute of Oil and Gas Law and Taxation, South West Legal Foundation pp. 201–361 cited in M. M. Olisa, 1997, *Nigerian Petroleum Law and Practice* p. 82.
14 MacLean, J. A. 1992 'The Canadian Association of Petroleum Landmen Operating Procedures: An Overview of the Revisions'. *Alberta Law Review*, Petroleum Law Edition, Vol. XXX, No. 1, p. 138.
15 'Nigerian Oil and Gas Industry' JOA, 2000, Article 2.5.2.
16 Ibid. Article 2.6.2.
17 Ibid. Article 2.7.
18 Ibid.
19 Ibid. Article 6.2.1.
20 Ibid. Article 6.2.4.
21 Ibid. Article 6.5.1.
22 Ibid. Article 6.5.2.
23 Ibid. Article 6.5.4.

14 MOU and JV operations

Introduction

Oil and gas exploration and production has long been a global business. Investors in the industry, especially the IOCs, have ventured into extremely challenging terrains to explore for oil and gas and one of the major enabling factors has been technological innovation which is derived from huge investments in research and development. Cutting edge technology greatly enhances exploration and production activities but it is often obtained at a premium. This adds to the cost elements of upstream activities and as a result deflates the expected profit accruing to IOCs. In entering into a Memorandum of Understanding (MOU) with an NOC, the IOCs are aware of the risks involved and for that reason expect reasonable returns on their investments. Oil and gas are the mainstay of Nigerian economy, accounting for about 80 per cent and 30 per cent of the nation's foreign earnings and of the GDP respectively. In consideration of the pivotal roles of oil and gas, the federal government is favourably disposed to creating an enabling environment as well as providing appropriate incentives for IOCs. Such incentives are designed to encourage the companies to commit funds and other related resources in the exploration and production business.

In specific terms, one would posit that it is the government objective to increase the national reserves beyond the 36 billion barrels mark and effectively cope with the vagaries of the global oil and gas market conditions. To achieve targeted reserve expansion and oil price fluctuations in the industry, the federal government introduced a package of incentives through an MOU with IOCs. The package of incentives is within industry parlance simply referred to as 'MOU'. A key element in the MOU relates to improved fiscal terms aimed primarily at ameliorating the impact of tax and royalty on the net earnings of the oil companies. The MOU is a legal document which places obligation on the parties to a JV agreement. In practice, the clauses contained therein are revered and upheld in utmost good faith by the parties.

Evolution of MOU

The global oil and gas industry experienced a glut in 1986 which resulted in the plummeting price of oil. Global oil demand shrunk thereby causing significant decrease in the earnings of oil companies. The decline in the ROI placed serious constraints on IOCs in terms of committing investment capital on new projects or embarking on de-bottlenecking programmes on existing projects. During the same period, non-OPEC production increased, further aggravating the declining price conditions. Observers noted that this singular event and the escalation of production from non-OPEC producers deflated the market share of OPEC from 80 per cent (attained in 1977) to 54 per cent in 1986. The consequences of the 1986 oil glut were grave and IOCs in Nigeria reacted by curtailing expenditure in the upstream sector. The scaling down of activities in the sector led to rapid depletion of existing reserves in the country. The prevailing circumstances rapidly eroded the cushioning effect of the $2.0/bbl fiscal notional margin introduced in 1983, which in practical terms became unrealisable. Furthermore, the accounting process which relied on the fiscal or posted price extracted from the Official Selling Price (OSP) derived from the OPEC pricing mechanism became rather untenable. The consequence of the cut back in expenditure in the upstream sector and the reliance on non-market related OSP was the dwindling flow of revenue into the national treasury.

The federal government acted swiftly to forestall further distress of the earnings from oil and introduced an incentive package to mitigate the impact of the oil glut. The incentive package so introduced is the MOU. In order to ensure the effective implementation of the MOU, the OSP and the market price were harmonised such that the market price was adopted as the applicable price. The adoption of this approach to a reasonable extent guaranteed an acceptable profit margin in the business. The MOU was initiated and executed in 1986 and was programmed to last for five years in the first instance.[1] The profit margin of the IOCs was adjusted four times between 1977 and 1991 and these adjustments were in response to changing price conditions in the global oil and gas market. Prior to the introduction of the MOU fiscal incentives applied as contained in the Petroleum Profit Tax Act (PPTA 1959) and subsequent amendments to the original Act. PPTA set a tax level of 85 per cent on taxable profits of each company. In order to mitigate the effects of taxes and/or critical expenditure on the profits of IOCs in the industry, allowance was made for certain deductibles to be applied as follows:

- all royalties paid on chargeable oil produced and disposed of;
- interest on borrowed funds other than money loaned to the company by subsidiary of the company;
- debts deemed unrecoverable and therefore written off;
- survey expenses incurred in preparation for drilling;

- rents paid for the period other than rental paid under prospecting licence and oil mining lease;
- intangible drilling expenses;
- any expenses attributable to chargeable oil but not including expenses incurred from the transportation business of the company;
- any other expenses allowed under the PPTA.

Prohibited deductions

Under the provisions of the PPTA the following deductions are prohibited for the purpose of arriving at an adjusted profit:[2]

- money spent on improvements rather than on categorical repairs;
- expenditure which cannot be ascertained to have been incurred purely for petroleum operations;
- any sum recoverable from an insurance policy or contract of indemnity;
- any rent or cost of repairs to a premises or part of premises not incurred for the purpose of the operations;
- depreciation of any premises, buildings, structures, works of permanent nature, plant, machinery or fixtures;
- interest on money borrowed from an affiliate;
- any cost incurred for the purchase of information relating to the existence and extent of petroleum deposits;
- payment to any provident fund, pension, saving or other society scheme or fund not approved by the Joint Tax Board;
- any royalty or other sums deductible from assessable tax as tax offset under section 17.

PPT and royalties were computed based on revenues derived from application of the fiscal or posted price of crude oil. The fiscal price in turn was derived from a formula premised on the official selling price which is a non-market related index originated by OPEC. Ali,[3] op. cit. pp. 320–321, in analysing the evolution and trend of MOU portrayed posted price as follows:

$$\text{Posted price} = \frac{OSP - (M + 0.15TC)}{0.88}$$

where M = fiscal margin, TC = technical cost.

Fiscal margins grew over the years from eighty cents in 1977 to \$2.30 in 1991. The technical cost of production also appreciated from \$1.00 in 1977 to \$2.50 in 1991. Available information indicates that prior to 1986, fiscal margins were well below \$2/bbl. This low margin as expressed by the IOCs warranted cut back in investments. As a result, seismic acquisition mainly in the 2D and 3D categories declined from 25,000 km in 1979 to about 6,000 km in 1983 and rose to 10,000 km in 1985.

In the drilling area, the experience was similar as footage drilled shrank from a level of 750,000 feet in 1978 to an all time low of 150,000 feet in 1985. Footage drilled in the industry was slightly below 500,000 feet in 1976 but increased to about 750,000 feet in 1978. However, from 1978 the footage drilled sharply declined to a little less than 200,000 feet in 1985. The increase in footage drilled to about 600,000 feet in 1977 was as result of the government revision of the fiscal technical cost to $1.00/bbl and that of fiscal notional margin to $0.80/bbl. Relief to the IOCs arising from the adjustments was temporary and in 1979 the footage drilled took a downward turn and thereafter the worst drilled footage results were recorded in the history of the industry. Ali noted that seismic data increased from the low of 8,000 km coverage in 1976 to 20,000 km in 1978. The decline in the drilled footage emanated from the fact that applicable notional margin of $1.00 in 1977 proved inadequate to stimulate investment which was critical for the incremental escalation of drilling activities. In the face of declining exploration and production activities, the government increased the fiscal technical cost in 1978 to $1.10/bbl. However, the guaranteed margin remained unchanged at $0.80/bbl. Between 1977 and 1983 the incentive package was adjusted upwards four times, but these incremental adjustments proved insufficient to ameliorate a fast deteriorating situation in the industry. Within the time span of 1982 and 1983, the notional profit doubled to $1.60 and the fiscal technical portion increased by about 50 per cent to $1.60. This adjustment was quickly overtaken by the deteriorating market conditions and increase in technical costs. The situation would have been quite easily handled if there were known applicable rules of thumb which can provide clues as to what level of notional margin and technical cost incentives would be sufficient to stimulate and sustain upstream activities. The problem depended mainly on global market forces in the industry and the cost of research and development which determine the cost of technology.

This apart, it was observed that the reliance on the OPEC generated non-market related OSP was an impediment in the packaging of sustainable incentives for upstream activities in the industry. A close analysis of the OSP showed a wide variance from the market price and in some cases the variance was as wide as $5/bbl. The price differential between OSP and market realisable price reached a climax in 1985 beyond which near irreversible damage could have been inflicted on the operations in the industry. This experience served as the turning point and the MOU was initiated in 1986 to avert a catastrophe. The MOU introduced a mechanism which mitigated the disparity between OSP and the market price. In this regard OSP was replaced with Realisable Price (RP) which was determined through a netback pricing mechanism. This method is generally used in calculating the portion of revenue due to the federal government. The computed government revenue is referred to as the Revised Government Take (RGT). Ali in his analysis further indicated that RGT is the aggregate of PPT and royalty[3]

$$RGT = OP - (TR_x\, TC) - OT$$

where OP = Offset Price = $b \times RP$
TR = Tax Rate (i.e. 85 per cent)
TC = actual Technical Cost
OT = Tax Offsets
$b(factor) = k[TR_x\,(1\text{-}Roy) + Roy]$

$$k(factor) = \frac{1 - (M + 0.15TC)/RP}{0.88}$$

RP = Realisable Price as computed from the net-back

MOU and programme execution

The revisions carried out in the incentive package produced the expected results, especially as it related to infusing confidence in the minds of IOCs. The guaranteed margin along with the introduction of RP in the equation improved the earnings to a reasonable level. An index used in assessing the performance of the sector was the acquisition of seismic data. In 1986, seismic data acquired in 2D and 3D categories was estimated to be 9,650 km. In the light of the fiscal adjustments, seismic data acquired increased to 28,980 km and 93,850 km in 1987 and 1990 respectively. This was a significant improvement. A corollary of seismic data acquisition was the footage of exploratory and appraisal wells drilled in the sector between 1986 and 1992. During the period 1990–1992 over 0.5 million was drilled. The trend in the Nigerian petroleum industry was exceptional because globally, drilling and other exploration activities stagnated due to low and in some cases negative margins recorded in upstream operations. Whereas activities plummeted in the global oil and gas industry, activities in Nigeria stimulated by the MOU recorded impressive results.

Keen observers in the industry noted that the prevalent conditions at the global scene led to low demand for rigs. Consequently, rig prices dropped significantly. In this regard, IOCs in Nigeria took advantage of the new incentives and deployed more rigs (at low prices) and seismic parties to increase the drilling footage as well as acquire considerable amount of seismic data. The increased activities required additional funds; therefore the CAPEX increased between 1988 and 1992 recording expenditure of between $0.5 billion and $2.5 billion during the period under reference. The aggregate impact of the impressive results in seismic data acquisition and the footage drilled was an increase in the national crude oil reserve. As indicated earlier the effect of the erosion of the guaranteed notional margin introduced in 1983 led to significant stepping down of upstream activities which translated into slow replenishment of the existing reserves. In fact the rate of the depletion at some point exceeded the rate of infusion of crude oil into the existing reserves. It was observed therefore that actual reserves depreciated from a high of 15.25 billion barrels in 1984 to about 13.63 billion in 1987. The rate of decline was, however, reversed in 1988 due to additional adjustments in the notional margin and technical cost variables embodied in the MOU. In view of the timely response to the price adjustment modulators, IOCs were able to cope with the dynamic global market conditions and pursue programmes which resulted in an upward movement of hitherto depleting reserves from 13.63 billion to 17.2 billion barrels.

The MOU in practical terms served as a catalyst in creating the right atmosphere for the critical upstream activities to progress and achieve the fundamental government objectives of expanding the crude oil reserves and also generating revenues through sale of crude oil at a globally competitive price. A new round of negotiations took place between the IOCs and the government in 1991 and 2000 respectively. During this period, major revisions were made to the 1986 and 1991 MOUs. In consideration of this, therefore, subsequent discussions in this chapter will dwell on the provisions of the 2000 MOU. The ultimate objective would be to assess the incentives offered therein and how these could engender satisfactory returns for the mutual benefit of stakeholders.

Provisions of 2000 MOU

Some aspects of the 1986 and 1991 MOU incentives were touched on previously only as a basis of discussing the trend of events, especially in relation to the erosion of the guaranteed notional margin which negatively impacted operations in the industry. This section will examine in detail the provisions of the 2000 MOU. The underlying objective would be to portray a lucid picture of the primary intent of the incentives mechanism. Clause 2.1 of the 2000 MOU (hereafter referred to as the MOU) stated that as at 31 December 1985 the fiscal regime based Royalty computation on the Posted Price and Petroleum Profit Tax (PPT) on the higher of the actual proceeds or posted prices of the PPT Act CAP 354 LFN 1990. The company shall revert to the fiscal regime stated in the clause in the event that it defaults on its obligations in clause 3.2 in accordance with clause 4.3.[4]

Clause 2.2.1 of the MOU ensures that unless the provisions of Clauses 2.5 and 2.6 are modified, based on mutual consideration of the parties to the MOU, IOCs

> . . . were guaranteed a minimum Notional Margin of $2.50/bbl after Tax and Royalty on its equity crude. The IOC would also earn a minimum of $1.25/bbl, after Tax and Royalty, on the NNPC equity crude which it lifted under this memorandum. However, this minimum guaranteed notional margin shall be premised on the fact that the technical cost of

operations does not exceed the notional fiscal technical cost of operations which at present is \$4.0/bbl.[5]

The provisions of the MOU further indicate that a fiscal technical cost band of \$2 and \$4 is mutually established and the clause enunciated that in the event that the average capital investment cost (T2) in a particular calendar year exceeds \$2/bbl, then the guaranteed Notional Margin as contained in Clause 2.2 would be adjusted upwards to \$2.70/bbl in respect of the company's equity crude and \$1.35/bbl for the NNPC equity crude.

The Government Take

The Government Take (Royalty and PPT) concerning the JV operations between NNPC and the IOC for any fiscal year shall be the lower of the Government Take as enshrined in the Royalty and PPT Regulations of 31 December 1985 as amended; calculated by substitution of Posted Price (PP) with Official Selling Price (OSP) and the Revised Government Take calculated by substitution of the PP with the Tax Reference Price (TRP).[6]

Tax Reference Price

In the context of the MOU the TRP is to be calculated as follows:[7]

$$TRP = \frac{RP - (M + 0.15 \times FC)}{0.88}$$

where *RP* = Reliable Price calculated in accordance with Clauses 2.11, 2.12 and 2.13 to determine or mirror the crude oil market values of Nigerian export crude oil

M = \$2.50/bbl when actual capital investment is \$2.0/bbl or less
= \$2.70/bbl when actual capital investment costs (T2) is greater than \$2/bbl

FC = notional Fiscal Technical Cost of \$4/bbl.

Revised Royalty

Under the provisions of the MOU the Revised Royalty (Roy_{TRP}) can be computed by substituting the TRP for the PP as contained in the 1985 Royalty and PPT Regulations as amended. In this regard the computation for (Roy_{TRP}) would be as follows:[8]

$$Roy_{TRP} = RR \times TRP \times V$$

where *RR* = applicable Royalty Rate according to 31/12/85 Royalty and PPT Regulations, as amended

TRP = Tax Reference Price as determined in Clause 2.4.1
V = company's crude oil and condensate production.

Revised PPT

Consistent with the procedure in the preceding section, Revised PPT based on PPT_{TRP} can be obtained by substituting TRP for the PP in the 31 December 1985 Royalty and Regulations as amended. The computation would be:[9]

$$PPT_{TRP} = [(TRP \times V_S) - TRP_{TRP} - TC)] \times TR$$

where V_s = company's crude oil and condensate volume
Roy_{TRP} = Revised Royalty as determined in Clause 2.4.2
TC = deductions under sections 10,14 and 18 (excluding royalty) of the Petroleum Profits Tax Act 1959 and its subsequent amendments
TR = Applicable Tax Rate.

Revised Government Take (RGT)

Clause 2.4.4 of the MOU defined the parameters of RGT as:[10]

$$RGT = Roy_{TRP} + PPT_{TRP} + TIP$$

where Roy_{TRP} = Revised Royalty as determined in Clause 2.4.2
PPT_{TRP} = Revised PPT as determined in Clause 2.4.3
TIP = Tax Inversion Penalty as determined in Clause 2.7.

Realisable notional margin

It is important to indicate that the formulae provided in Clauses 2.4.1 and 2.4.2 have cascading effects in the sense that once the basic premises are established in the formulae, subsequent parameters such as notional margin or TRP and RGT form the key variables of the formula used to elicit or generate other indices in the MOU. Therefore in calculating the notional margin, reference is made to Realisable Price (RP), notional Fiscal Technical Cost (FC) and company profit. In situations where RP is less than \$13.48/bbl the MOU provides that the applicable guaranteed notional margin would be calculated as:[11]

$$M = \left(1 - \frac{FC}{RP}\right) \times (RP_1 a_1 + RP_2 a_2 + RP_3 a_3)$$

where M = applicable guaranteed Notional Margin for RP less than \$13.48/bbl

RP = Realisable Price
FC = notional Fiscal Technical Cost of \$4.00/bbl
a = company's percentage share of field profit.

For

Realisable price in the range		*Company share applicable to price range*	
		T2 ≤ 2.00	T2> 2.00
$0 < RP_1 < \$5/bbl$	$a_1 =$	0.300	0.365
$\$5/bbl \leq RP_2 < \$10/bbl$	$a_2 =$	0.285	0.288
$\$10/bbl \leq RP_3 < \$13.48/bbl$	$a_3 =$	0.107	0.083

For RP between \$13.48 and \$15/bbl,

$$M = M_{15} + (RP - 15) \times a$$

where M = applicable guaranteed Notional Margin for \$13.48/bbl ≤ RP < \$15/bbl
M_{15} = \$2.50/bbl when actual capital investment cost (T2) is \$2/bbl or less
= \$2.70/bbl when actual capital investment cost (T2) is greater than \$2/bbl
RP = Realisable Price
a = company's percentage share of field profit.

For

*Realisable price in the range**		*Company share applicable to price range*	
		T2≤ 2.00	T2> 2.00
	a =	0.116	0.132

* examples of procedures for technical cost calculation are contained in Appendix 1.

The MOU also provides for the calculation of guaranteed notional margin in situations where the RPs are greater than \$19/bbl but less than or equal to \$30/bbl. In this case the formula would be as follows:

$$M = M_{19} + (RP - 19) \times a$$

where M = applicable guaranteed notional margin for \$19/bbl ≤ RP < \$30/bbl

M_{19} = \$2.50/bbl when actual capital investment cost (T_2) is \$2/bbl or less
= \$2.70/bbl when actual capital investment cost (T_2) is greater than \$2/bbl
RP = Realisable Price
a = company's percentage share of field profit.

For:

Realisable price in the range		*Company share applicable to pice range*	
		T2≤ 2.00	T2> 2.00
\$19 < RP ≤ \$30/bbl	a =	0.116	0.132

The MOU also provides that in situations where the RP exceeds \$30/bbl the Honourable Minister shall advise on applicable margins. Such advice from the Minister would be warranted if RP exceeds \$30/bbl for a minimum period of 45 days continuously. However if the RP falls below \$30/bbl the margin will automatically revert to the levels specified in Clauses 2.2, 2.3, 2.5 and 2.6 as would be deemed necessary.

Tax inversion

The primary purpose of the MOU is to provide appropriate incentives which could stimulate IOCs to invest adequately and derive tangible results through increased crude production. Perhaps of greater significance is the need to carry out the operations in the upstream sector in a safe and cost efficient manner. In order to engender cost efficiency the MOU provided for a Tax Inversion Penalty (TIP) of 35 per cent as contained in Clauses 2.7.1 and 2.7.2. T1 is adequately defined in Appendix 1 Paragraph 1. To the extent that T1 is less than \$1.70/bbl for any given calendar year the applicable TIP as applied in the RGT formula (Clause 2.4) shall be calculated as follows:[12]

$$TIP = (TR - TIR) \times (T1 - LTIT) \times V$$

where *LTIT* = Low Tax Inversion Threshold = \$1.70/bbl
TR = applicable Tax Rate
TIR = Tax Inversion Rate
T1 = sum of all costs and incomes approved by operating committee (Appendix 1 Paragraph 1)
V = company's crude oil and condensate production.

While adhering to the definition in Appendix 1, a situation may arise where

T1 is greater than \$2.30/bbl for a calendar year for companies with average production of 175,000 b/day or \$3.0/bbl for companies producing below 175,000 b/day in the same calendar year. The TIP as used in the RGT formula (Clause 2.4.4) shall be as follows:[13]

$$TIP = (TR - TIR) \times (T1 - UTIT) \times V$$

such that

UTIT	=	Upper Trigger Point for Tax Inversion for T1 = \$2.30/bbl for companies producing above an average of 175,000 b/d in the same calendar year or \$3.00/bbl for companies producing below 175,000 b/d in the same calendar year. To the extent that V, as defined herein, is adversely affected by circumstances outside the control of the company, UTIT shall be adjusted to negate such adverse impacts. The procedure is set out in Appendix C
TR	=	Tax Inversion Rate = 35 per cent
T1	=	as defined in Appendix 1 (para. 1)
V	=	company's crude oil and condensate production.

Imperative conditions

The introduction of MOU is mutually beneficial to both the federal government and the companies. However, the government considers the provisions in the MOU sufficiently attractive and for that reason expects the IOCs to meet certain conditions within the context of the MOU. Some of the conditions are as follows:

- notice/emergency volumes;
- interim payments;
- adjustment of interim payments;
- payment of Royalty and PPT;
- execution of programmes.

Notice and emergency volumes

Under Clause 3.1.1 the company (i.e. JV partner) accepts to lift on a monthly basis such volumes of crude which NNPC is unable to lift out of its equity crude oil. This provision notwithstanding, the volume of crude lifted on behalf of NNPC shall not exceed 50 per cent of the volume allocated to NNPC (by way of equity crude entitlement) in a particular month. The 50 per cent volume specification shall be limited to a maximum of three full cargoes each amounting to 920,000 barrels. In pursuance of the provisions of the clause, NNPC is required to give the company 15 days prior notice requesting the company to lift its unlifted crude. The notice shall also specify the crude stream to be lifted. In the event that 15 days prior notice is not given, the volume in question may be offered to the company as emergency crude. Any volume far and above the notice volume may also be offered to the company as emergency volume. The MOU further provides that in situations where the commercial production quota increased by more than 15 per cent, NNPC would be free to revise the notice volume as well as the conditions associated with the volume. It is to be noted, however, that the company is not under any obligation to lift emergency volumes but shall in the spirit of the JV endeavour to meet NNPC's requirement for lifting such volumes.[14]

Interim payments

In lifting the notice volume the company is expected to make interim payments to NNPC in connection with each lifting of crude oil for NNPC's equity crude. Such payments shall be made within 30 days of the Bill of Lading or any credit period granted by NNPC. The payment for each lifting shall be at the notional transfer cost enunciated in the MOU.

Adjustment of interim payments

To ensure that the lifting obligations are performed properly, the MOU (Clause 3.1.4) provides that at the end of each calendar year but not later than 31 March of the succeeding year, the interim payments shall be adjusted in relation to the transfer costs reported during the year. All sums due to NNPC shall be remitted in dollars, pounds etc. or any currency mutually accepted by the parties.

Payment of Royalty and PPT

Pursuant to the provisions of the MOU the company is expected to pay Royalty and PPT on the notice volumes lifted out of NNPC's equity crude oil. Such payment shall be made to appropriate government authorities.

Execution of programmes

Budgets are based on work programmes and other operations packaged and submitted to NNPC for consideration. The MOU is generally expected to be accepted in utmost good faith and as such all provisions contained therein are to be abided by. The incentives in the MOU are provided with the understanding that each party will fulfil its obligations pursuant to the implementation of the MOU. In this regard, especially in the context of the incentives, the company undertakes to execute all programme schedules as agreed barring any unforeseen negative intervening factors or circumstances. Based on mutual agreement the work programme may be revised periodically taking into consideration economic, capital and operational or other manifest exigencies.[15]

Non-performance

In the event that the company is unable to perform its obligation and fails to lift all or part of the notice volume pursuant to Clause 3.1.1 the company will be required to pay NNPC 2 per cent of the average RP for each barrel not lifted. These payments shall be made taking into consideration the provisions of Clause 3.1.4. However, if the total unlifted volume is less than 5 per cent of the notice volume communicated by NNPC or where a particular month's unlifted notice volume is lifted within 5 days of the beginning of the succeeding month, or as result of force majeure as stipulated in Clause 6, the provisions of Clause 4.1 referenced herein shall not be applicable. The imposition of port restrictions on crude oil, imports, tariffs barriers or other tangible restriction of trade is capable of affecting disposal of NNPC crude. In such circumstances, especially if NNPC is aware of the aforementioned conditions, the parties are obliged to meet and proffer an equitable solution.[16]

Force majeure

Clause 6 of the MOU states:

> . . . no failure or omission to carry out or to observe any of the terms, provisions or conditions of this memorandum shall, except as is herein expressly provided to the contrary, give rise to any claim by any one party . . . if such failure or omission arises from any cause reasonably beyond the control of the party. Such cause may be, but not limited to acts of God, breakdown of vessels or machinery and equipment, civil unrest, strikes, lock outs, wars etc. . . .

Arbitration

Oil and gas operations are world-class in nature and as such, parties to JVs involved in the business, as much as practicable, exhibit a high level of professionalism. In this regard all parties to oil and gas JV agreements, taking cognisance of the critical nature of the operations, endeavour to hold sacrosanct the provisions of the JV agreements. In the Nigerian context the operating committee constituted by the parties to the JV play an important role to ensure that disputes arising between parties or between non-operators and operators are resolved amicably. Clause 8 of the 2000 MOU states in part:

> . . . parties agree that if any difference or dispute arises between them concerning the interpretation or performance of this Agreement and if they fail to settle such differences or dispute upon exercise of due diligence

> then the party or parties may serve on the other or others a demand for arbitration. Within 30 days of such demand being served, each party involved . . . shall appoint an impartial arbitrator. . . .

As much as possible all disagreements are handled in such a manner that choice of arbitration for dispute resolution comes into the picture after all available avenues have been exhausted. Arbitration, of necessity, is a last resort. However, if the need arises, especially when all conceivable mechanisms have failed to produce a mutually acceptable solution, then arbitration comes into play.[17]

References

1 Ali, O. R. 'JV Investments and MOU Incentives: an Appraisal in the Nigerian Petroleum Business' (edited by V. E. Eremosele 1997), p. 317. Advent Communications Limited, Lagos.
2 Olisa, M.M. *Nigerian Petroleum Law and Practice*, 1997, Fountain Books, Ibadan , pp. 186–187.
3 Ali, O. R., op. cit., pp. 320–321.
4 Nigerian Oil Industry Memorandum of Understanding 2000, Clause 2.2.1. (all clauses refer to 2000 MOU).
5 Ibid. Clause 2.2.2.
6 Ibid. Clause 2.4; subsection 2.4.1–2.7.2.
7 Ibid. Clause 2.4.1.
8 Ibid. Clause 2.4.2.
9 Ibid. Clause 2.4.3.
10 Ibid. Clause 2.4.4.
11 Ibid. Clauses 2.5.1–2.5.2.
12 Ibid. Clause 2.7.1.
13 Ibid. Clause 2.7.2.
14 Ibid. Clauses 3.7.1–3.7.3.
15 Ibid. Clause 3.2.
16 Ibid. Clauses 4.1–4.2.
17 Ibid. Clause 8.

Further reading

Petroleum Profit Tax (Amendment) Act No. 15, 1973.
Petroleum Profit Tax (Amendment) Act No. 22, 1970.
Petroleum Profit Tax (Amendment) Act No. 55, 1977.
Petroleum Profit Tax (Amendment) Act No. 2, 1979.
Petroleum Profit Tax (Amendment) Act No. 3, 1979.
Petroleum Profit Tax Act CAP 354, 1990.
Company Income Tax Act CAP 60, 1990.
Petroleum Drilling and Production Amendment Regulation (L.N. No. 69), 1969.
Petroleum Drilling and Production Amendment Regulation, 1973.
Petroleum Drilling and Production Amendment Regulation, 1979.
Mineral Oils (Safety) Regulation CAP 350, 1990.

15 The Niger Delta

Introduction

The Niger Delta region has been associated with oil production activities for over eight decades but the presence of oil was first registered only in 1956 with the discovery at Oloibiri. Since then the region has increasingly grown to be the hub of Nigerian oil exploration and production activities. The Niger Delta is characteristically swampy, mainly inhabited by fishermen and peasant farmers, and in the early days (i.e. 1950s–1960s) the level of activities was low as the scope for production was limited. For this reason, major ecological disruptions were not recorded. With advancement in technology and expansion of entrepreneurial zeal, the ranks of the producing companies increased thereby resulting in a corresponding increase in the impact of these activities on the environment. It is important to state that early exploration activities, especially in the 1960s and 1970s, involved seismic activities on land, swamps, shallow waters and in the mangrove forest. Such activities involved seismic shots which were associated with vibrations capable of causing cracks in weak houses, and the companies were often confronted by host communities to pay compensation for damaged private buildings, community halls, schools, hospitals and church buildings. In some areas economic and tree crops destroyed by seismic crew activities were also paid for by the companies. These incidents, in most cases, were settled amicably although complications were often experienced when impostors also filed claims for farm land which had already been settled. In the swampy mangroves, no definite claims would be filed except in cases where fishing villages were involved. Such activities destroyed extensive areas of mangrove forests and the scars of damage remain for years, after which other trees grow to replace vegetation destroyed as a result of chemical toxicity.

Incidence of oil spillage was a rare event in the early oil and gas exploration periods. This was partly because the scope of operation and indeed the volume of crude oil handled were low. Also at that time indigenes in host communities believed that oil exploration activities and equipment stationed by the companies belonged to the government. The moral fabric in the Nigerian society was sacrosanct and host communities accepted responsibility for the safety of equipment of oil producing companies. The attitude was maintained only to the extent that the environment and the livelihood of the citizens were not jeopardised. It was a dominant belief also that government property should be protected in order to avoid the community leaders from being invited by law enforcement agents to account for any damage to oil assets. For this reason Niger Delta communities consciously cautioned youths against tampering with SPDC, Mobil, Agip, NAOC or other oil company equipment and installations. Things were quite different then, and youths kept their distance from the installations. However, this state of harmony between the oil companies and the communities was disrupted by the 1967 civil war. During that period most of the MNCs left the country, leaving behind much equipment including cars, trailers, barges, tug boats etc. Some of these items were looted by individuals and rebel forces, thereby breaking the long standing culture of communal protection of equipment and facilities of oil and gas producing companies.

The civil war era

One may admit that during the 1960s and 1970s minor oil spillages might have occurred. It would be fair to say that they did not constitute major environmental hazards. However, during the civil war, the Niger Delta region was a war theatre and initially under the control of the Biafran Army. Around 1969 Federal troops penetrated the region and took possession of some of the areas, especially Bonny, Okrika, Andoni and Opobo axis. In order to dislocate the activities of the advancing federal troops, regiments of the Biafran army bombed some oil installations on the Bonny River and other riverine locations, causing huge spillages which covered the mangrove forests, coastlines, creeks and beaches. The old Port Harcourt refinery was bombed and went up in flames. This incident caused significant pollution in the Bonny River including the adjoining creeks of Okrika, Ogu, Bolo and Wakama-Ama. Essentially these actions marked the beginning of major pollution activities in the Niger Delta region. The damage occasioned by hostilities was not accounted for by either side, neither was any compensation paid to the affected communities.

Current pollution activities

Since the 1980s, incidents of oil pollution have increased both on land and sea. Some of the spills associated with genuine drilling and production activities are usually reported by the affected companies. However, occasional spills which occur as a result of rupture of corroded pipes, accidental well blow outs, and spills occurring as a result of replacement of nuts and bolts have not been adequately recorded. In recent times, however, oil pollution incidents have become more frequent as youths of the Niger Delta now consider theft of petroleum products and indeed crude oil a source of livelihood.

For this reason, varieties of gadgets have been acquired and fabricated to drill holes in product and crude oil pipelines.

Product line vandalism

Vandalism is a common term used in the Nigerian oil and gas industry to characterise a deliberate act of sabotage targeted at oil and gas plants. NNPC currently has approximately 4,950 km of pipeline which link about 21 depots across the country. Incidents of pipeline vandalism have continued to increase. In 2002 a total of 516 cases of vandalism were reported. In the same year 26 ruptures occurred in the pipeline network. In 2003, however, the incidents of vandalism increased to 779 while ruptures increased to 48. Compared to 2002, the incidents of vandalism and ruptures in 2003 posted an increase of 51 per cent and 85 per cent respectively. Although NNPC and law enforcement agents have applied all known deterrent mechanisms, the vandalism process escalated in 2005, when the total number of incidents increased to 2,258. Incidents further increased dramatically in 2006. In a press brief in January 2007 the group managing director of NNPC, Funsho Kupolokun, categorised pipeline vandalism into three axis, namely Atlas Cove–Mosimi, Abuja–Suleja and Port Harcourt–Aba–Enugu–Markurdi. Available facts indicate that 1,650 line breaks were recorded between January and September in the Port Harcourt axis. This compares to a total of 600 in 2003. On the aggregate 2,625 line breaks were inflicted by vandals on the products distribution network. While the holes drilled in the pipes compromised their integrity, they also created room for contamination of the products in the pipelines. More importantly, the acts of sabotage led to loss of large volumes of products. In 2002 NNPC lost 308,725 litres of products,[1] while in 2003 a total of 402,927 litres were estimated to have been lost. In monetary terms, NNPC lost approximately ₦7.7 billion and ₦13 billion in 2002 and 2003 respectively as a result of criminal activities on the network of pipelines which connects the product depots. These incidents have on many occasions crippled the activities of the depots which receive products from the refineries. Often artificial scarcity of products has been linked to vandalism of pipelines.

Crude oil pipeline vandalism

Initial vandalism activities were restricted to products pipelines which provided a ready source of income for petroleum product hawkers. In recent times vandalism has been extended to crude oil pipelines and flow stations. These activities can be classified into two major categories, namely theft motivated and purely sabotage-related incidents. In theft motivated incidents, the perpetrators are interested in siphoning products from the pipeline for economic gains. However, in situations where the pipe break is sabotage-related, the perpetrators are simply interested in causing disruptions to the operations of the refineries receiving crude oil through the pipelines. Several

Figure 15.1 Fire explosion and pollution resulting from pipeline vandalism.
Source: Pipelines and Products Marketing Company Limited (PPMC), 2007

Figure 15.2 Pipeline vandalisation gadgets.
Source: Pipelines and Products Marketing Company Limited (PPMC), 2007

incidents of crude oil pipeline vandalism have been reported in recent years. Significant among these is the March 2003 destruction of the Escravos–Warri pipeline which supplies crude oil to the 115,000 b/d Kaduna refinery. This act of sabotage crippled the refining activities thereby causing products scarcity in the region. The contract for the rehabilitation of the damaged pipeline was awarded to Bill Finger Goss in 2003 and was only completed in January 2005. In situations where the pipeline belongs to a major producer it also brings to a halt temporarily the production activities of the company. In either case, the resultant effect is pollution because oil spills into the creeks, waters, swamp and land, inflicting colossal damage to the ecosystem. The degree of the damage may vary but the aggregate effect is negative and destabilising.

Impact of pollution

Farming

As pointed out in the preceding sections, the consequences of crude oil and products spillage are extensive. Some of the incidents involving pipeline vandalism often lead to wildfire. When fire occurs, large quantities of hydrocarbon burn for days and on a number of occasions expert fire-fighting teams have been mobilised to bring the inferno under control. The large combustion of hydrocarbon releases significant amounts of carbon dioxide into the atmosphere and the soil of the affected area is also subjected to high temperatures, thereby destroying all the manure and other nutrients. Such soil becomes unsuitable for farming purposes. In this regard, the occupations of the inhabitants of the affected area are seriously jeopardised: their incomes drop due to low agricultural yield and as a result they cannot afford medication to combat respiratory diseases. The children of affected families drop out of school as a result of decline or outright cessation of family income, which exacerbates poverty in the affected communities as well as creating economic disparity between urban families and rural families, especially those in oil producing areas.

Fishing

The aggregate impact of pollution on fishing activities in the Niger Delta is colossal. Traditionally fishing in the Niger Delta takes place in the creeks, small rivers and the ocean. Different types of fishing gear are used in the area ranging from single line hooks, multiple line hooks, cast nets, float nets or subwater barrier nets. All these activities require a clean and stable environment in order for fishermen to catch reasonable quantities of fish both for food and revenues. Pollution activities have destabilised fishing activities in two major ways. When oil spillage occurs, the slick covers the surface of the water thereby making it impossible for fishermen to cast nets or position float nets. The nets get completely entangled in the dark tar, preventing them from trapping fish. Fishing in the Niger Delta is also significantly affected by the flared gas. Float nets are used mainly at night and the basic principle underlying their use is that the sky and the waters must be dark so that fish stray into the nets and get trapped. In a situation where the sky is illuminated by gas flare the track of the net becomes discernible to the fish, enabling them to avoid it. The gas flares frequently explode and produce loud noise which scares the fish from the creeks and other fishing grounds. These activities adversely affect the economic wellbeing of fishermen and, as in the case of farmers, pollution activities create hardship and incomes decline. The environmental conditions arising from oil pollution create poor living conditions which prevent them from being able to afford the basic needs of life as well as provide education for their children.

Mangrove trees

The swampy vegetation in the blackish water zones of the Niger Delta is dominated by mangrove trees. These trees usually have roots that grow upwards out of the mud, the tips of which are exposed to air and serve as oxygen absorption channels for the root system. When oil spillage occurs the tips of the roots become coated with oil, the oxygen supply to the root system fails and the trees die. Mangrove forests are both feeding and breeding grounds for varieties of fish, so when the trees die the area becomes unsuitable for breeding and food provided by decaying leaves and pollens ceases to exist. The mangrove trees are often cut and sold for domestic use. This cutting is organised and allows for growth of younger trees, which also helps the regeneration of the forest. However, when pollution occurs, both young and old trees die in large quantities, thereby creating an ecological disaster. In many incidents of pollution the cycle of mangrove forest destruction occurs on a continuous basis because, apart from the impact of pollution, the forest suffers serious disruptions occasioned by drilling activities.

Figure 15.3 Mangrove forest in the Niger Delta before oil spill (Rivers State, Nigeria).
Source: Author

Figure 15.4 Mangrove forest in the Niger Delta after oil spill (Rivers State, Nigeria).
Source: Author

In order for the rigs to be properly positioned for drilling in the swamps, large areas of mangrove trees have to be destroyed to provide suitable space for the drilling platforms to be erected. All these activities cause serious disruptions to the fishing industry and indeed the incomes of the Niger Delta people. No definite estimate has been made of the losses suffered by fishermen in the Niger Delta but one can conclude, based on the magnitude of the spillages and vast areas of terrain affected, that the loss to the inhabitants is colossal.

Unemployment

The Niger Delta as currently constituted is a large area of about 70,000 km^2, 80 per cent of which is swampy in nature, and the two occupations that strive to survive in the area are peasant farming and fishing. The existence of oil deposits means several oil companies, namely Shell, Mobil Producing, ChevronTexaco, NAOC, Total, ConocoPhillips etc. operate in the region. Apart from the oil companies the area fails to attract major manufacturing industries, cottage industries, construction companies and other labour intensive commercial activities.

The families in the Niger Delta experience a vicious cycle which culminates in a poverty trap. As pointed out earlier, they earn low incomes because their farm lands and fishing grounds are regularly polluted. When pollution occurs the soil performs poorly and results in low farm yield. In the fishing sector pollution kills the fish and destroys the spawning grounds thereby reducing the fish harvest. These incidents sadly translate into low incomes for both the farmers and fishermen.

Fathers as family heads are obliged to put their children in school. This can only be done if the family income can go beyond subsistence and a part saved for education. In the majority of cases encountered in the Niger Delta, the families lack savings of any kind, which makes education for the siblings impossible. Beyond this, one can contend that the Niger Delta terrain is not attractive, therefore schools located in the typically wet swampy terrains fail to attract qualified and committed teachers. For this reason, the quality of education provided is low. In view of these circumstances the youths remain idle and frustrated. Such unoccupied minds are easily recruited to engage in criminal activities. Youths in the Niger Delta have over the period become restive and in the process caused chaos in oil producing communities. Incidents of closure of access roads to flow stations, destruction of oil installations, property and other vices have been recorded and these activities cause forced closure of operational facilities of oil companies resulting in the decline of oil production. The situation in the area continues to deteriorate and in January 2006, a militant group attacked a rig and took four expatriates hostage. This incident drew international attention and it took the intervention of Bayelsa State, South-South leadership, Ijaw National Congress and other socio-cultural groups (numbering 25 in all), to persuade the militants to release the expatriates. Shell experienced a similar incident when three expatriates were captured at Ogodobiri in Bomadi L.G.A. and ₦25 million ransom was demanded as a precondition for the release of the hostages. Similar incidents continue to occur and in June 2006, about five Daewoo staff were taken hostage only to be released days later. In August 2006 another member of the Shell staff was taken hostage and died in a shoot-out between law enforcement agents and Niger Delta militia. The level of youth agitation is definitely on the increase and hostage taking has shifted from the oil platforms to the streets and restaurants of Warri and Port Harcourt where expatriate staff of oil companies are harassed and abducted. It seems that this trend is inimical to oil and gas exploration and production activities and it is therefore imperative that the problem be addressed with a view to engendering a sustainable conducive business environment in the region.

Marginalisation

The plight of the people has been attributed to orchestrated marginalisation. On several occasions, both in private and in different public forums, people of the Niger Delta have unequivocally stated that they are marginalised. Several reasons and events are adduced as evidence of group marginalisation experienced by the indigenes. These include the following:

- political manipulation by the majority tribes;
- lack of infrastructure (roads, bridges, canals etc);
- pollution of environment;
- massive exploitation of oil and gas resources;
- unfavourable revenue allocation formula;
- underrepresentation in the federal civil service;
- underrepresentation in the armed forces;
- poor educational and health facilities;
- high youth unemployment;
- underrepresentation in federal government Parastatal Boards.

These issues have become points of contention and serve as unpleasant reminders of the backwardness of the region. Representatives of the region have repeatedly argued that the Niger Delta terrain is peculiar, requiring special resources for development. In view of the swampy nature of the terrain, the construction of roads requires sand filling and special site preparation, and these and subsequent activities demand more funds than road construction projects in other parts of the Federation. Multistorey buildings also require special piling and foundations, which means expenditure for development in these areas, even for minor projects, turns out to be almost triple the cost of similar projects in other parts of the country. In this regard the government and people of the Niger Delta contend that allocation of revenues to states on an equal basis without taking into consideration the peculiarities of the region is not justifiable. They argue that a region that accounts for over 90 per cent of the revenues of the Federation should not only be appreciated but also be developed through the provision of adequate infrastructure and employment generation facilities.

Indigenes of the Niger Delta also contend that the area is marginalised because its contributions to the National Treasury are not considered in the distribution of political offices. It is further argued that all the regions of the Federation have at one time or another contributed to the leadership of the country. However, the Niger Delta was deprived of political leadership until 2007 when it was given the opportunity to provide a Vice President under the People's Democratic Party (PDP). Furthermore, the area has experienced several incidents of pollution which have caused serious environmental degradation. These incidents have inflicted extensive economic hardship on the people. It may be noted that modest compensations have in the past been paid to the people. This may well be so, but it is also known that the culprits (the oil companies) manipulate the affected communities and offer compensation which to all intents and purposes cannot fully recompense for the hardship and health hazards suffered by the people. These issues are held dear to the hearts of the Niger Delta people. In 2004 the federal government convened a National Constitutional Conference to examine the existing constitution with a view to effecting amendments to relevant sections. The primary objective was to engender constitutional equity which would guarantee a stable polity. Several key issues were discussed among which was revenue allocation. The South-South representatives addressed the issue of revenue allocation and demanded an upward review of the derivation allocation to 50 per cent of the Federation account. This was not accepted by the Revenue Allocation Committee and 17 per cent was recommended. The recommendation fell far short of the demand of the South-South delegation and the representatives walked out of the conference as a mark of disapproval. It is important to say at this juncture that the agitations of the Niger Delta have occasionally been symbolically mitigated by the federal government through one programme or another. In an effort to mitigate the hardships and setbacks in development, the federal government, through an Act of the National Assembly, established the Niger Delta Development Commission (NDDC). The purpose of the commission is to develop a blueprint and execute programmes which would ameliorate the problems experienced by the region. An effort will be made to examine the Act establishing the commission with a view to determining its structure, objectives, strategies and funding mechanisms. Its achievements will also be assessed to determine the impact on the region.

The Niger Delta States

The Niger Delta is an integral part of the southern tip of the tributaries of the River Niger which evacuates into the Atlantic Ocean. At the early stage, only the six states of Akwa Ibom, Bayelsa, Cross River, Delta, Edo and Rivers were classified as the constituent components of the region. In recent years, however, the region has been expanded to include the states of Abia, Imo and Ondo. The Deltaic wetlands are interspersed with creeks, brackish water and mangrove trees so do not present themselves as easy terrains for commercial ventures or general development. The difficulty of the terrain was first encountered by the early explorers who had a daunting experience forging their way into the adjoining hinterland to locate the choice territories in which to establish their administrative machinery. These peculiarities of the terrain prompted the British government to set up the Henry Willink's Commission charged with the responsibility of mapping out strategies to develop the area. One of the recommendations of the report, submitted in 1958, was that the Niger Delta should be accorded special attention in order to achieve meaningful development in furtherance of the needs of the region. In 1960 the federal government established the Niger Delta Board, but it was poorly funded and remained inactive for several years. The outbreak of the civil war in 1966 further incapacitated the agency thereby preventing it from making a meaningful contribution to the development of the area. Shortly after the war the Niger Delta Development Basin was created alongside other development basins established across the country.

However, in the 1960s, when oil exports increased, the economic impact of the region became prominent. Based on agitations from the people, a Presidential task force was inaugurated which was allocated 1.5 per cent of the Federation account for the exclusive development of the Niger Delta. The activities of the task force lacked focus; therefore the objectives for which it was established remained unachieved. In 1993 the Oil Mineral Producing Area Development Commission (OMPADEC) was created to further pursue the objectives of the Niger Delta Development Board and the Presidential task force. Although OMPADEC was well structured, with an executive chairman and directors, it did not inherit any framework or blueprint that would aid vigorous pursuit of its objectives. The absence of a development master plan, proper office space and other related facilities adversely affected its inauguration and its activities were further seriously impacted by the military Generals and their associates who influenced the award of contracts. These contracts were awarded not based on competence but on the source of the sponsorship. Most beneficiaries of OMPADEC contracts acted unethically and abandoned the contracts without completing them. These events further aggravated the plight of the Niger Delta people who were yearning for positive change in the environment and living conditions. Youth unemployment became more pronounced and their pent-up energies were sadly channelled to criminal and mischievous activities. Violence increased in the area and oil exploration and production activities continued to be interrupted. These interruptions significantly affected federal government revenues and also eroded the confidence of oil producing companies as it related to the safety of their employees and installations.

In the light of these events the federal government was compelled to provide an alternative which would assuage the setbacks of the region. The democratically elected government of Chief Olusegun Obasanjo which came into existence in May 1999, considered it necessary to bring into existence through an Act of the National Assembly a commission which would focus on the developmental needs of the region. This feeling led to the establishment of the Niger Delta Development Commission (NDDC). The National Assembly in 1999 passed into law the NDDC Act, No. 2, 1999, and the commission was configured to execute several functions. Details of the structure, objectives, powers, funding etc. will be discussed in subsequent sections.

Establishment of the NDDC

As pointed out above, the NDDC was established through Act No. 2, 1999 of the National Assembly. The basic component parts of the Commission can be delineated as follows:[2]

- the governing board
- structure of the Commission

- funding of the Commission
- functions of the Commission.

Governing board

The Commission with its head office in Port Harcourt was duly registered with the CAC (Corporate Affairs Commision); therefore it exists as a corporate entity with perpetual succession and a common seal. In this regard, the Commission may sue and be sued in its corporate name by individuals or other corporate organisations. The membership of the Commission comprises:

- Abia State
- Akwa Ibom State
- Bayelsa State
- Cross River State
- Delta State
- Edo State
- Imo State
- Ondo State
- Rivers State.

The board of the Commission is constituted through nomination of one individual from each of the member states. In addition the following are members of the board:

- three persons representing non-mineral producing states and members have to be drawn from the geo-political zones which are not represented on the commission;
- one representative jointly nominated by the oil producing companies;
- the managing director of the Commission;
- two executive directors;
- one representative of the Ministry of Environment.

The chairman and other members of the Commission are to be appointed by the President of the Federal Republic subject to confirmation by the Senate in mutual consultation with the House of Representatives. The Act provides that all appointed members shall be men and women of proven integrity and would serve the commission in a part-time capacity. It further provides that directors are to be appointed for a period of four years in the first instance and may be reappointed for an additional period of four years or more based on satisfactory performance. The Commission also comprises an advisory committee made up of the governors of the oil producing states and two other members appointed by the President. The primary function of the committee is to advise the board and monitor the activities of the Commission in order to enhance the achievement of its objectives.

Funding of the Commission

The Act establishing the Commission outlined the funding mechanism as follows:

- 15 per cent of the total monthly statutory allocations due to the States of the Commission from the Federation account;
- 3 per cent of the total annual budget of oil producing companies;
- 50 per cent of the ecological fund of each member state;
- donations and grants from Federal and State governments or any other local or foreign organisations;
- legitimate gifts, loans, grants-in-aid etc.;
- proceeds from assets.

Structure of the Commission

The Commission is for administrative purposes divided into the following directorates:

- administration and human resources;
- utilities, infrastructure development and waterways;
- environmental protection and control;
- community and rural development;
- education, health and social services;
- commercial and industrial development;
- agriculture and fisheries;
- finance and supply;
- legal services;
- planning, research, statistics and management information systems;
- project monitoring and supervision.

Functions of the Commissions

The Commission was mandated to carry out the following:[3]

- formulate policies and guidelines for the development of the Niger Delta;
- conceive, plan and implement development projects and programmes;
- develop a master plan to promote physical development;
- identify and document factors inhibiting development in the Niger Delta;
- monitor and report on projects funded by the Commission;
- engage in the mitigation of ecological problems caused by oil exploration and production activities;

- interface with oil producing companies and service companies on matters of pollution and environmental degradation.

The Niger Delta subregion is characteristically multicultural which accounts for the difficulty in reaching concensus on major issues affecting the region.

References

1 PPMC Operations Report 2005.
2 Niger Delta Development Commission Mission, Vision and Strategy 2006. www.nddcconline.org
3 Niger Delta Development Commission Act, No. 2, 1999. www.waado.org/Niger Delta

Further reading*

Alagoa, E. J. *History of the Niger Delta.* Onyoma Research Publications, Lagos, 2005.

The Price of Oil: Corporate Responsibility and Rights Violation in Oil Producing Communities, Human Rights Watch, 1999.

Akpobibibo, O. 'Sustainable Development as a Strategy for Conflict Prevention: The Case of the Niger Delta'. www.ogele.org

Wolfson, F. 'Trade and politics in the Niger Delta 1830–85'. www.links.jstor.org

Douglas, O. Kemidi, V. et al. 'Alienation and Militancy in the Niger Delta: A Response to CISS on Petroleum'. *Politics and Democracy in Nigeria*, www.fpif.org

Peters, S. W. 'Conservation and the Development of the Niger Delta'. www.onlinenigeria.com

Omeje, K. 'The State, Conflict and Evolving Politics in the Niger Delta' *Review of African Political Economy*, Vol. 31 No. 101, 2004.

Dafinone, D. O. 'Niger Delta, Today, Yesterday and Tomorrow'. www.nigerdeltacongress.com

* All entries are Federal Government Publications, Lagos, Nigeria

16 Environmental pollution

Introduction

Oil and gas production has enormous financial benefits but associated with these benefits are major hazards or consequences. One such hazard is environmental pollution. Active oil and gas exploration in Nigeria derived its roots from early 1908; however, active production of crude oil only started in 1956 and early production activities were low; therefore they were not associated with major oil spills. Over the years exploration and production activities in the Niger Delta have escalated and consequently oil spillage in the region has become a common occurrence. The primary purpose of this chapter therefore is to examine critically the pollution incidents and evaluate their impact on the environment with special attention being focused on the Niger Delta, which is actively involved in oil production. In order to put the issue in the right perspective, an attempt will be made to discuss pollution episodes in other producing regions of the world. The aim of an historical approach will be to establish if a particular pattern exists that links oil pollution incidents in various oil producing regions of the world.

Origin of oil spills – global view

In the recent past and even now, large volumes of oil are being transported around the world and in addition, oil is often transferred from one vessel to another. During these processes minor oil spillages occur as well as many small spillages, as a result of discharge of oil-contaminated ballast water from tankers. Although the spillage from an individual tanker may be considered small and perhaps inconsequential, a number of spills accumulate to constitute a threat to the environment. In 1964 the increasing trend of pollution associated with ballast water provoked public outcries which forced oil companies and transporters to explore ways of preventing such incidents. Although several preventive measures have been taken, the practice continues on the high seas. Small spills also occur as a result of tanker to tanker transfers of oil, tanker to refinery discharge, adjoining areas of refineries, ruptured pipes and storage facilities.[1] The aggregate effects of these activities cause considerable damage to the environment and the ecosystem.

Although oil pollution has from inception been a source of concern, the quantities arising from spillages especially in the 1960s caused limited impact on the environment. In the early 1930s, the largest category of tankers carried about 20,000 MT of oil. In view of technological development in recent years, however, the sizes of tankers have increased significantly; therefore a single haulage of 300,000–310,000 MT of crude oil is a common occurrence. The large haulage provides economic advantages in terms of freight cost but only to the extent that the cargo is safely delivered. The Very Large Crude Carriers (VLCCs) and the Ultra Large Crude Carriers (ULCCs) are difficult to manoeuvre and take a long time to change direction or stop in order to avoid icebergs, on-coming vessels or other hazards. Consequently accidents involving super-tankers discharge large volumes of crude oil into the sea thereby causing serious environmental problems. Various degrees of oil spills occurred globally from 1967 to 1994 involving 230 million to 240 million gallons. In recent times the *Exxon Valdez* incident seems to have taken a prominent place in the contemporary history of oil spillage episodes. The vessel, carrying 50 million gallons of crude oil, went aground on a reef near Valdez thereby piercing the hull. Approximately 10.8 million gallons of oil spilled into the Gulf of Alaska.[2] Available records show that about 240 million gallons of crude oil were discharged into the Persian Gulf in 1991 as a result of missiles fired during the Gulf war. Beyond these incidents other major spills have been recorded. Subsequent sections of this chapter will attempt to provide a lucid account of such episodes.

Major oil spills

As shown in Table 16.1 several major oil spills have occurred around the world. This section will attempt to provide brief accounts of some of these incidents and assess their impacts on the environment and the ecosystem.

Torrey Canyon, *England 1967*

The first marine pollution, which involved *Torrey Canyon* (118,285 DWT), occurred around 1967 off the coast of England. The vessel, which at the time of the incident was carrying crude oil from Kuwait, got trapped on the Seven Stones in the Scilly Isles. In an attempt to beat the tide at Milford Haven the captain made a costly decision to navigate between the Seven Stones. The adventure turned sour and the vessel ran aground on the Stones. The accident broke open the hull of the vessel and 35 million gallons of crude leaked into the sea causing catastrophic damage to the beaches of Cornwall and Brittany. The spill rapidly spread on the water surface aided by strong winds. Although incidents of oil spill had been experienced previously they were very small quantities which could easily be quarantined using booms and other containment devices. In 1967 clean-up technology was at its rudimentary stage; therefore the British government was not quite prepared to remedy the damage caused by the spill. Sadly the impact of the event on the environment was considered secondary to saving the life of the tanker. The flames rose hundreds of feet into the sky, discharged several tons of soot and the southeastern beaches of England were threatened with massive pollution. The public put pressure on the government and a clean-up team was constituted to prevent further damage to the beaches and adjoining environment. An international team of scientists and engineers were also assembled and their deliberations advocated the use of powdered chalk for the clean-up process. Technically, powdered chalk binds the oil into sizeable particles which sink to the bottom of the sea. On the whole, 30,000 MT of chalk was applied on the assumption that it would control about 20,000 MT of oil as powdered chalk is less toxic and injurious to existing flora and fauna.

Table 16.1 Major global oil spills

S/N	*Name and place*	*Year*	*Cause*	*Millions of gallons*
1	Terminals and Tankers, Persian Gulf	1991	War	240.0
2	Ixtoc-1 oil well, off Mexico	1979	Blowout	140.0
3	Nowruz Field, Arabia	1980	Operations	80.0
4	Fergana Valley, Uzbekistan	1992	Operations	80.0
5	*Castillo de Bellver*, off South Africa	1983	Fire	78.5
6	*Amoco Cadiz*, off NW France	1978	Grounding	68.7
7	*Atlantic Express* and *Aegean Captain*, off Trinidad and Tobago	1979	Collision	48.8
8	Well, 480 miles SE of Tripoli, Lybia	1980	Operation	42.0
9	*Irene's Serenade*, Greece	1980	Grounding	36.6
10	*Torrey Canyon*, off SW UK	1967	Grounding	35.6
11	*Sea Star*, off Oman	1972	Collision	34.0
12	Storage tanks, Shuaybah, Kuwait	1981	Operation	31.2
13	*Urquiola*, off N Spain	1976	Grounding	29.0
14	*Hawaiian Patriot*, N Pacific	1977	Fire	29.0
15	*Braer*, Shetland Islands	1993	Grounding	25.0
16	*Sea Empress*, Wales	1996	Grounding	24.0
17	Pipeline, Usinsk, Russia	1994	Burst pipe	23.0
18	*Exxon Valdez*, Alaska	1989	Grounding	10.8

Source: International Oil Spill Statistics, 1994.

Florida barge spill – 1969

On 15 September 1969 a supply tug-boat pulled out of Tiverton, Rhode Island and sailed in the direction of Cape Cod Canal. During the journey the towline snapped and the barge drifted and eventually went aground on submerged boulders near West Falmouth Harbor. The barge ruptured and discharged 175,000 gallons of refined products into the sea bordering the shores of Buzzard Bay. In view of its closeness to a vibrant economic and tourist zone of the US this incident turned out to be the best investigated and studied pollution episode in US history. The dark-coloured slick stretched several miles and aided by strong winds moved toward the shore line. Clean-up teams laid booms in an effort to quarantine the spill but the efforts of the clean-up crew only had a modest effect on the movement and impact of the oil. Aquatic life was adversely affected and within days the Wild Harbor beach and adjoining areas were littered with dead lobsters, scallops, fish and other organisms. The affected area was a key fishing ground of the shell fish industry and an initial estimate of the loss to the industry was $250,000. This estimate did not capture the volume of dead birds, lobsters, and fish affected by the incident. In subsequent days researchers trawled the bottom of the New Silver beach and discovered that over 70 per cent of the benthonic animals had died and more were still dying. This incident triggered an elaborate long-term study of the spill, an extension of which continues to date in various forms. The subtidal mash mud is the habitat of a variety of aquatic life; therefore large concentrations of oil in the area inflict high mortality on the organisms. Considering the relatively small size of the spill officials of the clean-up company predicted a full recovery of aquatic life within a short period. This notwithstanding, traces of spill resurfaced about six months after the incident and further contaminated areas of the subtidal line of the adjoining beaches. The effects of the spill were indeed far reaching with regard to the decrease in the population level of the species, in population growth, in species diversity, changes in population of predators and reproductive rates. The overall effect of the spill on the marshes and the aquatic community was to say the least, severe.

Amoco Cadiz – *1978*

The tanker *Amoco Cadiz* (216,000 DWT), with a crude oil cargo, encountered fierce winds which drove it ashore at Brittany, France in March 1978. *Amoco Cadiz* was a state-of-the-art vessel but lost its steering gear a few nautical miles off the coast. A tug-boat tried in vain to tow the vessel out from shore amidst strong winds and waves. The vessel drifted and hit a submerged rock which pierced the hull, causing 68 million gallons of Iranian crude oil to be discharged into the sea. There were procedural lapses during the accident. An investigation showed that the captain of the vessel waited about two hours before a distress call was issued. The oil washed ashore covering a stretch of approximately 400 miles of the French coast including key fishing harbours and a popular tourist area. All the fishing boats were covered in layers of oil; seafood such as crabs, lobsters, oysters and fish were destroyed in large numbers. Adult finfish were severely poisoned and thousands of birds died as a result of the pollution. In addition snails, periwinkles, crabs and other invertebrates on the oily beach suffered decimation. The incident dealt a serious blow to the fishing industry as 40 per cent of the total fish stock of France was depleted from the area. The severe impact of the incident resulted in public agitation for definite preventive and remediation mechanisms to be put in place by the government. Amoco in response to the spill set aside $2 million to examine the effects and the French government also provided money for remediation activities.

The *Amoco Cadiz* oil spill ushered in an era of extensive investigation which for the first time would try and determine what happened to the oil itself. A variety of techniques was adopted to determine both the concentration and distribution of the oil in the water column. Sampling stations were engaged, shorelines chemically monitored and aerial photographs of the shoreline taken, all in an effort to provide an accurate account of the incident and to facilitate the restoration of the ecosystem. The findings of the extensive analysis showed that 28 per cent of the oil evaporated while 27 per cent was deposited along the shoreline. Furthermore, 8 per cent was deposited in subtidal sediments and 4 per cent was believed to have been degraded by microbes. The analytical approach notwithstanding, 20 per cent of the 68 million gallons could not be accounted for.

IXTOC-1, Mexico – 1979

In June 1979 the IXTOC-1 well near the city of Carmen in the Gulf of Mexico blew out and caught fire.[3] This explosion involved 140 million gallons of oil and the magnitude of the spill was second only to the massive spills which occurred in the Gulf as a result of the 1st Gulf War. The IXTOC-1 well located offshore in the Gulf of Mexico blew out in the early hours of the morning. Initially it was estimated that only 400,000 gallons were lost. As the fire raged the NOC – Petroleum Mexicanos (PEMEX) – claimed that 50 per cent of the spill had burned off while 30 per cent evaporated. This was only an optimistic view from a thoroughly agitated company. As it turned out no crude oil burned, only the associated natural gas. It was not until after nine days that the magnitude of the incidence became obvious as a slick measuring 100 miles long and approximately 8 miles wide was moving in a westward direction. PEMEX made frantic attempts to cap the well but failed. Alternatively, two wells were drilled at approximate locations to the explosion site to release the pressure, but to no avail. The slick, aided by fierce winds, travelled as far as the Florida and Texas coastlines. In order to control the damage, skimmers were deployed but crude oil recovery efforts were seriously hampered by strong winds. PEMEX injected 1,000 ball-size lead and steel objects submerged in a heavy solution into the errant well and this attempt reduced the flow of crude oil by about 50 per cent, but about 300,000 gallons/day continued to spill into the water. PEMEX, in a further effort to extinguish the fire, lowered a 310 ton steel cone over the well thereby reducing the flow rate to about 74,000 gallons per day. From mid-October 1979 the oil on the water continued to burn until March 1980 when three concrete devices sealed the gusher and disabled the well. Beyond the ecological damage, the US Coast Guard spent $8.5 million on clean-up activities while PEMEX incurred $132 million and $87 million on fire containment activities and production loss respectively.

Castillo de Bellver, *1983*

On 6 August 1983 the Spanish tanker *Castillo de Bellver* went ablaze 70 miles off the coast of Cape Town. In a period of seven hours the intensity of the fire caused a massive explosion breaking the vessel into sections in shallow waters. As a preventive measure the marine officials and fire fighting agents towed the vessel further away from shore where it sank in 2,743 m of water. Oil from the sunken ship continued to spill out until January 1984, discharging over 78 million gallons into the sea. It was estimated that approximately 10 per cent of the oil burned off, 40 per cent evaporated and the balance remained as slick. The oil remained at sea, however, and over 1,400 birds were stained with it. Clean-up efforts commenced immediately with about 60,000 gallons of dispersant sprayed on the slick. The burning oil discharged large amounts of soot into the atmosphere, causing black rain in the following days which discoloured the white wool of sheep and settled on wheat fields. Concerns over the spill were high in view of the fact that the affected area was a breeding ground as well as a nursery for commercially important fish and lobsters. The clean-up efforts proved effective and the slick remained mainly offshore and caused limited coastal and ecological damage.

Exxon Valdez, Alaska – *1989*

The *Exxon Valdez* spill is by far the most widely discussed oil spill in recent times. On 24 March 1989 a 211,469 DWT tanker of American origin loaded with 50 million gallons of Alaskan crude oil in the Valdez region set out on a journey for a destination outside the region. Consistent with known practice the vessel was guided through the hazardous waters by a pilot. Thereafter the captain of the vessel took charge, and along the course of the journey some error occurred in the charting and manoeuvring processes and the vessel went aground on a reef. The hull of the vessel ruptured and discharged 11 million gallons of crude oil into the sea. The slick extended to Prince William Sound, Gulf of Alaska, Kodiak Island and adjoining areas. The spill, which was the largest in US history, occurred in a quiet natural environment known for the existence of a variety of wild and aquatic life.

The impact of the spill was the immediate killing of sea otters, fish, lobsters, birds and other wildlife. More than that, it caused severe occupational dislocation among native Alaskans. The catastrophic spill received wide economic, social and scientific assessments which also involved an extensive study of the impact on humans as well as on a wide range of organisms. The Federal and State governments and the Exxon Corporation mobilised to curtail the impact of the disaster. Programmes were set up to clean up the environment and restore subsistence hunting and fishing. Provision was also made to capture and rehabilitate wildlife with a view to mitigating the impact on marine life within the Prince William Sound region. The immediate objective of the remediation team was to minimise the spread of the oil slick and reduce the mortality of fish, aquatic life and wildlife in general. Following the spill, over a hundred studies were commissioned by the federal government and the State of Alaska under the National Resources Damage Assessment Regulation. The regulation essentially outlined a process of compensation to the public for damage inflicted on natural resources. In September 1991 Exxon and the US government entered a Criminal Plea Agreement. Consequently, Exxon was ordered to pay a fine of $100 million. A civil settlement of $900 million to be disbursed over a period of eleven years was also brokered.

Oil spill in the Niger Delta

Oil spill occurs in various forms and degrees depending on the cause and source of the spillage. Environmental experts in the industry often classify spillages as either minor, medium, major or of the scale of a disaster. Spills involving quantities of the range of 40 barrels on land or inland waters are often classified as minor. Spillages of the order of 300 barrels in inland waters or 2,600 barrels on land or offshore are classified as medium in scale. Major incidents involve quantities in excess of 2,000 barrels on land or inland waters.[4] The last category of spillage which often involves well-blow falls in the category of a disaster. As the name implies it involves massive discharge of crude oil on land, in the inland waters or on the high seas. Such incidents arise from oil-well blow-out, pipeline rupture or storage tank failure. Blowouts are usually uncontrolled initially and involve discharge of high volumes of crude oil into the environment. In the Nigerian context spillages have occurred as a result of ruptures, system failure (as in the case of 1998 Idoho blow-out) and outright sabotage. Some experts have attempted to assign percentages to the various causes, but such analysis in my view is rather a theoretical exercise due to the lack of accurate information on the causes of spillages. The Niger Delta is very swampy and the rivers contain brackish water. Salt water has high corrosion potential; therefore some of the pipelines which have existed for over 20 years are susceptible to corrosion.

Under such circumstances they can easily rupture and cause a large spillage. The incidents of pollution are increasing on a yearly basis. Between 1976 and 2005 a total of 8,768 spills were reported. Further analysis showed that between 1989 and 2005 a total of 6,491 incidents occurred. This implies that about 70 per cent of all spillages occurred in the past 15 years. Perhaps more significant is the fact that spillage incidents are increasing even as more cutting edge technology is being applied in the industry. The increase in the number of spillages simply indicates that they are certainly not as a result of equipment failure in the exploration and production cycle. The sources of these incidents could be ruptures, or sabotage. In recent times acts of sabotage perpetrated by disgruntled elements have accounted for a large percentage of the spillages.

Between 1976 and 2005, nearly 9,000 incidents of spillage occurred and these incidents involved approximately 2,662,900 barrels of crude oil (Table 16.2). Available records did not show quantities recovered between 1976 and 1988. However, between 1989 and 2005 a total of 1,038,947 barrels were spilled as a result of various causes and out of this volume 94,600 barrels were recovered. The recovered quantity only represents 9 per cent of the total quantity spilled and the low recovery rate can be attributed to a number of factors. One could postulate that low quantities of oil are recovered at sea or on land due to delayed response from the appropriate intervention agencies. Another reason could be lack of appropriate technology in the pollution monitoring and containment agencies such as DPR, Ministry of the Environment and Clean Nigeria Associates.[5] Regardless of the reasons, one can assert that the occurrence of the incidents is inexcusable. If the 9 per cent recovery rate is a typical performance index in the industry then it can be inferred that out of the 2,662,900 barrels spilled between 1976 and 2005 only 239,661 barrels could have been recovered. This further implies that 2,423,239 barrels were unaccounted for and presumed to have been absorbed into the ecosystem (i.e. coastlines, swamps, mangrove forests etc.).[6] The volume of oil discharged into the Niger Delta environment is not only staggering but capable of decimating a wide colony of aquatic life and fauna. Most sadly, the inhabitants of the region eat the contaminated fish and other seafood and in the process hydrocarbon substances find their way into the human blood circulation system. These individuals then suffer from different ailments which are traceable to the environmental degradation orchestrated by the oil exploration and production activities in the region.

As indicated earlier, large volumes of crude oil were spilled between 1976 and 2005 into the Niger Delta environment. Some of the spillages may be minor but available data isolated some as major incidents. In 1978 GOCON (Gulf Oil Company of Nigeria – now Chevron) was involved in a spill of about 300,000 barrels at Escravos. Shell was also involved in a spill of over 100,000 barrels from the Forcados terminal. Similar incidents were recorded at Texaco Funiwa-5 facility and the Abudu pipeline blow-out involving about

Table 16.2 Niger Delta oil spill data 1976–2005

Year	*No. of spills*	*Quantity spilled (barrels)*	*Cumulative vol. spilled*	*Quantity recovered (barrels)*	*Spilled vol. recovery rate (%)*
1976	128	26,157	26,157	N/A	N/A
1977	104	32,879	59,036	"	"
1978	154	48,294	107,330	"	"
1979	157	694,170	801,500	"	"
1980	241	600,511	1,402,011	"	"
1981	238	42,722	1,444,733	"	"
1982	252	42,841	1,487,574	"	"
1983	173	48,351	1,535,925	"	"
1984	151	40,209	1,576,134	"	"
1985	187	11,876	1,588,010	"	"
1986	155	12,905	1,600,915	"	"
1987	129	13,866	1,614,781	"	"
1988	208	9,172	1,623,953	"	"
1989	195	7,628	1,631,581	1,883	24.69
1990	160	14,940	1,646,521	2,286	15.30
1991	201	106,827	1,753,348	1,544	1.45
1992	378	51,187	1,804,535	1,476	2.88
1993	428	9,752	1,814,287	2,357	24.17
1994	515	30,282	1,844,569	688	2.27
1995	417	63,677	1,908,246	3,109	4.88
1996	430	46,353	1,954,599	806	1.74
1997	339	81,727	2,036,326	53,781	65.81
1998	399	99,885	2,136,211	5,954	5.96
1999	225	16,903	2,153,114	294	1.74
2000	637	84,071	2,237,185	1,456	1.73
2001	412	120,976	2,358,161	76	0.06
2002	446	241,617	2,599,778	864	0.36
2003	609	35,284	2,635,062	120	0.34
2004	543	17,104	2,652,166	16,843	98.47
2005	496	10,734	2,662,900	1,069	9.96

Source: Department of Petroleum Resources 2006.

18,000 barrels of crude oil.[7] The Idoho oil spill in January 1998 from a Mobil Offshore facility which involved 40,000 barrels was a major environmental setback and the 1998 Jesse fire incident, in Idjerhe, Warri, Delta State, involving scavenging of petroleum products, in which an estimated 1,000 people died, is indeed an example of a major disaster associated with pollution in the region. These incidents collectively have caused serious environmental problems for the oil producing region and severe consequences for the inhabitants.[8] The economic and social dislocations caused by the unfortunate incidents are serious and culminate in creating a poverty web which has engulfed the fishermen and peasant farmers in the area.

References

1 Burger, J. *Oil Spills*. Rutgers University Press, New Brunswick, NJ, 1997, p. 27.
2 International Oil Spill Statistics, 1994 (Cutler Information Corporation).
3 Burger, J. op. cit.
4 Nwilo, P.C. and Badejo, O. T. 'Impact of Oil Spillage Along the Nigerian Coast', *Soil Sediment and Water*, October 2001.
5 DPR Statistical Data 2006.
6 Niger Delta Environmental Survey 1997. Environment and Socio-economic Characteristics. Environmental Resources Manager Limited.
7 Nwilo, P. C. and Badejo, O. T. op. cit., p. 2.
8 'Water Pollution as a World Problem', Report at Conference held at the University College of Wales, Aberystwyth, July 1970, pp. 53–83.

17 Shipping and cabotage practice

Introduction

Ships have for many years been relied upon as an important medium for achieving commercial activities in the contemporary maritime industry. Ships account for carrying over 80 per cent of 5.1 billion tons of cargoes involved in international trade, and shipping is generally considered an efficient system for the movement of goods and at drastically reduced freight costs. In 1980 global freight charges were about 6.64 per cent of the value of goods transported but dropped to 5.24 per cent in 1997 and further declined in later years. Kitte-Powell[1] indicated that the efficiency of shipping has been highly accentuated, making it possible for a 20-foot container to be transported for less than ten cents per mile. This rate is far lower than the cost of transporting a 20-foot container by road or rail. The shipping industry has benefited from emerging technologies in the areas of computers, communication systems and engineering innovations. Bulk shipping technology has experienced little change but the process is generally considered economical, efficient and competitive. On the other hand, container shipping has undergone dramatic technological evolution and has projected containerisation as the preferred means of cargo transportation. In 1950 both dry and tanker cargo amounted to 460 million tons. In 1988, however, the total amount transported across the globe increased by a factor of 10 to approximately 5,070 million tons. The contents of Tables 17.1 and 17.2 indicate that bulk and tanker commercial activities have grown at an average rate of 3 per cent per annum.

Global fleet

Global marine trade is currently executed by an international fleet of about 25,000 commercial ships. The cargo volumes in the various international routes have changed over the period. As shown in Table 17.2, tanker cargo increased from 130 million tons in 1950 to 1,945 million tons in 1998. These cargoes are transported through some major routes, namely Middle East–Asia, Middle East–Europe, Intra-American and Middle East–the Americas.

On the aggregate the Middle East–Asia route accounts for 530 million tons of petroleum cargo which represents 26 per cent of total bulk liquid cargo.

Origin of tanker transportation

Tanker transportation is a vital link in the movement of crude oil and petroleum products from their various services to the consumer in far-away destinations. In 1998 over 279 million DWT of oil tankers operated in the world fleet. This segment of the shipping industry accounted for about 30 per cent of all seaborne cargoes transported by tankers. In view of the volume of cargo moved through tankers, it is widely recognised as a crucial sector for a broad spectrum of global consumers, oil suppliers as well as traders. Although tankers are important in liquid bulk cargo movement, they have also become the epicentre of international focus in view of the environmental consequences associated with their operations. Pollution incidents in November 2002 off the Brittany coast of France, *Exxon Valdez*, and many other incidents create nostalgic feelings among the governments and environmentalists.

The US based standard oil in the 19th century pioneered oil trading through the transportation of refined oil to Europe utilising a fleet of small purpose-built ships. In the 1880s the Nobel brothers built a fleet of ships in Europe to transport Russian oil until the Bolshevik revolution. Crude oil trade did not flourish at the initial stage due to the dominance of coal as a major source of energy. Shortly after World War I, crude oil transportation commenced from Mexico and Venezuela to the United States. The increase in crude oil trade triggered expansion of refinery capacity in the primary consumption centres in Europe. During World War II the demand for petroleum products to sustain the operations of Allied Forces increased. This warranted the construction of high volume T-2 tankers in US shipyards to supply fuel to the fleet of the Allied Forces. At the end of World War II the T-2 tankers found little use in the military and were therefore sold at low prices to entrepreneurs. The sale of the vessels provided an opportunity for private entrepreneurs to lay the foundation for fleet expansion. As far back as 1886 the Gluckauf, a prototype of a modern oil tanker, was introduced. However, the more specialised and modern tankers were introduced into the market only in the 1940s and 1950s.

In the post-World War II era the profile of the Middle East as an oil producing region increased. Crude oil was therefore regularly transported from the region to major consuming centres. In this regard the demand for modern tankers increased in order to transport crude oil to the US and Japan. The sizes of the tankers at the early stage were small thereby warranting several voyages to satisfy a particular demand. In an effort to meet the demand for crude oil the companies increased their fleets to meet charter demand. Tanker demand is essentially a derived demand; therefore fluctuation in oil demand impacts on the demand for tankers. High demand for crude oil essentially translates into an increase in the demand for tankers. In a situation of limited supply, long haul voyages put greater pressure on the tanker market, considering the fact that such journeys require extended voyage days to meet a particular demand. Major oil companies, especially the Seven Sisters (Chevron, Texaco, Exxon, Standard Oil, Mobil, Gulf, Shell and BP) dominated the tanker business until 1952. Initially they treated transportation (shipping) as core business.[2] Marine transportation divisions formed part of the corporate hierarchy of major oil companies. In 1953 traditional ship owners entered the tanker market followed by the government-owned shipping companies.[3]

In the 1960s the ownership of vessels among Multi-National Oil Companies dwindled, paving the way for strict entrepreneurial ownership. Shipping suffered a setback in the 1973 Yom Kippur war as a result of the oil crisis. None the less, from 1975 cargo volume increased by 75 per cent rising from 3,000 million to 5,256 million MT.[4] This increase in cargo volume brought with it a corresponding increase of 2.5 billion tonnes of imports in the Pacific and Atlantic regions. The increase in cargo tonnage can be attributed to a combination of factors hinged on economic development and trade liberalisation among nations. GATT initiative, financial re-engineering and deregulation contributed significantly to the elimination of trade barriers. The situation ushered in an era of free flow of goods between nations, which boosted shipping activities in the industry. The growth of the Asian economies has also aided the expansion of the tanker sector which heavily engaged in the transportation of crude oil to satisfy the growing demand of refineries. In the 1960s ships were restricted to 65,000 DWT imposed by the Suez Canal. In subsequent years ship owners took advantage of economies of scale and constructed large vessels to lower the unit cost of transportation. The construction of *Universe Apollo* in 1957 and *Universe Iran* in 1969 recorded major landmarks and signalled the revolution in the sizes of tankers. Tusiani, in a study, indicated that although the first 200,000 DWT tanker was introduced only in 1966, the number of such vessels or Ultra Large Crude Oil Carriers (ULCCs) increased to 366 in a short period (Table 17.1). Within the same period 525 of such vessels were under construction or on order. The demand for crude oil remains strong; therefore it can be posited that the demand for tankers will continue to increase in the near future.

Yom Kippur war and sea transport

The Yom Kippur war of 1973 introduced a dramatic turn in the oil and tanker market. The Arab embargo of oil shipment to the United States and the Netherlands created an artificial glut in the tanker market. During this period the price of oil increased from $3.65 to $10 per barrel. Arab nations also expropriated American and European investments in their countries. Despite these incidents, as the demand for oil increased the tanker market steadily regained momentum. The stabilisation of the tanker market was again disrupted between 1979 and 1980 when hostilities resumed between

Iran and Iraq. The shortfall arising from the cutback in the production of Iran and Iraq were replaced by suppliers from non-OPEC countries, namely Russia, Mexico and the North Sea. These sources sustained the global oil demand thereby sustaining the global tanker market but the events of the period were a source of concern for tanker owners. It brought the golden transport era to an abrupt end and created a tanker market glut in the shipping industry in the 1980s.

Table 17.1 Major vessel categories in the world's ocean-going cargo ships

Cargo type	*Category*	*Typical size*	*Number in world fleet*
Dry bulk	Handy size	27000 dwt	2,700
	Handymax	43,000 dwt	1,000
	Panamax	69,000 dwt	950
	Capesize	150,000 dwt	550
Tanker	Product tanker	45,000 dwt	1,400
	Aframax	90,000 dwt	700
	Suezmax	140,000 dwt	300
	VLCC	280,000 dwt	450
General cargo	Container	2000 TEU	800
	Container feeder	2000 TEU	1,800
	Ro/Ro	2000 TEU	900
	Semi container	1,000 TEU	2,000

Source: J. Steel et al., eds. *Encyclopedia of Marine Science*, Academic Press 2001.

Table 17.2 World sea-borne dry cargo and tanker trade volume (million tons 1950–1998)

	Dry cargo	*Tanker cargo*	*Total*	*% change*
1950	330	130	460	
1960	540	744	1,284	
1970	1,65	1,440	2,605	
1986	2,122	1,263	3,385	3
1987	2,178	1,283	3,461	2
1988	2,308	1,367	3,675	6
1989	2,400	1,460	3,860	5
1990	2,451	1,526	3,860	3
1991	2,537	1,573	4,110	3
1992	2,573	1,648	4,221	3
1993	2,625	1,714	4,339	3
1994	2,735	1,771	4,506	4
1995	2,891	1,796	4,687	4
1996	2,989	1,870	4,859	4
1997	3,163	1,944	5,107	5
1998	3,125	1,945	5,070	–1

Source: J. Steel et al., eds. *Encyclopedia of Marine Science*, Academic Press 2001.

Imperatives for shipping

OPEC has become a key variable in the global energy equation. However, early experiences of the oil producers in developing countries were fraught with frustrations arising from manipulations of the MNCs. In an effort to mitigate the dominance of the multinationals, some oil producing countries (Iraq, Iran, Saudi Arabia, Kuwait and Venezuela) held a conference in Iraq (10–14 September 1960) with the primary motive of forming an association that would coordinate and unify oil related policies. In addition, the conference sought to provide a common voice for producers from the developing countries. This became necessary in the face of unilateral fixing of the price of crude oil in the global market by the MNCs.

In the light of emerging events, OPEC found it necessary to encourage NOCs to actively participate in the upstream and downstream sectors of their respective petroleum industries. This among other things was aimed at maximising revenues accruing from oil and also to allow NOCs to have greater control over their natural resources. Some members identified with OPEC's position on the need to participate fully in both the upstream and downstream sectors of the various oil industries. Shipping was considered a vital component of the downstream chain which had the capacity to provide a sustainable source of revenue. The establishment of national shipping companies was seen as a strategic step required in transforming the oil producing countries into major economic powers.[5] The primary motive of establishing national shipping companies was clear and novel. This notwithstanding, the development of the shipping business in these countries was slow. Inexperience in maritime operations, finance and lack of managerial skills for the operation of a modern fleet accounted for the inability of some members to achieve this laudable and economically beneficial objective.

Shipping business in NOCs

Beyond experience and managerial skills, the attractive revenues derived from the sale of crude oil in the international market served as a distraction in the pursuit of the perceived consensus of establishing national shipping companies for the crude oil transportation business. Failure of some NOCs, such as Nigeria, to establish a viable shipping company, especially in the 1980s when prices of second-hand ships of all categories were depressed (Drewry 1982), turned out to be forfeiture of significant opportunities and revenues. It is important to state that some oil exporters (Kuwait, Saudi Arabia, Malaysia and Indonesia) who regarded ship owning as purely an avenue for financial transaction and cross-trading, are currently in the league of successful shipping countries in the world. The international shipping business was well defined when NOCs executed their strategies to penetrate the market. At the initial stage very little success was achieved in penetrating the global shipping business. However, the vantage position of the producing countries as the suppliers of oil to the international market played a key role in opening up the global crude oil and products shipping business to aggressive NOCs. Some of those listed below have taken giant strides in shipping and are currently globally recognised as successful shipping companies:

- Vela International Marine – Saudi Arabia
- Kuwait Oil Tankers Company – Kuwait
- Malaysia International Shipping Company, (MISC) – Malaysia
- Pertamina Shipping – Indonesia.

Vela International Marine – Saudi Arabia

The Vela International Marine Company was established in 1984 with four second-hand ships by the Saudi Arabia Oil Company (Aramco) to meet the emerging need for crude oil transportation. Vela has grown from a four ship company to become one of the largest crude oil and products vessel owner-operator in the world. The company currently has a fleet of 23 (VLCCs) and four tankers which transport crude to North America, Europe and Asia. Vela has a history of commitment to accident free voyages and quality maintenance. Saudi Aramco operates terminals which handle 4,000 tankers per annum. It has a team that works to ensure that the company delivers crude oil products on time in a cost-effective manner. All its services are executed without compromising integrity and safety.

Figure 17.1 Very large crude oil carrier.
Source: Daewoo Shipbuilding and Marine Engineering Limited, 2006

Kuwait Oil Tanker Company (KOTC)

KOTC is a subsidiary of Kuwait Petroleum Corporation (KPC). KOTC was formed in April 1957 by a group of Kuwaiti pioneer investors who had a vision for the important role sea transportation would play in the global crude oil business. As crude oil gained prominence in the global energy mix, transportation of the commodity-refined and liquefied products developed into an important dimension of the overall marketing chain. In view of the progressive strategic role of transportation in the Kuwait oil business and in an effort to bring oil operations from the well to the consumer under one company, the government in 1976 acquired 100 per cent equity in KOTC thereby making it a wholly government owned company.

When the Kuwait Petroleum Corporation (KPC) was formed in 1980, KOTC assumed the responsibility for the transportation sector of the corporation. Prior to the emergence of oil as a major revenue earner, sea carriage was the main source of income for Kuwait. Skills in marine transportation acquired in the late 1950s have become an integral part of the shipping business in the Kuwait petroleum industry. KOTC has, from the receipt of its first tanker, visualised an expansionist programme in shipping that would position it as one of the major global players in crude oil transportation. The fleet of KOTC has over the years increased to 16 petroleum product tankers, six crude oil tankers, six liquefied gas tankers, two ship fuel boats and several supply and tow boats.

Malaysia International Shipping Company (MISC)

MISC was incorporated in 1968 as a JV between the Malaysian government and other private investors. It is principally engaged in ship owning, ship operating and other shipping related activities. Realising the importance of shipping in the downstream chain, Petronas empowered MISC to acquire modern vessels for the marine transportation business. It currently enjoys 62.4 per cent equity holding in the shipping company. MISC currently operates a fleet of about 100 vessels with a combined tonnage of 8.22 million DWT. These vessels operate from seven core business units which include:

- LNG tanker services
- petroleum tanker services
- chemical tanker services
- dry bulk carrier services
- integrated liner-logistic services
- offshore services
- heavy engineering services.

The company has grown to be a global player in shipping and developed a reputation for excellent shipping services. It owns 17 LNG tankers which account for 11 per cent of the global LNG fleet. Furthermore, MISC has placed an order for six additional LNG vessels which were scheduled for delivery during the period 2004–2007. In the financial year ending March 2004, it generated $2.00 billion revenue and posted a Profit Before Tax (PBT) of $612 million. MISC has shareholder's funds amounting to $2.897 billion and assets of the company are estimated to be $5.88 billion. The development of marine transportation in the downstream by MISC has contributed immensely to the fortunes of Petronas. This success as in the case of Vela (Saudi Arabia) and KOTC (Kuwait) has justified OPEC philisophy which encouraged NOCs to participate actively in the oil and gas industry with a view to operating world-class integrated oil and gas companies.

Pertamina Shipping

The shipping activities of Pertamina were transformed into a shipping division in 1959 with the acquisition of two 3,200 DWT vessels through the Bare Boat Hire Purchase (BBHP) system from PT Caltex. In March 1990, Presidential Decree No. 11 established a shipping Habour and Communication Directorate which transformed into Pertamina Downstream Directorate. This later transformed into Pertamina Shipping Company. The company's vision is to be an outstanding, developed and well respected shipping company. Some of its activities include professional logistics services for oil, gas, petro-chemical and other refinery products. At present, Pertamina shipping owns approximately 32 vessels of different types and also operates over 50 vessels which range from bulk lighter to very large crude carriers (VLCCs). As part of the vessel replacement programme the company ordered 12 vessels including two VLCCs. Plans had been concluded to acquire additional vessels by the year 2008.

Nigerian National Petroleum Corporation (NNPC)

OPEC has consistently held the view that transportation is an extension of downstream investment which like any other would provide OPEC producers an opportunity to expand their economic infrastructure. In response to this philosophy Pertamina of Indonesia and Petronas of Malaysia commenced Marine Transportation in 1959 and 1968 respectively. Kuwait Petroleum Corporation and Aramco also introduced shipping in their downstream chain in 1976 and 1984 respectively. These companies have not only made significant progress in shipping but have also improved their earnings in the downstream operations. Although NNPC aggressively pursued development of the upstream sector, the same objectives in the downstream sector can only be described as partially achieved. The vital missing link in the chain is 'Shipping'. Three decades after joining OPEC (1971) NNPC is yet to transform into an international integrated oil and gas company. As at 2006, foreign companies moved 100 per cent of Nigerian crude oil. This practice is inimical to the revenue generation and mobilisation policies of the federal government. Under the democratic dispensation the government has undertaken policy review to take advantage of the huge tonnages available in the Nigerian shipping market. In this regard therefore, NNPC in 2007 formed JV shipping companies with Daewoo Ship Building and Marine Engineering (DSME) and Hyundai Consortium to transport crude oil and LNG respectively.

World oil demand and supply

Global oil demand in 2005 was approximately 83.3 mmbd. This represented 2.3 mmbd or a 3.5 per cent increase over the level attained in the same period in 2004. The US and China orchestrated the increase in demand thereby contributing 27 per cent and 35 per cent respectively. China led in quarterly demand, except in the second quarter of 2004, with transportation, petrochemicals and power being its major consuming sectors. Cold weather, the switch to fuel oil in OECD countries and regional crises stimulated demand in 2005. Demand increased in North America, Europe and the Pacific by about 0.39 mmbd, 0.17 mmbd and 0.14 mmbd respectively and this accounted for a 1.5 per cent increase over 2004. A relatively mild northern summer and seasonal quarterly contraction of OECD demand for heating oil were some of the factors that constrained oil demand in 2004. OPEC crude oil supply in 2005 was 29.3 mmbd, an increase of 1.7 mmbd or 6.8 per cent over the preceding year. Demand for crude oil has remained firm since the first quarter of 2004 and global demand in the fourth quarter of 2006 was envisaged to rise to about 86 mmbd. On the whole, the call on OPEC crude was expected to increase to 29.3 mmbd in the fourth quarter of 2006 in response to the global industrial growth especially in China, India and other expanding economies in East Asia.

Unutilised opportunities

The fact that the Nigerian petroleum industry provides a wide variety of investment opportunities in shipping business has been adequately stated. To buttress this point further, it is important to capture in empirical terms one such opportunity. In the area of crude oil production, daily production increased from 5,100 b/d to about 2.4 million b/d between 1958 and 2005. During the period under review a total of 23 billion barrels of crude oil were produced. It is discomforting to note, however, that NNPC or its affiliates did not participate in the transportation of 3.04 billion MT of the crude oil available in the Nigerian market. This is clearly an unutilised opportunity which has a high potential for revenue generation. Out of the 3.04 billion MT of crude oil the JV agreement guaranteed NNPC exclusive rights over 11,398 billion barrels or 1.52 billion MT of equity crude oil which could have been shipped by its transport company. It is clear that the failure to establish a shipping company has deprived NNPC of huge revenues.

It is interesting to note, however, that the opportunity still exists for NNPC to take advantage of the huge tonnages. It is estimated that an NNPC Shipping Company formed in collaboration with a reputable shipping company can commence business by lifting 10 per cent, 25 per cent and 30 per cent respectively of Nigerian equity crude over a period of 10 years (2005–2014). Higher lifting is possible in view of NNPC's vantage position as the custodian of the national crude oil pool. If 10 per cent of the equity crude is allocated, it is estimated that 12.98 million MT of crude oil will be available annually to the proposed NNPC Shipping Company. On the other hand, if the allocations are increased to 25 per cent and 30 per cent over the same period, 32.45 million MT and 38.94 million MT respectively, would be available for shipment during the same period. These volumes are sufficient to sustain the business. The current situation in Iraq, the Niger Delta crisis in Nigeria and other trouble spots globally continue to create uncertainty in the oil transportation business. Charter rates would continue to increase if demand for crude continues to grow. In this regard NNPC stands to benefit from its transport company which would naturally be readily available to transport crude oil to its customers globally, especially in situations when tanker owners place voyage route restrictions on their vessels.

Shipping opportunities in Nigeria

Shipping tonnages in the Nigerian petroleum industry are robust, the current production being 2.4 mmbd. Consistent with the equity interest in the JV operations, NNPC controls about 60 per cent of all crude oil produced in the country. Its share of the crude oil produced annually averages about 525 million barrels or 70 million MT. Furthermore, NNPC imports refined products to satisfy domestic consumption. For instance, in 2003, 7.32 million MT of products were imported by PPMC while in 2004 about 5 million MT were imported. The tonnage available from Nigerian crude oil exports is sufficient to create a giant captive market for a fully-fledged NNPC shipping company. The recently enacted Coastal and Inland Shipping (Cabotage) Act 2003 (the Act), provides an NNPC shipping company with an opportunity to participate in the coastal transportation of refined products to Warri, Port Harcourt, Calabar and the oil rigs. It is important to indicate at this juncture that current oil reserves are estimated to be 36 billion barrels. This is expected to increase to 40 billion barrels in 2010. At the current rate of production it can be assumed that approximately 70 million MT of Nigerian equity crude tonnages will be available annually for the next 40 years. It is also important to indicate, however, that in the light of the aggressive Deep Water exploration, reserves are expected to further increase, thereby expanding the life span of the crude oil transportation business.

The pioneer LNG plant has come on stream deploying over eighteen (18) LNG vessels. The tonnages from this business are committed to Bonny Gas Transport (BGT). However, new LNG projects such as Brass LNG,

OK-LNG and others which are at various stages of development will emerge in the near future thereby consolidating the horizon of the shipping business. Nigeria is primarily considered a gas province with estimated reserves of 187 TCF which will last well over 120 years. All these opportunities in crude export, products import, LNG tonnages and coastal shipping tonnages arising from the Cabotage Law positively indicate that an NNPC shipping business can be sustained. This assertion is further premised on the fact that NNPC has full control over the mode of purchase and freighting of Nigerian crude oil. The current FOB sales arrangement has zero 'value addition' in freight revenues. The practice is inconsistent with the dominant crude oil sales policies of OPEC and non-member countries. The ultimate sales policy which supports national interest (value addition and local content enhancement) is CIF sales. NNPC can sell crude oil CIF in collaboration with a reputable JV partner without the risk of creditors impounding their vessels or encountering pollution problems as widely feared in the past.

Demand for shipping service

Crude oil production in Nigeria has experienced a steady increase since 1958 when oil was first discovered at Oloibiri in Bayelsa State. Whereas crude oil production was 5,100 b/d in 1958, it increased to about 2.4 mmbd in 2006. It is further projected that it will increase to 4.5 mmbd in 2010 and government policies are geared towards achieving this objective. Recently, the federal government awarded 24 MFs to indigenous companies. All these activities will contribute towards the expansion of the crude oil reserve base to well over 40 billion barrels in 2010. Reserves expansion further guarantees an elongated life span for the shipping business. In view of the positive results being achieved in the upstream sector it can be posited that adequate crude oil tonnages will be available for many years to come. Nigeria's recent JV agreement with São Tomé and Príncipe in the JDZ will also contribute toward the expansion of the national reserves. In definite terms it is safe to assume that the equity crude oil in the JV operations in the JDZ will be available to an NNPC owned or affiliate shipping company. Such a company will also be in a position to penetrate the crude oil transportation business in neighbouring African countries.

Global cabotage practice

Maritime nations globally have historically enacted laws designed to promote and protect their merchant transportation. For many years developed countries depended on a robust maritime sector as a foundation for military and economic security. In the modern maritime industry over 90 per cent of international trade is transacted through marine transportation and nations world-wide have adopted laws that prohibit domestic or cabotage shipping in order to promote and protect the local maritime industry. The Act stipulates that ships engaged in coastal trade must be in-country built, owned and operated by nationals. This practice is adopted by the United States, Brazil and to a certain extent India. Several other countries in various forms introduced coastal trade restrictions in their respective maritime industries in order to promote indigenous entrepreneurial participation in the sector. In recent years the maritime industry has experienced significant growth especially with the elimination of trade barriers between a group of nations who are signatories to the General Arrangement on Trade and Tariffs (GATT). The volume of shipping and the related maritime activities of some national coastal lines have increased and past policies and guidelines are no longer adequate to regulate these activities. There is an emerging trend among nations to introduce new laws, regulations and interpretation that delineate activities which are covered by the Act. Maritime nations have established maritime boundaries and Exclusive Economic Zones (EEZs) which guarantee jurisdiction over resources within 200 miles of a conservation zone, and all sea-bed mineral resources on the continental shelf.

Advocates in the maritime industry have proposed the expansion of the Act in order to limit unfair foreign competition in the maritime sector. Although the proposal for restriction of foreign competition in coastal trade has merit, the introduction of such measures can subject some industrial sectors to negative economic consequences. According to the Office of Technology Assessment general cabotage activities in coastal limits involve the following areas:[6]

- commercial fisheries
- oil and gas exploration
  - mobile drilling rigs
  - service vessels/supply boats
  - anchor handling boats
  - launch barges/crane barges
  - production platforms
  - seismic survey boats
- commercial cruise vessels
- marine mining vessels
- dredging vessels
- waste disposal vessels
- ice breaking vessels
- offshore lightering
- diving vessels.

Cabotage in Nigeria

The Act is aimed at restricting the use of foreign vessels in domestic coastal trade. The primary objective is to promote the development of indigenous tonnage and to establish a cabotage vessel financing fund. The coastal trade for the purposes of the law involves the carriage of goods by vessel(s) or any mode of transport above Nigerian waters from one location to the other. The goods carried or transported may be for the purpose of exploration, exploitation of minerals and other non-living mineral resources.

Restriction of vessels in domestic coastal trade

The Act further provides as follows:[7]

> A vessel other than a vessel wholly owned and manned by a Nigerian citizen, built and registered in Nigeria shall not engage in domestic coastal carriage of cargoes or passengers.
>
> Vessels, tug boats or barges determined to be foreign owned shall not participate in the carriage of materials or supply services to oil rigs, platforms and installations. Such vessels, boats or barges are also excluded from carrying petroleum products to rigs, platforms etc.

The oil and gas sector accounts for 95 per cent of the coastal trade in Nigeria, and this trade is currently dominated by foreign vessels. The purpose of the Act therefore is to create opportunities for indigenous participation in coastal marine transportation. The services provided in the subsector are not highly specialised; therefore it provides a suitable entry point for indigenous participation. However, the Act exempts crude oil export, products import/export, LPG, LNG and NGL exports from any restriction imposed by the Act.

Furthermore the Act also provides opportunities in the provision of Floating Production, Storage and Offload (FPSO) and Floating Storage and Offload (FSO) services for oil producing companies. Currently about five FPSOs are in operation at Okono,Yoho, Abo, EA and Ukpokiti. Four additional FPSOs are expected to be introduced into the system in the next five years. In the area of FSO about eight are in operation at Escravos, Amenam and Ima. The Deep Offshore operations are expanding rapidly; therefore it can be expected that more such vessels and facilities will be required. A wide range of vessels, namely crew boats, barges, fuel tankers, diving vessels, tug boats, cable and pipeline layers etc., will be required in the provision of services in the upstream sector. The new law creates an array of opportunities for indigenous entrepreneurs to provide varieties of goods and services to key players in the oil and gas industry.

References

1 Kite-Powell, H.L. 'Marine Policy: Shipping and Ports' in Encyclopedia of Marine Science, Academic Press, 2001, pp. 2768–76.
2 Rodrigue, Jean-Paul, *International Oil and Transportation* http/www.people.hofstra.edu/geostrans
3 Kumar, S.N. *Tanker Transportation*, Maritime Academy, 2006, p. 4.
4 Stopford, M. 'Defining the Future of Shipping Markets'. ITIC Forum, 2000.

5 OPEC Bulletin, April 1981.
6 US Congress, Office of Technology Assessment, and Competition in Coastal Seas: An evaluation background paper, 1989.
7 Federal Republic of Nigeria Official Gazette, July 2003.

Further reading

Containerisation Year Book (2000). National Magazine Co. Ltd, London.

Huber, M. (2001) *Tanker Operations: Hand Book for the Person-in-Charge* (4th edn). Cornell Maritime Press, Centreville MD.

Kuvussano, M. (1996) 'Price Risk Modeling of Different Size Vessels in the Tanker Industry', *Logistics and Transport Review* 32: 161–76.

Glen, D. and Martin, B. (2002) 'The Tanker Market: Current Structure and Economic Analysi', in C. Grammenos (ed.), *The Hand book of Maritime Economics and Business*. LLP, London, pp.251–279.

National Research Council (2002) *Oil in the Sea: Inputs, Fates and Effects.* The National Academies Press, Washington DC.

Tusiani, M. D. (1996) *The Petroleum Shipping Industry: A Nontechnical Overview Vol. II*. Pennwell, Tulsa OK.

UNCTAD (2001) *Review of Marine Transportation, 2001.* United Nations, New York.

Zannestos, Z. (1966) *The Theory of Oil Tankership Rates*. MIT Press, Boston MA.

18 Privatisation and liberalisation

Introduction

The term privatisation made its incursion into the Nigerian economy only in recent times, especially the 1980s. It generally refers to the sale or lease of assets in which the government has a majority interest. Privatisation can also involve the contracting out of hitherto publicly provided services. In many developing countries, and in particular Nigeria, policies are formulated to stimulate the private sector to take over the provision of various goods and services. This practice is not new; however, the wide range of public sector activities packaged for privatisation, the level of accompanying privatisation policies and the various techniques applied in achieving relevant objectives differentiate current privatisation efforts from previous attempts.

The growing interest in the general concept of privatisation in industrialised countries derived its roots from the 1970s. In the 1960s and early 1970s rapid public sector expansion was seen as not only a catalyst and contributor to economic growth but an essential prerequisite for the attainment of social and political stability. For this and other related reasons, the expanding role of the public sector was seen as a necessary development and for that reason continued unchallenged. This situation changed dramatically when the aggregate effects of the oil price shock in 1973 triggered a marked deterioration in macro-economic performance, especially in developing economies. The effect of the shock was profound and economies experienced snail speed recovery. Hemming in his analysis indicated that the poor macro-economic performance was attributed mainly to the large public sector enterprises which by virtue of their heavy reliance on the government deprived the economy of the flexibility required to achieve necessary adjustments.[1] In addition, the efficiency and effectiveness of public enterprises and services were subjected for the first time to elaborate scrutiny. Most countries, especially the US and Britain, experienced a fresh political movement which demanded the curtailment of their government's involvement in public sector activities. Potential public office holders used reduction of the size of the government as a major political campaign tool.

Privatisation was regularly recommended for developing countries in which key elements in the commercial sector and indeed the industrial sector are dominated by public enterprises. Most public sector enterprises in developing countries which performed poorly on a year-to-year basis depended on various forms of grant and subsidies from the government for survival. In recent times the macro-economic problems of these countries have expanded exponentially thereby warranting the adoption of strategies to curtail the drain on the government. Primarily, privatisation has been identified as a suitable process of relieving the government of the burden of overdependence of public enterprises. Considering the size of the public sector in many developing countries, the potential macro-economic change associated with privatisation programmes is substantial. This derives from the fact that the State is the single largest employer of labour. It often plays the role of a default employer of the last resort for post secondary institutions. In addition the State is looked upon for provision of healthcare, education and housing.

In developing countries, subsidies extended to consumers of petroleum products distort the actual price of the commodity at the pump. Participation of the government in ownership of banks and insurance companies determines both interest rates and insurance premiums. The broad scope of the government-related activities in public enterprises was often seen as an avenue of extending the government presence to the masses especially as it relates to social equity and income redistribution. In this regard, therefore, it is often contended that the introduction of privatisation, apart from having the potential for creating economic efficiency, will serve as an instrument for the depoliticisation of issues of social equality and income redistribution. Although privatisation streamlines economic activities, the primary question remains as to what extent it can go to in engendering efficiency in governance. Privatisation essentially is adopted by developing countries as a tool for implementing fiscal and austerity measures aimed at mitigation of spiralling inflation and deficit generated by inefficiently run public sector companies. Perhaps of greater concern are the short and long-term negative externalities that might be associated with such programmes in the context of seeming economic gains. The value of privatisation notwithstanding, it might be essential to know the scope and sequence of its implementation.

The mode of implementation of privatisation therefore is essentially determined by the initial configuration of the public sector. However, Jones and Mansoor[2] are of the view that the pattern of evolution of the public sector enterprises is consequential especially in relation to the nature of vested interests associated with them. The establishment of public sector enterprises has motivating factors which are premised on political and ideological dispositions. Bienen and Waterbury opine that public sector enterprises are founded in most cases based on, among others, the following political and ideological exigencies:[3]

- infrastructure development to promote commerce and industry.

- collection of monopoly rents on minerals;
- mitigate capitalist monopolistic practices;
- assurance of economic sovereignty in a foreign dominated economy;
- nationalisation of dominant foreign or indigenous private enterprises (banks);
- establishment of equity through income redistribution, job creation and regional development;
- national security through reinforcement of the defence sector.

The preceding notwithstanding, it is clear that current privatisation efforts are mostly propelled by pragmatic reaction to many years of unfruitful involvement by the government in public sector enterprises. In this regard therefore it can be contended that genuine ideological departures which question the propriety of government intervention in the economy are infrequent and it is essential to indicate that certain types of government interventions may be easier to dismantle than others. In this regard one would posit that enterprises established to promote private sector growth and development are easier to relinquish to private participations. However, enterprises created as a result of bias against private capital expansion would certainly be more difficult to hand over. Privatisation in developing economies is uniquely difficult because even when public enterprises are adjudged to be underperforming, public opposition against hand-over of such enterprises to private ownership remains high. A typical example is the low capacity utilisation of Nigerian refineries which has gone as low as 40 per cent. The performance of the Port Harcourt, Warri and Kaduna refineries have been erratic and unreliable and the government has spent huge sums of money to rehabilitate them, without success. In this regard, the government considered privatisation as the rational path to the efficient utilisation of the assets, but the general public and workers' opposition to this is rather robust. One may admit that it is possible for new political parties and the governments in developing countries to initiate a clean break from past policies and obligations. However, it is essential to bear in mind that the origins and ideological underpinnings for State intervention to a large extent determine the nature of the privatisation programme. Privatisation in practical terms has constraints and possibilities which are governed by the magnitude of the gains or losses experienced by the economic sectors and the political coalitions.

Privatisation in industrialised States and LDCs

In less developed countries (LDCs), privatisation which is engendered by the need for a broader process of structural adjustment will manifest itself in a determinate sequence. In this regard an initial effort would be made to reform public enterprise performance. Expectedly, when these measures fail to produce efficiency, liquidation and privatisation of certain enterprises are anticipated. In such cases policy makers are always of the view that the public enterprises are beyond redemption.[4] In Mexico, however, the restructuring of public enterprises to perform efficiently operated in tandem with the price fixation process and this model is often the approach adopted by developed economies. In general one is inclined to state that privatisation programmes in developing countries are instigated by negative incentives such as the need for the reduction of deficits and rates of inflation. On the other hand, in privatisation in industrialised countries such as Britain and France, revenue generation from sale of public assets has been a major consideration. In LDCs privatisation aspires to eliminate costly income redistribution programmes regardless of the consequences of alienation of traditional sources of political support.

In the case of Britain and France, privatisation provided an attractive avenue for popular capitalism to flourish and in the process promote middle class political support. Between 1975 and 1985 deficits ranged between 3.5 and 5.6 per cent of GDP but in the 1980s and beyond the figure escalated, due mainly to huge external debt servicing obligations. Under such circumstances budget deficit reduction was imperative in order to improve public finance and mitigate spiralling inflation, both of which are crucial for the maintenance of a stable economy. As valuable as privatisation may seem in developed countries, the degree of success in its implementation in developing countries has been low. One distinguishing factor between developed countries and LDCs which embark on privatisation is the relative size of the middle-class income strata. The middle-income strata in LDCs are relatively thin and this accounts for the difficulties encountered by privatisation programmes targeted at small shareholders.

In Britain and France, which have outlets for private bank loans, the level of participation is much higher. In this regard, both countries have, through privatisation, added new investors to their respective private sectors. Among civil servants and other middle-income workers in LDCs, lack of suitable collateral stands as an obstacle in their efforts to secure bank loans. More important is the fact that banks in most cases do not consider small share purchases in a privatised enterprise to be a lucrative commitment of bank funds. They fear that such loans may stall if the privatised enterprise fails to post impressive returns. In this regard, the general approach is non-committal, which ultimately deprives potential investors from acquiring shares in what might turn out to be a successful privatised entity. Sadly, these financial bottlenecks facilitate the transfer of middle-class privileges to the super rich, thereby steering the overall income distribution further in favour of the wealthy. These observations are manifest in the Nigerian privatisation exercise such that some of the major companies, namely NICON Hilton, Sheraton Hotel, NAFCON, NITEL, Eleme Petrochemical company etc., have been acquired by a few companies owned by a few individuals through selective tendering processes. The general public, comprising mainly of middle-income earners, did not benefit from the exercise because they lacked the capital to join the mini-consortia formed to acquire the businesses.

Politics of privatisation

Politics of privatisation derives from the perceived attempt of the government to secure some future gains both for the general public and the private sector capitalists who have, prior to the privatisation effort, competed with the sector. The gains of the process are often weighed against the emergent costs of those who lost employment and influence.[5] From a political context it can be argued that beneficiaries of the status quo would under normal circumstances make attempts to destabilise the implementation programmes. The question has often been asked as to what persuades leaders and policy makers to show devotion to the reform process. It has been observed in the recent past that structural adjustment, or for that matter privatisation programmes, are often prescribed by the IMF, the World Bank, collateral donor agencies or commercial banks. The experience of some countries which have undertaken structural adjustment and privatisation programmes show that the final result turns out to be the opposite of what is expected. In this regard there is a general feeling that the IMF, World Bank and donor organisations are insensitive to the hardships inflicted by the programmes. It is important to note that most reform programmes undertaken in Tanzania (1980) were instigated by donors. The same is true of the structural adjustment programme of 1986 in Guinea linked to a World Bank loan.[6] Whereas the preceding observations may be true, Hyden on a different parlance expressed the view that the need for reform independent of external forces is abundantly clear.[7]

For instance, in 1980 following the plummeting of oil prices, Algeria undertook an elaborate assessment of its development strategy and the role of government intervention. The result of the re-evaluation process provided further impetus for the acceleration of deconcentration of PSEs. It also led to deregulation and the emboldening of the private sector to register greater participation in the takeover of PSEs. In a similar manner, India around 1980 embarked on liberalisation and deregulation programmes without external catalysts orchestrating the transformation process. The factors impinging on the various privatisation programmes are numerous. This notwithstanding, Bienen is of the view that translation of a new awareness into a comprehensive policy constitutes the main political challenge for policy makers and technocrats alike. In a truly pragmatic milieu the implementation of a privatisation programme will be determined by the following:

- the magnitude of government deficits and associated fiscal crisis;
- the origins and structural state of the public sector enterprises;
- the characteristics of the key coalitions;
- the quest for power by the incumbent leadership and the desire to control both economic resources and patronage.

Budget deficits

PSEs in developing countries have historically underperformed and as a result contributed to a significant portion of government deficits. It is interesting to note that at the end of the 1970s the debt burden of PSEs in many LDCs increased significantly and in Brazil it ranged between 66 and 76 per cent of the overall government deficit.[8] Underperforming PSEs are highly leveraged and debt servicing constitutes a significant portion of their operating deficit. Privatisation creates a fast solution for PSE deficits to exit the public accounts in the sense that once the enterprise is transferred, the government ceases to extend subsidy to the PSEs. The quick exit of the government from the burden of the PSE is not without cost. Privatisation is time-consuming and entails expensive processes which must be undertaken in order to attract suitable buyers. Furthermore, the process may require government absorption of the outstanding debt of the enterprise. In this regard it is important to indicate that the primary concern of the government in such situations is to terminate the dependence of the ailing enterprises on the government purse as quickly as possible. Under this circumstance, one would say that revenue generation is rather secondary. During the military era, in some LDCs public enterprises were undervalued and hastily privatised in order to avoid the opposition by interest groups. This was the experience of Sudan in 1972 and Bangladesh in the 1980s.[9] Privatisation has both positive impacts and negative externalities which have far reaching consequences.

Bienen opines that privatisation programmes have discernible linkages with employment embargo and salary freeze, reduction in budgets and social services, de-indexing of wages and reduction of consumer subsidies. These actions collectively inflict immediate hardship on the affected personnel. Such effects on individuals arising from the above factors tend to be more profound than the direct impact of privatisation. Sometimes certain actions are executed preparatory to the implementation of privatisation and these actions would in most cases entail job cuts. Between 1981 and 1983 Mexico lost an aggregate of 1.5 million jobs in the formal economy prior to the introduction of the privatisation programme. LDCs in most cases lack the technocrats to implement privatisation programmes and this shortfall in specialised manpower requirement accounts for the poor performance of some privatisation programmes.

Even when the required specialists are available their time and input would be needed on more pressing state matters such as raising money to service foreign debt. Under this circumstance most LDCs concentrate on selling minority state shares or outright liquidation. Experts are of the view that a strong correlation exists between the extremely poor performance of PSEs and the degree of difficulty in disposing of them through the privatisation process. Consistent with research findings Nellis expressed the view that poor performances of PSEs are in the majority of cases linked with bureaucratic and political interference. It is important to note also that investment capital in Africa is very scarce and the capital markets are at low levels of development. In such cases where indigenous capital is not readily available foreign ownership would prevail. When this occurs citizens might be inclined to refer to privatisation as foreign ownership. Such stigmatisation invokes sentiments which suppress the political will to implement the privatisation programme.[10] Granted, privatisation is a desirable and instrumental exercise, but nonetheless, the intervening factors, if not adequately addressed, can impose enduring obstacles in the implementation process.

Ideological imperatives

In the 1970s, when ideological orientations commenced, Socialism, Welfarism and Marxism held sway in Africa. During that period the majority of PSEs were conceived and established primarily in the context of such ideologies. Therefore if a particular ideology promoted income redistribution, the PSEs functioned to accomplish that objective even at the expense of efficient operation of the enterprises. In recent years, especially with the collapse of the Soviet Union, some LDCs have eschewed Socialist programmes and progressively moved toward market economy ideologies. Some of the Tiger economies of Asia, Brazil, Côte d'Ivoire, Nigeria and Pakistan (except during the Bhuto era) fell into the categories of countries that have embraced capitalist ideology. Such countries would not find it difficult to privatise. However, countries which emphasised the role of the state in wealth redistribution and social justice would see the privatisation process as anti-ethical to their ideological convictions. Adoption of privatisation, it was believed, would call into question their commitment to the party or group ideology. Tanzania, India, Algeria and Egypt found themselves in this category in the 1970s and 1980s. This situation has changed as the majority of LDCs globally have restructured their political, ideological and bureaucratic processes to the extent that privatisation has become widely accepted and implemented. Nigeria, although non-aligned in ideology, established PSEs which served redistribution and social justice agenda. Since 1999 privatisation in Nigeria has assumed an elaborate scope, details of which will be discussed in subsequent sections.

Influential coalitions

Influential coalitions interacted in definite patterns and served as major constraints on the government in their efforts to embark on transformation programmes. It can be recalled that most LDCs adopting principles of economic development initiated Import Substitution Industrialisation (ISI) programmes. These programmes had benefits that appealed to a cross-section of interest groups – namely professionals, organised labour, technocrats and captains of industry. These groups interacted in definite patterns dictated by their vested interests and ultimately crystallised into formidable coalitions which influenced State-orchestrated programmes. The formation of these coalitions paid less attention to the greater majority of stakeholders involved in agriculture, cottage industries and the informal sector. Bienen and Waterbury in their analysis indicated that anti-ISI coalitions in India, Mexico, Turkey, Algeria and Tanzania sustained momentum and attained stability. These coalitions perceive privatisation and related structural adjustment programmes as destabilising factors and as such developed mechanisms to resist the implementation of such transformation programmes. On the other hand, the farmers, small scale industrialists and operators in the informal sector who can benefit from privatisation programmes failed to muster support mechanisms because they lacked cohesion. Incumbent leadership in the government which intends to enlist their support to push through privatisation programmes achieved little success because they lacked a network which could easily be mobilised for action. They further contend that in the ISI paradigm, organised labour and other lower and middle-income groups associated with the coalition consciously traded political support in exchange for the government commitment to protect their standard of living, resource redistribution, and provide free healthcare and consumer subsidies. The social pacts on the part of the government which attracted political trade-off from the members of the coalition tend to be costly. Consequently, successful structural adjustment programmes tend to drastically alter the self-serving pacts in order to create operational efficiency in the economy. Governments engaged in structural adjustment and privatisation programmes tread cautiously in order not to push labour too far. For this reason the government provides channels for dialogue with organised labour to renegotiate social pacts and secure their tacit endorsement of certain austerity measures. This approach was extensively used by the Nigerian government in its efforts to increase the price of petroleum. Although labour did not completely endorse the increase in price, it negotiated social palliatives which were designed to cushion the effects of the price increase.

Perpetration of power

Perpetration of power is a major constraint against privatisation programmes in LDCs. Numerous scholars have portrayed public enterprises in various forms. Sheperd expressed the view that PSEs are simply an expression of political power and their ability to serve distinct public interests.[11] Baklanoff in his analysis stated that the state seeks to maximise its utility (i.e. perquisites, income and power). Therefore the promotion of the welfare of the masses through PSEs is a rather secondary objective. Leaders had resources (funds, status and jobs) – including a wide array of activities which they manipulated and distributed to individuals and groups based on political support – placed at their disposal by private enterprises. The administrative institutions of LDCs were considered weak; therefore politicians relied on PSEs as avenues to control the various coalitions of the polity.

It is customary for leaders to articulate strategies to limit the scope of government involvement in the economy. However, such political pronouncements and implementation agendas are down-played when it is realised that resultant actions would lead to loss of control of political resources.[12] A typical example was the case of Turgut Osal of Turkey who, as Prime Minister, pledged to curtail State intervention in the economy but failed to move ahead with the corresponding reforms for fear of losing control of State resources which were considered crucial in maintaining his narrow electoral majority.[13] Similarly Herbst opined that the elaborate introduction of PSEs in Africa was instigated by the need to acquire greater authority over the electorate rather than for purposes of responding to ideological convictions or market failure.[14]

Country experiences

Privatisation was applied as a tool for engendering greater efficiency in the PSEs in Britain. Between 1979 and 1987 the United Kingdom raised a total of £12.0 billion (Table 18.1b) through a carefully crafted privatisation programme. British Telecom and British Gas were by far the biggest PSEs that were privatised. Sales of a majority of shares in British Telecom generated £4.1 billion while a similar exercise in British Gas yielded £1.8 billion in the first instance and was expected to generate additional funds which would amount to a total of £5.1 billion. The privatisation team in LDCs effected sale of the PSEs by adopting one of two methods, or in some cases a combination of both, namely offers for sale at fixed price and the sale of equity by tender. Offers of sale at fixed price was by far the most common approach. Both methods presented problems because most tenders failed to reach the reserve prices. Perhaps more significant was the fact that offers were oversubscribed. It is important to indicate at this point that determining the market price of a public sector enterprise is a difficult task due to lack of reliable data on or comparators against which the estimate can be benchmarked. In this regard selling by tender was considered a logical route for achieving the privatisation programme.

This notwithstanding, the process has been regarded as a complex sales approach that fails to attract the participation of small investors who are deliberately targeted for involvement by the UK government privatisation policy. Offers for sale in practical terms have encountered all the problems identified earlier. Hemming and Mansoor stated that in an offer for sale in 1982 the small electronics firm Amersham International was oversubscribed 25 times. Although the exercise yielded £64 million, subsequent trading led to an immediate discount of over 30 per cent on the sale and 65,000 shareholders traded their shares to larger organisations at a high profit.[15] Mayer and Meadowcroft in a similar study also established an existence of oversubscription and subsequent discount of about 26 per cent on a range of sales which were effected between 1979 and 1985.[16]

Table 18.1a France: major privatisations, November 1986 to 4 June 1987

Date		*Enterprise*	*Proceeds (FF billion)*
November	1986	St. Gobain	8.2
January	1987	Paribas	12.6
April	1987	Credit Commercial de France	2.0
May	1987	Compagnie Generale d'Electricite	10.0
May	1987	Havas	2.5
June	1987	Societe Generale	17.0
June	1987	TFI (television station)	4.5

Source: International Library of Critical Writings in Economics, 2000.

Table 18.1b United Kingdom: privatisation of major public enterprises, February 1981 to January 1987

Date		*Enterprise*	*Proceeds (£ million)*
February	1981	British Aerospace	43
October	1981	Cable and Wireless	181
November	1983	Britoil	627
December	1983	Cable and Wireless	263
June	1984	Enterprise Oil	382
July	1984	Jaguar cars	297
November	1984	British Telecom	4,090
May	1985	British Aerospace	346
August	1985	Britoil	426
December	1985	Cable and Wireless	571
December	1986	British Gas	1,796
January	1987	British Airways	415

Source: International Library of Critical Writings in Economics, 2000.

The French government also undertook a five-year privatisation programme in mid-1986. Sixty-five enterprises and subsidiaries which included banks, insurance companies, and financial holding groups were packaged in the plan. The programme was successfully implemented and about FF 55 billion was raised in the early rounds (Table 18.1a). Other countries, namely Chile 1970 and 1983, Germany 1984, Italy 1982, Japan 1985, Malaysia 1985 and Turkey 1984, implemented privatisation programmes in response to the need to create competition and efficiency, and to minimise reliance of PSEs on the public treasury through subsidies, administered prices and outright grants.

Fiscal impact of privatisation

Privatisation in some cases has been preceded by a fanfare, especially where it served as a political instrument. The physical sale of assets is tangible and the transfer of ownership is perceptible. However, the fiscal impact of privatisation is latent. In practice the *Government Finance Statistics Manual* classifies proceeds from the sale of assets to the private sector as either loan repayment or capital revenue. Therefore the sale of existing government fixed assets (such as property) or intangible assets (such as mineral rights) have their proceeds recorded in the books as capital revenue. On the other hand, if the government sells part or all of its interest in a public enterprise, the transaction is treated as sale of equity and the proceeds are considered as loan repayment. In the event of asset sales by the government, fiscal impact is felt through the reduction of the deficit (i.e. difference between total expenditure and total revenue). The amount of reduction would in normal cases be equal to the financial value of the sales.

However, it should be borne in mind that the remittance made to the government in the form of profit by a public enterprise would be foregone and as a result the reduction in the government deficit would be lower. The sale of the government asset in practical terms would lead to a once and for all reduction of the government deficit unless the sale proceeds are lower than income that would have accrued to the government. In a situation where the liabilities of the company are higher than the asset value, outright liquidation might be the solution. In the Nigerian context, NAFCON fell into this category and was therefore liquidated. It is important to indicate that the overall deficit in the government serves as a barometer for a number of activities in the economy. A deficit measures the difference between the government revenues and expenditures; therefore changes in the overall deficit followed by adjustments generally signify changes in the government fiscal position and indicate whether 'demand management policy has been more expansionary or contractory'.[17]

Privatisation and liberalisation in Nigeria

Privatisation

Nigeria's pattern of development and the policy initiatives adopted in the 1970s and 1980s are quite similar to measures taken in Kenya, Uganda, Ghana and Côte d'Ivoire. As frequently espoused by experts of development economics, Nigeria embarked on the establishment of Public Sector Enterprises with a view to providing basic goods and services for the citizens. Only a few private sector companies existed, namely UTC, Uniliver, PS, Kingsway, Nestlé, Cadbury etc., and these were essentially non-indigenous. By virtue of their private ownership, the government was constrained in making public welfare related demands on them so the practical alternative was the establishment of PSEs such as NAFCON, Niger Dock Shipyard, Kaduna Steel Rolling Mill, Federal Palace Hotel, Ajaokuta Steel Mill, ALSCON, etc. These companies were in most cases of international standard and well equipped. It was the expectation of the government that the PSEs would operate effectively, generate modest profit and at the same time provide goods and services to the public. Under the military regimes, establishment of PSEs was seen as popularity programmes which must be embarked upon to win the support of the citizenry. The early military governments in the 1960s and 1970s frowned at profligacy, therefore heads of PSEs and their lieutenants conducted affairs with some measure of accountability. In later years several regime changes occurred and the government both at the State and Federal levels became unduly interested in the affairs of the PSEs, appointing the CEOs and members of the boards of directors. Such appointments were products of intense lobbying and appointed members represented interest groups. In order to satisfy the demands of political sponsors and supporters appointees in turn made some demands on the PSEs. The CEOs were expected to oblige all or a part of any request made by directors, but compliance with such demands created distortions in the cash flow and projected revenues. Boards of PSEs, in carrying out their statutory duties of approving contracts, created outlets for their surrogates to benefit. This process went on for years, became deeply entrenched and finally led to the destruction of systems and processes in the enterprises. The revenue generation capacity declined and the integrity of the primary equipment and other production facilities deteriorated. Ultimately the revenues generated became far too low to pay salaries, overhead costs and other obligations of the companies.

During the military era the presiding Generals believed that closure of Federal or State PSEs might be interpreted as non-performance on the part of the government. It was also considered that dismissal of staff would negate the government employment creation agenda. For these and other related reasons, Federal and State governments extended subventions to the ailing companies to rehabilitate their plants, provide goods and services and generate profit. These gestures from the government were always eagerly expected by the beneficiary companies, but their performance consistently fell far below government expectations. It is important to emphasise at this point that the problems of the companies were not exclusively from senior government officials, directors or the military class. The malpractices of the board and executive levels trickled down to the finance and accounts departments, procurement, maintenance etc. These departments devised techniques to siphon funds from the system through invoice inflation, purchase of interior parts and inflation of maintenance contracts. All these malpractices exacerbated the precarious conditions of the PSEs. Malpractice in the systems continued for many years and some of the enterprises experienced complete collapse. A notable example was the National Fertilizer Company of Nigeria (NAFCON) which was constructed at Onne to produce different grades of fertilizers. The company was constructed by world-class contractors and notable among them was Kellogg.

Kellogg operated the plant for a number of years during which the plant ran efficiently and the urea and other grades of fertilizer produced were sold both in the local and international markets. At the end of the management contract period, the plant was handed over to Nigerian leadership with several layers of other Nigerians understudying the expatriate staff. Upon completion of the understudy years Nigerian engineers and technicians took over the various positions in the company. The new team compromised the systems and processes in the running of the plant and as a result the performance level declined. More importantly, it was observed in the case of NAFCON that there was outright corruption both at the top and middle-level management. These activities drastically affected the liquidity of the company and the debt burden of the company increased exponentially. Under the poor financial conditions TAM was skipped and the integrity of the plant deteriorated, and ultimately production activities ground to a halt.

The failure of NAFCON caused serious problems for the agricultural sector as fertilizer could no longer be procured in sufficient quantities locally to apply on farms. NAFCON relied on natural gas as its primary feed stock. This feed stock is available in large quantities; therefore the failure of the company which was running on subsidised gas was rather inexcusable. The closure of the company was an embarrassment and it created unnecessary recourse to the use of scarce foreign reserves for the importation of fertilizers sold at subsidised rates to local farmers.

The collapse of NAFCON was the subject of several federal government panels of inquiry which indicted past and serving chief executives of the company. Although the company stopped production, the federal government was saddled with the responsibility of paying staff salaries. This continued from 1999 to 2005 and this dependence on the government along with similar cases in other PSEs, warranted the government decision to liquidate the company. The Bureau for Public Enterprises (BPE) scheduled NAFCON for privatisation. This proved abortive in view of the fact that the company had about ₦20.0 billion of debt. Potential investors were uncomfortable with the level of indebtedness of the company and as a result withdrew from the privatisation process. The attempt at privatising NAFCON by BPE started around 2001 and dragged on until 2005 when it was liquidated at the cost of about $220 million. There are several cases of PSEs which underperformed and were subsequently privatised. Niger Dock, Nigeria Ports Authority, NICON Hotel, Eleme Petrochemicals Company, Benue Cement Company, NITEL, Ajaokuta Steel Company, ALSCON etc. are the government PSEs that have been privatised as a result of poor performance and heavy reliance on the government for subventions.

The global political polarisation along capitalist and Communist lines has subsided following the collapse of the Soviet Union. LDCs in most cases towed the middle path of socialism or welfare-ism which advocated market economy as well as State subsidy of essential services – education, health, housing etc. – for the benefit of the masses. In the case of Nigeria the government is genuinely interested in the masses and would to a practicable extent provide public services for the benefit of the citizenry. More importantly, it is keen on providing infrastructure for private-sector driven economic and industrial development. There is the general belief among technocrats that a robust private sector will stimulate the economy and create jobs. In this regard the government considers it inexpedient to support non-profit yielding PSEs. Rather, current government policy is to provide infrastructure, power, roads and the enabling environment for private businesses to grow.

Liberalisation

Commercial activities in the Nigerian economy experienced various degrees of interruption between 1990 and 1999 due to scarcity of petroleum products. The Obasanjo administration, upon resumption of office in May 1999, set up a 34-man special committee on the review of petroleum products supply and distribution. The committee, in a period of three months, examined the various problems plaguing the industry, namely low refinery capacity utilisation, scarcity of products, hoarding, mishandling of products, problems of cross-border leakages and waste of valuable man-hours at petrol stations. The committee, which comprised industry stakeholders, professionals, labour leaders etc., recommended that the downstream sector of the oil and gas industry should be liberalised. Liberalisation in this context implies the abolition of the government monopoly (through NNPC) of the supply and distribution of petroleum products. This implied therefore that private individuals who have the means and technical knowledge should be allowed access to refining and products distribution in the sector. The entry of new players on a level playing field would engender competition.

The committee recommended that to effectively implement the policy two conditions must be met, namely:

- the government must ensure availability of products in every part of the country at whatever cost prior to the implementation of the policy; and
- the government should undertake a nationwide tour for an enlightenment campaign on the policy in order to allow Nigerians to understand the nature of the proposed liberalisation and the associated benefits.

The government did embark on an extensive enlightenment campaign and 33 States of the Federation were visited where citizens were informed about the policy and its benefits. In addition the National Assembly, industry professional bodies, labour leaders, student organisations and media houses were visited and consulted on the matter. The two conditions were met and on 1 January 2002 the government officially announced the commencement of the liberalisation policy along with a new products price – petrol ₦26.00/litre, diesel ₦26.00/litre and kerosene (₦24.00/litre). Partial deregulation has taken effect and marketing companies have adjusted prices beyond the initial levels to about ₦70.00/litre to reflect the cost of crude oil in the international market. In order to ensure that the liberalisation policy did not lead to unrealistic price increases, the Petroleum Products Prices Regulatory Agency (PPPRA) was established. The members included:

- refiners
- marketers and retailers
- mass media
- labour
- transporters
- Nigeria Association of Chambers of Commerce
- Manufacturers Association of Nigeria (MAN)
- Nigeria Employers Consultative Association
- airline operators, and
- consumers of products.

The mandate of the commission was to liaise with the 19 stakeholders and establish from time to time a fair and affordable price of products. In implementing the liberalisation policy of the government PPPRA identified the advantages of the policy as follows:

- establishment of private refineries;
- construction of new product storage facilities by major and independent marketers;
- introduction of competition and improvement of operational practices;
- opportunity for private importation of high volumes of products;
- stability of product supply;
- competitive pricing of products at filling stations;
- promotion of investments and flow of FDI;
- macro-economic growth through the emergence of new industries and resuscitation of dormant plants;
- job creation through new industries and poverty alleviation.

References

1 Hemming, R. and Ali Mansoor (1988). 'Privatisation and Public Enterprises'. *Privatisation in Developing Countries*, Vol. II, 2000, 1988, p. 167.

2 Jones, L. P. and Mansoor, E. S. *Public Enterprises in Less Developed Countries* (Cambridge: Cambridge University Press. Adopted from *Privatisation in Developing Countries*, Vol. II, p. 619).

3 Bienen, H. and Waterbury, J., 'The Political Economy of Privatisation in Developing Countries'. *World Development*, Vol. 17, No. 5, 1989, pp. 617–32.
4 UNDP, Subregional Meeting on the Role of the Private Sector in Economic Development, draft report (Lagos, Nigeria: UNDP May–June 1988).
5 Bienen and Waterbury, op. cit., p. 623.
6 Hyden, G. 'Business and Development in Sub-Saharan Africa'. USFI Reports, No. 25 (1986).
7 Wilson, E. (1988). 'Privatisation in Africa. Domestic Origins, Current Status and Future Scenarios'. *A Journal of Opinion*, Vol. 16, No. 2, pp. 29–34.
8 Bienen and Waterbury, op. cit., p. 624.
9 Lock, K. (1988). The Privatisation Transaction and its Long Term Effects. A Case Study of the Textile Industry in Bangladesh, unpublished. Cambridge, MA: Harvard University, Centre for Business and The government.
10 Cowan, L. G. (1987). A Global Review of Privatisation in Hanke S.H. (ed.), *Privatisation and Development*. Institute of Contemporary Studies, San Francisco, CA.
11 Shepherd, W.G. (1976). *Public Enterprises: Economic Analysis of Theory and Practice* (Lexington, MA).
12 Bienen, H. and Waterbury, J., op. cit., p. 628.
13 Herbst, J. P. (1988). Power and Privatisation in Africa, paper presented at the Privatisation Working Conference, Princeton University, Princeton, NJ.
14 Baklanoff, E. N. The Dependent Entrepreneurial State, Public Enterprises and External Debt in Latin America. UFSI Reports No. 7 (1986).
15 Hemming and Ali Mansoor. *Privatisation and Public Enterprises*. IMF Occasional Paper, No. 56, Washington, DC: International Monetary Fund, 1–22.
16 Mayer, C. P. and Meadowcroft, S. A. 'Selling Public Assets: Techniques and Financial Implications'. *Fiscal Studies* (Oxford), Vol. 6, Nov.1985, pp. 42–45.
17 Heller, P. S, Haas, R. D. and Mansoor, A. S. *A Review of Fiscal Impulse Measure*. Occasional Paper No. 44 (Washington: IMF 1986).

19 Investment opportunities

Investment imperatives

Major oil producing companies in the Nigerian petroleum industry continue to register impressive crude oil discoveries, especially in the Deep Offshore. Similar discoveries are being made in the gas subsector while the oil services sector is also expanding rapidly. Successive governments have placed a high premium on the socio-economic development of the country and investment remains a major catalyst for development in all sectors of the Nigerian economy. For this reason the government has made deliberate efforts to encourage genuine investors to take advantage of the available opportunities. Over the years several companies, namely Shell, ExxonMobil, ChevronTexaco, Total, Agip, ConocoPhillips etc., have taken advantage of the enormous investment opportunities in the Nigerian petroleum industry. Their sojourn has lasted for many years with attractive financial benefits and the investment horizon in the industry continues to expand, especially with the progression of exploration and production activities into the Deep Offshore.

Nigeria's journey to full development, while progressing, is beset with several challenges in the areas of employment creation, capacity building, NCD, power generation, education, health etc. These challenges can only be overcome through the generation of revenues. Such revenues will primarily manifest through new investments in the oil and gas sector and other sectors of the economy. It is against this background that the NIPC and other collaborating organisations made it a priority to organise forums in order to encourage and invite potential investors to explore and take advantage of the various investment opportunities in the Nigerian petroleum industry.

Oil and gas sector liberalisation

Liberalisation in the industry involved opening up the downstream sector of the petroleum industry for healthy competition. The upstream sector of the industry was from the outset liberalised and exposed to foreign competition. At present MNCs in the upstream sector include Shell, ExxonMobil, ChevronTexaco, Agip, Total, ConocoPhillips etc. On the other hand, the downstream sector of the industry was regulated by the government. For instance, all refining activities in Nigeria were up until 2008 undertaken by the the government-owned NNPC. In the early stages the refineries were in good condition and the petroleum products supply was not a problem. However, in 1999 the integrity of the plants deteriorated and capacity utilisation (445,000 barrels/day) declined to approximately 34 per cent. This situation caused serious dislocations in products supply thereby warranting government importation of approximately 12 million litres of petroleum products daily. The importation exercise of the government involved a huge subsidy which impacted negatively on other sectors of the national economy. In order to stem this trend, the government in 2003 liberalised the downstream sector to achieve the following:

- induce competition;
- eliminate bureaucracy arising from the monopoly enjoyed by the NNPC;
- set the stage for privatisation of the sector;
- encourage refiners to be export oriented;
- promote investment and create a vibrant industry;
- eliminate incessant petroleum products scarcity.

In pursuance of these objectives, the government in 2002 issued licences to 18 companies to construct private refineries. In this regard the liberalisation process created additional investment opportunities in the downstream sector.

Joint ventures

The need to develop, build capacity and transfer technology as well as attract the huge investment capital was an underlying reason for the establishment of JV operations between the NNPC and the MNCs. Nigeria commenced active involvement in JV operations with MNCs in the upstream sector in 1971. JVs provide NNPC opportunities for participation in a highly dynamic sector and in the area of funding NNPC bears 60 per cent of all funding requirements of oil and gas investments in the JV relationships. Seven JVs are currently in operation, as follows:

- NNPC/Shell/Elf/Agip
- NNPC/Mobil
- NNPC/Chevron
- NNPC/Agip/Phillips
- NNPC/ElF
- NNPC/ChevronTexaco
- NNPC/Panocean.

In 2001/2002 the federal government conducted a licensing round for the allocation of oil Blocks in the Deep Offshore with other licensing rounds being conducted in 2005 and 2006. The Nigerian oil and gas terrain is broad with many other Blocks yet to be allocated in the Deep Offshore. Therefore the opportunity exists for forward looking investors to win Blocks and join other multinationals who have reaped attractive returns from their investments. The Gulf of Guinea has also opened up vast investment opportunities with an estimated 15 billion barrels of oil locked up in the area.

Production Sharing Contract (PSC)

A PSC is an agreement between a NOC (NNPC) and a foreign oil company. The contractor in this case is not a co-owner (with the NOC) of the petroleum licence or lease. The contractor, guided by preliminary information on the block, enters into a contractual agreement which covers a period of 10–30 years with NNPC and provides funds for exploration, development and production. However, the contract guarantees the contractor the right to recoup all expenses incurred through 'Cost Oil'. Cost Oil is the amount of available crude oil allocated to the contractor to enable the company to recover all costs as specified in the contract. The contractor also gets a share of the 'Profit Oil'. There are currently nine existing PSCs and 16 new PSCs operating in the industry.

Service Contract (SC)

A SC is an agreement between NNPC and a foreign oil company. A service contractor has neither a working interest nor an ownership claim in the licence or the lease. In this case the contractor accepts to carry out on behalf of NNPC exploration and development of the oil field. Contrary to the provisions of a PSC a service contractor does not have an automatic right to produce oil discovered. However, the company may carry out production activities if NNPC grants it the option to do so. A service contractor is responsible for all costs of exploration and development work and the expenses incurred by the contractor are recouped through cost oil and profit oil. At present only Agip Energy has an SC with NNPC.

Marginal Field parameters

In the Nigerian context, an MF is 'any field which has reserves booked and reported continually to the DPR and has remained unproduced for a period of 10 years'. Specifically MFs have the following characteristics:

- fields not produced due to marginal economics;
- fields bearing oil but not produced for over ten years;
- fields with crude characteristics different from current crude stream by virtue of high viscosity and low API gravity;

- fields not produced as a result of low ranking in the portfolio of hydrocarbon fields;
- fields scheduled for farm-out due to portfolio ranking.

MF development

Petroleum exploration, development and production in Nigeria as in other oil producing countries are highly competitive and capital intensive. Over the years the major oil companies have produced many fields. However, some of the fields were not produced because the volume of crude oil in the reservoirs do not satisfy the economic criteria of the companies. Naturally, fields with crude oil in the range of 10 million to 60 million barrels could not satisfy the economic criteria for production. In this regard, therefore, the federal government, through the DPR, recovered the MFs from the major producers with a view to providing opportunities for indigenous companies to acquire fields and produce crude oil and gas for the overall benefit of the country.

Criteria for evaluation

In an effort to encourage participation in the upstream sector, the government in the past discretionarily allocated oil fields to indigenous companies. Most of these allocations were not utilised, thereby negating the government's effort to engender domestic entrepreneurial empowerment. In order to avoid the pitfalls of the past, clear criteria were established to ensure that beneficiaries were qualified to execute the programmes. The set criteria include the following:

- company details;
- evidence of the company's technical and managerial capability in exploration and production;
- payment of a signature deposit;
- statement of HSE policy;
- NCD plan stating how the company intends to create jobs, utilise local materials in fabrication, capacity building etc.

Partnering opportunities

In 2002 the federal government awarded 24 MFs with crude reserves ranging from 10 million to 60 million barrels, but most of the beneficiaries lack the capacity to raise capital to execute the programmes. More importantly, in view of problems of low capitalisation, the Nigerian banks (which are traditionally risk averse) are more disposed to short-term lending and rebuff financing long-term projects in the upstream sector of the industry. Recent collaboration between NNPC and local banks is expected to provide initially $350 million for indigenous contractors in the upstream. MF operators are yet to benefit from the scheme.

Some of the beneficiaries desperately explored partnering opportunities with foreign companies in order to achieve the objectives of producing MFs. These opportunities deserve consideration as successful partnerships arising from such collaborative efforts will lead to mutually beneficial and financially rewarding relationships. Collaborative efforts in the upstream sector have a history of success and this achievement can be repeated in MF development and production.

Gas monetisation–utilisation

Nigeria Liquefied Natural Gas (NLNG)

LNG production is one key area of the monetisation of gas (conversion of gas into a steady revenue stream). The NNPC (49%) in collaboration with major investors, namely Shell (25.6%), Total (15%) and ENI (10.4%) established the NLNG in 1989. The plant which was constructed in Trains, has progressed to the sixth Train and shareholders have mobilised funds in excess of $8.0 billion to produce LNG. FID was made on the sixth Train and came on stream in 2007. NLNG has successfully exported LNG cargoes to off-takers in Europe (Enel of Italy, Gas natural of Spain, Botas of Turkey, Gaz de France and Trans-Gas of Portugal). Upon completion the six Trains will produce 21.5 million MT of LNG and 5.0 million MT of LPG per annum respectively. This level of production will be sustained by a feed stock of 2.8 billion scf of natural gas per day. The implementation of the LNG programme which cost over $8 billion was funded through equity contribution of stakeholders and syndicated loans from a consortium of international banks.

Gas-to-liquid (GTL)

Nigeria is endowed with abundant reserves of associated and non-associated gas estimated to be in excess of 187 trillion scf. This represents over one third of Africa's total gas reserves which are estimated to last for about 120 years. It is important to indicate also that Nigeria ranks tenth globally in gas reserves. Until recently, 2 billion scf (75 per cent) of gas produced (i.e. approximately 400,000 b/d equivalent) was flared daily. Currently, approximately 43 per cent of gas produced (i.e. 230,000 b/d) is flared daily. There is an urgent need to stem gas flaring in the oil producing fields. In view of the UN initiatives for global environmental protection, the federal government has introduced a policy which makes it unlawful for any oil and gas producing company to flare gas beyond 2009. The zero flare programme creates an opportunity for investors who have the appropriate technology to extract gas liquids to produce ethane, propane, butane, pentane etc. which have a wide range of commercial and industrial applications.

Fertilizer production

Nigeria has a total land area of 923,770 km^2 with over 34 per cent of the land mass as arable. The price of fertilizer is high; therefore the utilisation of the commodity in Nigeria is low and in some cases is not available for farmers to purchase required quantities. The National Fertilizer Company was established in 1986 to produce different grades of fertilizer and urea/ammonia for export. NAFCON, like other government parastatals, collapsed and the objective of fertilizer production was defeated. Aggregate demand for fertilizer is 6.8 million MT while total production was estimated to be 1.25 million MT, so Nigeria is currently a net importer of fertilizers. The West African subregion also requires large quantities of fertilizers for their agricultural programmes and there is also ample opportunity to export urea/ammonia to buyers in Europe and other markets globally. Natural gas, the primary feed stock for fertilizer plant, is available in large volumes in Nigeria; there is therefore an investment opportunity in fertilizer and urea production to satisfy national, subregional and export demands. In this regard, the government is ready to support genuine investors to secure gas and construct fertilizer plants and there is an adequate infrastructure (roads, seaports etc.) and a pool of experienced personnel to operate the plants.

Other gas monetisation programmes

Beyond the NLNG plant, other gas monetisation projects are evolving. These include the following:

- The WAGP project. This is a subregional JV project being executed by NNPC, Ghana, Benin, Togo, ChevronTexaco and Shell. It is estimated to cost $500 million.
- NNPC/ChevronTexaco GTL project estimated to cost over $1 billion and was expected to be completed in 2007.
- Independent Power Producers. Under this gas monetisation scheme a number of power generation facilities are expected to be executed:
    - NNPC/Agip 450MW power station at Kwale with a price tag of $240 million was commissioned in 2005;
    - NNPC/Mobil 350MW power station at Bonny is estimated to cost $285 million.
- Brass LNG is estimated to cost $3 billion and will come on stream in 2010.
- OK-LNG is estimated to cost over $3 billion and will come on stream in 2010.
- Trans-Sahara gas pipeline to Algeria and Europe is also being vigorously pursued.
- Investment opportunities also exist in the areas of LPG extraction, compressed natural gas for automobiles, petrochemicals etc.

Funding of oil and gas projects

Between 1980 and 2005 several projects were embarked upon by the NNPC to catalyse the industrialisation process. Accordingly, within the period under reference, the following oil and gas projects were executed:

- Port Harcourt Refining Company Limited;
- Warri Refining and Petrochemical Company Limited;
- Kaduna Refining and Petrochemical Company Limited;
- Eleme Petrochemical Company Limited;
- Butanisation Project;
- Bonny Export Terminal;
- Pipeline Phases I–III linking all the products depots;
- NLNG (Trains 1–6);
- Brass LNG;
- OK-LNG.

Brass LNG, OK-LNG, WAGP and the Trans-Sahara gas pipeline are currently at various stages of execution. It is important to indicate at this juncture that the three refineries (Port Harcourt, Warri and Kaduna), the Butanisation Project, Eleme Petrochemical Company, Bonny Export Terminal and Pipeline Phases I–III were funded directly by the federal government. On the other hand, NLNG, Brass LNG and OK-LNG are JV projects between NNPC and MNCs. This dichotomy is essential in view of the sharp contrast in the funding mechanisms of these categories of projects.

Related investment opportunities

Investment opportunities also exist in seismic data acquisition, drilling operations, civil engineering works, casing and tubings manufacturing, construction of logistics boats and manufacturing. In the downstream sector opportunities exist in gas treatment, drilling, chemicals production, transportation and marketing of products. Compressed natural gas production for vehicular use, LPG and NGL liquid extraction are also potential areas of investment. Subsequent discussions will attempt to identify various financing options which are amenable to the execution of oil and gas projects. The funding options will be examined with the primary objective of determining their propensities to accommodate local entrepreneurial needs and ability to develop and achieve federal government oil and gas investment objectives.

The funding gap

Consistent with the experiences of most developing countries, Nigeria is constantly in need of development funds. The earnings from oil over the past five decades notwithstanding, investment capital is in seriously short supply. This accounts for the stagnation of oil and gas infrastructure development and indigenous participation in the highly capital intensive upstream sector. It is important to note at this point that the local content development policy is designed to grow indigenous participation in the upstream sector incrementally through deliberate assignment of certain less technical jobs (e.g. conceptual designs, front end engineering, fabrication etc.) for indigenous contractors. The envisaged growth requires reliable sources of funding. It also calls for liberalisation of loan requirements and a paradigm shift on the part of banks from short-term funding of commercial activities to high revenue yielding activities in the oil and gas sector. Banking institutions should recognise the expediency of the NCD policy and support oil and gas investments through robust syndicated loans. This will provide a suitable avenue for Nigerian banks to capture a significant proportion of the investment opportunities in the upstream sector.

Funding options

Funding of oil and gas investments is often achieved through a variety of financing options. These, among others, include bank loans, stocks, bonds, promissory notes, JV contracts, PSCs, SCs and Foreign Direct Investment (FDI).

Bank loans option

The Nigerian banking sector has registered a high growth rate in the past two decades. This success apart, the capitalisation of the banking industry is low in relation to over $12 billion funding requirements of the oil and gas sector. The federal government funds 60 per cent of this amount through cash calls while the balance of 40 per cent is funded by the JV companies. Oil and gas investments often have long maturity periods spanning about three years or more and most banks hesitate to part-finance such projects in view of the long gestation period and the perceived risk elements associated with the projects. In 2003 the federal government awarded 24 MFs to Nigerian entrepreneurs. These fields will require funding and the participation of local banks in this could usher in a new set of investment activities in the industry, but these banks have failed to provide the required funds for the development of the MFs arguing that the beneficiaries lack the necessary collateral to secure loans. Total expenditure on the NLNG project to date has exceeded $7.5 billion, but so far six Nigerian banks have syndicated $160 million which accounts for only 2.1 per cent of the total. The project is a valuable opportunity which has not been exploited by the local banks and there is need for Nigerian financial institutions to support oil and gas projects in order to stem the huge capital flight associated with the offshore financing of these investments.

Stocks option

The capital market in developed economies is a major source of funds for oil and gas projects. Although the current capitalisation of the market is estimated to be ₦12 trillion (about $70 billion), the Nigerian Stock Exchange (NSE) is yet to float the stocks of upstream sector companies. This makes it difficult for companies to raise capital locally for oil and gas projects. At present, indigenous contractors rely heavily on internal cash flow and overdraft facilities to fund oil and gas investments, and considering the high interest rates (approximately 21%) associated with overdrafts, this funding option diminishes the profitability index of oil and gas investments. The limitations of the capital market in this regard inhibit the execution of oil and gas projects, which ultimately affects economic expansion and the overall industrialisation process. The absence of the upstream sector in the stock market is in my view a critical problem and the expert attention of bankers is required in order to generate innovative solutions that will allow upstream sector companies to be quoted.

Bond issuance

The bond market in Nigeria is localised and certainly at the infant stage. Although the federal government and some State governments have floated bonds, the financing option attracts low patronage from the general public and the organised private sector. Investment capital derived from the process is low; therefore bond issuance may for some time to come not be classified as a viable funding option for oil and gas projects.

JV option

Lack of expertise, technology and investment capital among Nigerians was an underlying reason for the establishment of JV operations between the NNPC and the MNCs. Although Nigeria initiated steps to participate in the operations of MNCs as far back as 1964, actual participation of the federal government only started in 1971. JV operations play dual roles as they provide opportunities for the NNPC to participate in a highly technical and dynamic sector and also provide alternative sources of funding for oil and gas investments. The operations which require collaborative funding make it possible for the government to fund the upstream activities in the face of serious financial demands in other economic sectors. JV operations are funded through 'cash calls': the parties to a JV contribute funds in the proportion of their equity interests (e.g. 60:40 ratio) to finance capital costs and other expenses of approved operations. In practice, the operator of a particular JV project will serve notice of cash requirements to a non-operator (i.e. a partner in the JV – NNPC) 15 days prior to the first day of the calendar month in which specific costs are to be funded by the partners. The partner is obliged to honour the request from the operator. JVs provide technical assistance and capital which facilitate oil and gas investments and promote socio-economic development in the host country. Major projects such as the Oso condensate project, Agbara, Obiafor/Obrikom gas re-injection project, Nigerian Liquefied Natural Gas (NLNG) etc. were funded through JVs. They are governed by agreements which define the relationships of the stakeholders. The rights, duties and obligations of the operators are usually clearly enunciated.

PSC

A PSC is an agreement between an NOC (NNPC) and a foreign oil company. It covers a period of 10–30 years. In a typical PSC the contractor is not a co-owner of the OPL and the OML but plays the role of a contractor under the supervision of NNPC. PSC is essentially a sole risk operation in which the contractor provides all the funds for exploration, development and production. Other expenses on facilities as well as operating expenses are also borne by the contractor.

However, the contract guarantees the contractor the right to recoup all expenses incurred through ‘cost oil’. Cost oil is the amount of available crude oil allocated to the contractor to enable the company to recover operating costs as specified in the contract. The discussion in this section can be summarised by cautiously stating that although PSC is a viable funding option for oil and gas investments, the cost components of the contractors can only be taken in good faith. PSC is currently the dominant funding option by the NNPC, especially in the Deep Offshore exploration activities.

SC

A Service Contract (SC) is an agreement between an NOC and a foreign oil company. In this case the contractor accepts to carry out on behalf of the NNPC exploration and development of the oil field. The contractor does not have an automatic right to produce any oil discovered. However, the company may carry out production activities if the NNPC grants it the option to do so. A service contractor is responsible for all costs of exploration and development work, but all expenses incurred are recouped through cost oil.

Equity and syndicated loan funding

The NLNG is a typical example of a project which relied on equity contributions and syndicated loans. The project is a success story derived from the commitment of the JV partners – NNPC 49%, Shell 25.6%, Total 15.0% and ENI 10.4%. NLNG Trains 1 and 2 (Base Project), which cost \$3.6 billion, were funded through equity contribution in proportion to shareholders’ interests. The expansion project (Train 3) cost \$1.8 billion but was funded through additional equity contribution by the partners and supplemented with revenues from charter fees from NLNG vessels and surplus funds from the Base Project. Trains 4 and 5 (NLNG Plus) cost $2.1 billion and were funded as follows:

- five international banks (BNP Paris, Citigroup, Credit Lyonnais, MCC and West LB) syndicated $800 million;
- the African Development Bank financed $100 million;
- six Nigerian banks (Citibank Nigeria, First Bank, FSB, Guaranty Trust, Union Bank and UBA) financed $160 million.

Bonny Gas Transport, which is a subsidiary of the NLNG, acquired vessels at different stages of the project. Under the Train 3 project, five vessels were acquired with syndicated loans estimated to be $470 million, with Credit Suisse of Boston as the leader. Under Trains 4 and 5 four vessels were also acquired at the cost of $460 million. This facility was arranged by ABN AMBRO Bank, Credit Lyonnais, Fortis, ING Bank, HVB Verein and West Bank LB.

The NLNG project has been globally acclaimed as a success story. The funding of the project reveals the magnitude of the revenues forgone by the Nigerian banks as a result of their inability to fully participate in the programme. This classic example should not only serve as food for thought but also serve as a challenge to the banking industry. Brass LNG and OK-LNG would require funding but the question remains as to whether or not Nigerian banks will take full advantage of the loan syndication opportunities at their disposal.

Direct government funding

In an effort to provide the necessary infrastructure for rapid industrialisation, the government embarked on the construction of the three refineries (Port Harcourt, Warri and Kaduna), Eleme Petrochemical Company, Bonny Export Terminal and the Butanisation projects etc. The federal government provided Letters of Credit, promissory notes, marketing agreements, trust agreements, deferral agreements etc. as guarantees for loans used for the execution of the projects.

As can be seen from the preceding section, the conditions for government guaranteed loans are quite elaborate. This notwithstanding, government direct intervention brought on stream major oil and gas projects which have contributed significantly to achieving national development objectives. Bonny Export Terminal, Pipelines Phase III and the Butanisation project were funded by the federal government under the special projects scheme.

The funding patterns of JV projects as well as the government projects have been discussed earlier. In the case of the NLNG JV all projects were completed on schedule due to the fluency of the funding mechanism.

However, the government projects lagged behind primarily as a result of delayed payments to contractors. This situation led to cost escalations arising from time extensions and increase in the exchange rate. In general JV projects enjoy better coordination and performed better both in terms of cost containment and meeting completion schedules.

Appendices

Appendix 1

Procedure on the calculation of technical cost per barrel with reference to the memorandum on incentives for encouraging investments in exploration and development activities and enhancing crude oil exports between the government of the Federal Republic of Nigeria, NNPC and the company

1 For the purpose of clauses 2.7 and 2.8 of the Memorandum, T1 is to be understood as the Operating Committee for the Joint Venture for any given calendar year under the following NNPC budget codes divided by the Joint Venture crude oil and condensate production for that year

NNPC Budget Code	Description
4000	Direct lifting costs
4001	Direct handling/ transportation costs
4002	Other direct production costs
4003	Fields overheads
4011	Environmental protection
4012	Safety
4022	Personnel and personnel amenities
4023	Material handling/ storage expense
4025	Other general and administrative expense
4029(*)	Fees received for services ancillary to petroleum operations

(*)Credit

2 The following expenses and costs reported in the corporate account for

the Joint Venture only during the period January/December of the year of lifting shall be included in calculating the actual production cost per barrel for the purpose of the transfer cost referred to in Clauses 3.1.3, 3.1.4 of the Memorandum.

2.1 Production Operating Expenses (T1)

- i Direct Production Expenses as per items 400 and 401 of Report No. 002 of Account Reporting Manual.
- ii Portion of Administrative and General Expenses allocated to production – refer to item 402 of Report No. 002 of Account Reporting Manual.
- iii Custom Duties and Gross Rentals allocated to Production – refer to items 4043 and 4045 of Report No. 002 of Account Reporting Manual.
- iv Extraordinary/prior year expenses/incomes – refer to item 405 of Report No. 002 of Account Reporting Manual.

2.2 Capital investment costs which qualify for treating as expenses for PPT calculation and chargeable to production costs (T2)

- i Exploration costs.
- ii Appraisal drilling cost (1st and 2nd wells)
- iii Intangible drilling and development costs
- iv Capital allowances – for the Joint Venture only, restricted to the capital allowances applicable direct to production. The tax allowance should be reconciled with the allowances claimed for the year under Section 15 of the PPT Act.

2.3 It is expected that the production operating expenses and capital investment costs above will be reconciled by the Company with reports Nos. 002 and 001 of the Accounts Reporting Manual respectively.

2.4 The recommended basis for allocation of expenses and costs in 2.1 ii, 2.1 iii and 2.1 iv above to production shall be

$$\frac{P}{E + P + X}$$

Where

P = The additions to the capital costs during the year and the operating expenses for the year reported in the corporate accounts of the Joint Venture. The expenses/costs must be directly related to the production function.

E = The additions to the capital costs during the year and the operating expenses for the year reported in the corporate accounts of the Joint Venture. The expenses/costs must be related to the exploration function.

X = Any function/company other than the production and exploration function which should share the expenses being allocated.

2.5 Production cost per barrel $= \dfrac{(a)(i)to(iv) + (b)(i)to(iv)}{\text{Annual production(barrels)}}$

2.6 The advice on cost per barrel forwarded to NNPC should be supported with a working paper.

Appendix 2

Worked examples for establishing the guaranteed notional margin for R.P. <$15/bbl – Clauses 2.5 & 2.6

	FOR T2 < \$2.00	*FOR T2 ≥ \$2.00*
1.	If RP = \$5 RP = \$5 a1 = 0.3 and FC = \$4.0/bbl Margin M = $\left(1-\frac{4.0}{5}\right)\times(5\times0.3)$ M = \$0.3	If RP = \$5 RP = \$5 a1 = 0.365 and FC = \$4.0/bbl Margin M = $\left(1-\frac{4.0}{5}\right)\times(5\times0.365)$ M = \$0.365
2.	If RP = \$9 RP1 = \$5 a1 = 0.3 RP2 = \$4 a2 = 0.285 Margin M = $\left(1-\frac{4.0}{9}\right)\times(5\times0.3+4\times0.285)$ M = \$1.467	If RP = \$9 RP1 = \$5 a1 = 0.365 RP2 = \$4 a2 = 0.28833 Margin M = $\left(1-\frac{4.0}{9}\right)\times(5\times0.365+4\times0.28833)$ M = \$1.655
3.	If RP = \$12 RP1 = \$5 a1 = 0.3 RP2 = \$5 a2 = 0.285 RP3 = \$2 a3 = 0.10744 FC = \$4.0/bbl Margin M = $\left(1-\frac{4.0}{12}\right)\times(5\times0.3+5\times0.285$ $+\,2.0\times.10744)$ M = \$2.093	If RP = \$12 RP1 = \$5 a1 = 0.365 RP2 = \$5 a2 = 0.28833 RP3 = \$2 a3 = 0.08286 FC = \$4.0/bbl Margin M = $\left(1-\frac{4.0}{12}\right)\times(5\times0.365+5\times0.28833$ $+\,2\times.08286)$ M = \$2.288

4. If RP = $14
   a = 0.116
   M15 = 2.5
   Margin M = 2.5 + (14 – 15) × 0.1160
   M = $2.384

   If RP = $14
   a = 0.1315
   M15 = 2.7
   Margin M = 2.7 + (14 – 15) × 0.1315
   M = $2.569

For RP > $19

RP	*FOR T2 < $2.00*	*FOR T2 >$2.00*
General Formula	$M = 2.5 + (RP - 19) \times 0.116$	$M = 2.7 + (RP - 19) \times 0.1315$
20	$M = 2.5 + (20 - 19) \times 0.116$ = $2.616	$M = 2.7 + (20 - 19) \times 0.1315$ = $2.832
22	$M = 2.50 + (22 - 19) \times 0.116$ = $2.848	$M = 2.7 + (22 - 19) \times 0.1315$ = $3.095
24	$M = 2.5 + (24 - 19) \times 0.116$ = $3.080	$M = 2.7 + (24 - 19) \times 0.1315$ = $3.358
26	$M = 2.5 + (26 - 19) \times 0.116$ = $3.312	$M = 2.7 + (26 - 19) \times 0.1315$ = $3.621

Appendix 2a

Calculation of realisable price

The market weighted FOB Nigeria Netback Value (NBV) for the purpose of calculating realisable price as defined in Clauses 2.4, 2.11, 2.12 and 2.13 of the Memorandum made the .... Day of .... 2000 shall be calculated from the data given below and in accordance with the worked example shown in Appendix 2c.

1 Delivery	Each export grade of crude oil will be deemed delivered to	
	U.S. Gulf Coast	60 per cent
	North West Europe (NEW)	20 per cent
	Mediterranean (MED)	20 per cent
2 Grades	The export grades of crude oil are (all standard export Gravity)	
	Bonny Light	37° API
	Forcados Blend	31° ,,
	Bonny Medium	26° ,,
	Qua Iboe Light	37° ,,
	Escravos Light	36° ,,
	Brass Blend	43° ,,
	Pennington Light	36° ,,
	Commingled Antan	32° ,,
3 Conversion Factors		
	Bonny Light	7.5060 barrels per metric tonne
	Forcados Blend	7.2396 ,, ,, ,,
	Bonny Medium	7.0160 ,, ,, ,,
	Qua Iboe Light	7.5060 ,, ,, ,,
	Escravos Light	7.4625 ,, ,, ,,
	Brass Blend	7.7741 ,, ,, ,,
	Pennington Light	7.4625 ,, ,, ,,
	Commingled Antan	7.2844 ,, ,, ,,

4 Yields The following refinery yields will be applied to each geographical area unless amended under the terms of Clause 2.11 of the Memorandum.

U.S. Gulf Coast (expressed as Volume)
(Same yields apply Summer and Winter)

	Bonny Light % Vol.	**Forcados Blend** % Vol.	**Bonny Medium** % Vol.
LPG Propane	2.30	2.40	2.50
LPG Normal Butane	2.30	2.40	2.50
Gasoline Regular	17.80	16.42	14.25
Gasoline Unleaded	17.80	16.43	14.25
Naphtha	12.30	8.30	5.20
Jet Kero	12.80	10.50	8.50
No. 2 Oil	22.40	29.55	36.70
Max. 1% S Fuel Oil	12.30	14.00	16.10
Refy Fuel Loss	—	—	—
TOTAL	**100.00**	**100.00**	**100.00**

NORTH WEST EUROPE (NEW) AND MEDITERRANEAN (MED)
(Expressed as weight %)

	Summer % wt.	**Bonny Light** Winter % wt.
Gasoline		
Premium	24.50	20.00
Regular	8.60	8.50
Jet Kerosene	10.00	8.50
Gasoil	23.10	34.50
Fuel Oil 1%	28.80	23.50
Refy Fuel/Loss	5.00	5.00
TOTAL	**100.00**	**100.00**
		Forcados Blend
Gasoline		
Premium	19.00	15.40
Regular	7.50	5.80
Jet Kerosene	8.80	8.80
Gasoil	29.00	36.30
Fuel Oil 1%	30.70	28.70
Refy Fuel/Loss	5.00	5.00
TOTAL	**100.00**	**100.00**

Bonny Medium

Gasoline		
Premium	8.70	7.00
Regular	3.60	3.00
Jet Kerosene	8.00	6.00
Gasoil	27.60	30.50
Fuel Oil 1%	47.50	48.90
Refy Fuel/Loss	4.60	4.60
TOTAL	**100.00**	**100.00**

Summer yields are to be used for the calculation of all prices for the months of April, May, June, July, August and September.
Winter yields will apply for the months of October, November, December, January, February and March.

5 Processing Fees

U.S. Gulf	–	$1.90 per barrel
NWE	–	$1.40 ,, ,,
MED	–	$1.30 ,, ,,

6 Valuation of Refined Products

Reference to the Platt's Oilgram Price Report quotations outlined in Attachment la to Appendix A will be made to value the refinery yields in accordance with Clause 2.12 of the Memorandum. In each case the average of the mid-range product prices for each quotation day for the period 1st to 20th day (inclusive) of the month in question prior to 31 March 2000 and for the period 1st to the last day of the month in question after 31 March 2000 will be used.

7 Freight

US Gulf Coast	–	LR 2 for Bonny Light and Forcados, one port loading and one port discharge.
	–	LR 1 for Bonny Medium, one port loading one port discharge.
NWE and MED	–	25% VLCC plus 75% LR 2 for Bonny Light and Forcados, one port loading and one port discharge
	–	LR 1 for Bonny Medium, one port loading and one port discharge.

Freight rates for the various ship sizes will be based on monthly assessments obtained from the London Tanker Brokers Panel.

8 Insurance and Outturn Loss

The following will be allowable deductions in the calculation of the realisable price.

Insurance	$0.03 per barrel
Outturn Loss	$0.05 per barrel

9 Method used in Calculating NBV

See Appendix 2C

10 Price Differentials between Bonny Light and Other Nigerian Light Crude Oil Grades

Bonny Light will be used as the reference for all other Nigerian crude oils except Forcados and Bonny Medium. Forcados is separately quoted in Platt's Oilgram Price Report while Bonny Medium will be taken as Bonny Light less $0.75 per barrel. The differential between Bonny Light and other Nigerian Light grades shall be as follows:

	Price Range 0<RP<$20/bbl	*Price Range* $20<RP<$25/bbl	*Price Range* $25<RP
Brass	6 cents	6 cents	6 cents
Escravos	–8.5 cents	–11.0 cents	–13.5 cents
Pennington	13 cents	13 cents	13 cents
Qua Iboe (QIT)	3 cents	7.5 cents	10 cents

11 Formula for Realisable Price

$$\text{Bonny Light (BL)} = \frac{\text{Bonny Light} - \$0.25/\text{bbl} + \text{NBV}}{2}$$

$$\text{Forcados Blend (FB)} = \frac{\text{FB} - \$0.25/\text{bbl} + \text{NBV}}{2}$$

$$\text{Bonny Medium (BM)} = \frac{\text{Bonny Light} - \$1.00/\text{bbl} + \text{NBV}}{2}$$

With NBV limited to a $0.40 tunnel around Bonny Light, Forcados and Bonny Medium.

Appendix 2b

Quotations used in realisable price calculations

Market	Product	Quotation	Source
USG	LPG	Propane	Gas Liquids – Mont Belvieu
	LPG	Normal Butane	Gas Liquids – Mont Belvieu
	Butane	Unl. 87	US Gulf Coast – Waterborne
	Gasoline Regular	Unl. 87	US Gulf Coast – Waterborne
	Gasoline Unleaded	Naphtha	US Gulf Coast – Waterborne
	Naphtha	Jet/Kero 54	US Gulf Coast – Waterborne
	Jet Kero	No. 2	US Gulf Coast – Waterborne
	No. 2 Oil	No. 6 1.0%S10.5 API	US Gulf Coast – Waterborne
	Max 1.0% S Fuel Oil		
NWE	Gasoline Premium	Prem Unl × 1.0225	Cargoes CIF NWE Basis ARA
	Gasoline Regular	Reg Unl × 0.925	Cargoes CIF NWE Basis ARA
	Jet Kerosene	Jet Kerosene	Cargoes CIF NWE Basis ARA
	Gasoil	Gasoil 0.2 × 0.996	Cargoes CIF NWE Basis ARA
	Fuel Oil 1.0%	1% Fuel Oil	Cargoes CIF NWE Basis ARA

Market	Product	Quotation	Source
MED	Gasoline Premium	Prem 0.15G/L × 0.98	Cargoes CIF Med Basis
	Gasoline Regular	Prem 0.15G/L × 0.903	Genoa/Lavera
	Jet Kerosene	Jet Kerosene	Cargoes CIF Med Basis
	Gasoil	Gasoil 0.2 × 0.996	Genoa/Lavera
	Fuel Oil 1.0%	1% Fuel Oil	Cargoes CIF Med Basis

For Mediterranean, when cargoes CIF are not quoted, use the FOB quotation.

Appendix 2c

Worked example of netback calculation to determine realisable price for Bonny Light Crude

Note

All figures calculated are rounded to 4 decimal places, that is rounding down if the 5th decimal place is 4 or less, otherwise rounding up.

Realisable Prices are based on the standard export gravities as per paragraph 2 of this Appendix. The final calculated realisable prices will be adjusted to take account of API variations as follows:

For each 0.1° API difference above or below the reference gravity, an adjustment of $0.003 per barrel will be added to or subtracted from the calculated realisable price.

<u>Section 1 – Calculation of NBV per Market</u>

(A) *U.S. Gulf Coast (USGC)*

	Winter Yields (Volume) Cents/Gallon		$/bbl. %
LPG Propane	29.8846 × .42 = 12.5515 × 2.30	=	0.2887
LPG Normal Butane	43.0962 × .42 = 18.1004 × 2.30	=	0.4163
Gasoline Unleaded	60.1827 × .42 = 25.2767 × 17.80	=	4.4993
Gasoline Unleaded	60.1827 × .42 = 25.2767 × 17.80	=	4.4993
Naphtha	60.8173 × .42 = 25.5433 × 12.30	=	3.1418
Jet Kerosene	66.7019 × .42 = 28.0148 × 12.80	=	3.5859
No. 2 H.O.	66.4135 × .42 = 27.8937 × 22.40	=	6.2482
Max 1.0% S.F.O.	($/bbl) = 11.5962 × 12.30	=	1.4263
	Gross Product Worth (GPW)		24.1058
	Conversion Factor		7.5060

From the above, following deductions to apply:

Processing Fee					1.900
	Flat Rate ($/MT)		LR 2 (WS Points)	Conversion Factor	
	10.62	×	125.8%	7.5060	1.7799
Outturn Loss					0.0500
Insurance					0.0300
	(a)		Netback – USGC		20.3459

(B) North West Europe (NWE)

		$/MT		Winter Yields (Weight) %		$/MT
Gasoline Premium	=	242.6923	×	20.00	=	48.5385
Gasoline Regular	=	210.3664	×	8.50	=	17.8811
Jet Kerosene	=	308.6538	×	8.50	=	26.2356
Gasoil	=	287.6346	×	34.50	=	99.2339
Fuel Oil 1%	=	93.0000	×	23.50	=	21.8550
Refinery Fuel and Loss	=			5.00		
		Gross Product Worth (GPW) –USD/MT			=	213.7441
		Conversion Factor 7.5060				
		Gross Product Worth (GPW) –USD/BBL			=	28.4764

From the above, following deductions to apply:

	Flat Rate ($/MT)		Vessel Class (WS points)	Conversion factor	
Processing Fee					1.4000
Freight					
LR2 – 75%	8.79	×	130.4%	7.5060	1.1453
VLCC – 25%	8.79	×	90.0%	7.5060	0.2635
Outturn Loss					0.0500
Insurance					0.0300
		(b) Netback – NWE			25.5876

(C) Mediterranean (MED)

USD/MT		USD/MT		Winter Yields (Weight) %		
Gasoline Premium	=	238.0000	×	20.00	=	47.6000
Gasoline Regular	=	219.1980	×	8.50	=	18.6318
Jet Kerosene	=	307.5385	×	8.50	=	26.1408
Gasoil	=	294.3846	×	34.50	=	01.5627
Fuel Oil 1%	=	98.3077	×	23.50	=	23.1023
Refinery Fuel and Loss	=			5.00		
		Gross Product Worth (GPW) –USD/MT			=	217.0376
		Conversion Factor 7.5060				
		Gross Product Worth (GPW) –USD/BBL			=	28.9152

From the above, following deductions to apply:

	Flat Rate ($/MT)		Vessel Class (WS points)	Conversion factor	
Processing Fee					1.3000
Freight					
LR2 – 75%	8.16	×	131.0%	7.5060	1.0681
VLCC – 25%	8.16	×	85.0%	7.5060	0.2310
Outturn Loss					0.0500
Insurance					0.0300
		(b) Netback – MED			26.2361

<u>Section 2 – Calculation of NBV</u>

	Market Contribution	Weighting		Netback	
(a)	USG	60%	×	20.3459	12.2075

(b)	NEW	20%	×	25.5876	5.1175
(c)	MED	20%	×	26.2361	5.2472
(d)				Initial NBV	22.5722

Section 3 – Calculation of Final NBV

	A.	Initial NBV		22.5722
	B.	Bonny Light Average		20 8712
	F1.	Greater of (B – 40c) and A	22.5722	
	F2.	Lesser of (B + 40c) and F1	21.2712	
Final NBV			21.2712	

NOTE:

NBV shall be adjusted only if its value is greater/lower than Bonny Light by more than 40 cents.

Section 4 – Calculation of Realisable Price

Crude Element	Bonny Light			20.8712
	Less 25c/bbl differential		0.2500	
		(i)	20.6212	
Product Element Final NBV		(ii)	21.2712	
	Average of (i) and (ii)		20.9462	
	Realisable Price for Bonny Light		20.9462	

Appendix 3

Procedure on tax inversion penalty adjustment

For the purposes of Clauses 2.7 and 2.7.2 of this Memorandum, where the impact of crude oil or condensate production lost as a result of circumstances outside the Company's control, causes the Company's T1 (as defined in Paragraph 1 of Appendix 1 to the Memorandum) to exceed the UTTT, such impact shall be excluded for the purposes of calculating the tax inversion penalty described in Clause 2.7.2, as provided herein below. The Company shall qualify for this protection only if the cumulative effect of the circumstances and situations referred to herein causes the Company's UTIT to be exceeded by more than five per cent per barrel.

Circumstances outside the Company's control, which result in the Company's production loss and which may necessitate an adjustment as envisaged above, shall arise from:

i Effects of quota restrictions which restrict the Company's production;
ii Production interruptions caused by sabotage and/or community disruptions.

If the Company's production for the relevant budget year is reduced as a result of circumstances which are outside the Company's control, then to the extent that the cumulative effect of such circumstances result in the escalation of T1 cost per barrel (EC), the UTIT shall be increased by the EC. The NNPC shall advise the Company of its minimum annual production volume in barrels (PM) for the relevant year based on Nigeria's anticipated OPEC quota for that year. This advice shall be issued by NNPC before 31 July of the preceding budget year. In the event that no advice is received from NNPC by this date then the previous year's PM shall apply. To the extent that NNPC advises the Company to reduce production below PM during the relevant year, the Company's production shall be deemed quota restricted.

Any such situation or circumstance shall be logged by the Company after its discovery and the Company shall notify NAPIMS on behalf of NNPC in writing of an occurrence of such situation within one week of the date of discovery by the Company, and shall facilitate NNPC a visit to the site at NNPCs earliest convenience. NNPC shall verify such occurrence within two months of receipt of the notice and shall thereafter convey its objection or acceptance in writing. If no objection is received within this period then such occurrence shall be deemed verified. The cumulative effect of such situations and circumstances on T1, as defined in paragraph 1 of Appendix 1 to the Memorandum, for the applicable budget year shall be determined by the Company and NNPC within two weeks of approval of the budget and programme performance of Company for the relevant year by the Operating Committee of the NNPC/Shell/Elf/NAOC Joint Venture. If such determination is not made within the said two weeks, the matter shall be referred to arbitration in accordance with Clause 8 of this Memorandum.

Appendix 4

Achievements of the Nigerian oil and gas industry

1908: Nigerian Bitumen Co. & British Colonial Petroleum commenced operations around Okitipupa.
1938: Shell D'Arcy (from 1979 called SPDC) granted exploration licence to prospect for oil throughout Nigeria.
1955: Mobil Oil Corporation started operations in Nigeria.
1956: First successful well drilled at Oloibiri by Shell D'Arcy.
1956: Changes name to Shell-BP Petroleum Development Company of Nigeria Limited.
1958: First shipment of oil from Nigeria.
1961: Shell's Bonny Terminal was commissioned.
1961: Texaco Overseas started operations in Nigeria.
1961: Gulf Oil Company started operations in Nigeria.
1962: Elf started operations in Nigeria (as Safrap).
1962: Nigeria Agip Oil Company started operations in Nigeria.
1963: Elf discovered Obagi field and Ubeta gas field.
1963: Gulf Oil discovered their first oil at Okan-1.
1965: Agip found its first oil at Ebocha.
1965: Gulf's first production.
1965: Phillips Oil Company started operations in Bendel State.
1966: Elf started production in Rivers State with 12,000b/d.
1967: Phillips drilled its first well (dry) at Osari-1.
1967: Phillips' first oil discovery at Gilli-Gilli-1.
1968: Mobil Producing Nigeria Limited (which changed its name to Unlimited in 1991) was formed.
1968: Gulf's terminal at Escravos was commissioned.
1970: Mobil started production from four wells at Idoho field.
1970: Agip started production.
1970: government started its Department of Petroleum Resources Inspectorate (DPR).
1971: Shell's Forcados terminal commissioned.
1971: Mobil's terminal at Qua-Iboe commissioned.
1973: First participation agreement; federal government acquires 35% shares in the companies.

1973: Ashland started PSC with the then NNOC (NNPC).
1973: Pan Ocean Corporation drilled its first discovery well at Ogharefe-1.
1974: Second participation agreement; federal government increases equity to 55%.
1974: Elf formally changed its name from 'Safrap'.
1974: Ashland's first oil discovery at Ossu-1.
1975: First oil lifting from Brass Terminal by Agip.
1975: DPR upgraded to Ministry of Petroleum Resources (MPR).
1976: MPE renamed Ministry of Petroleum Resources (MPR).
1976: Pan Ocean commenced production via Shell-BP's pipeline at a rate of 10,800b/d.
1977: Government established Nigeria National Petroleum Corporation (NNPC) by Decree 33 (NNOC & MPR extinguished).
1977: Government established Nigeria National Petroleum Corporation (NNPC) by Decree 33 (NNOC & MPR extinguished).
1979: Third participation agreement (through NNPC) increases equity to 60%.
1979: Fourth participation agreement; BP's shareholding nationalised, leaving NNPC with 80% equity and Shell 20% in the Joint Venture.
1979: Changed name from NNPC to Shell Petroleum Development Company of Nigeria (SPDC).
1984: Agreement consolidation NNPC/Shell Joint Venture.
1986: Signing of Memorandum of Understanding (MOU).
1989: Fifth participation agreement (NNPC=60%, Shell=30%, Elf=5%, Agip=5%).
1991: Signing of revised Memorandum of Understanding and Joint Venture Operating Agreement (JOA).
1993: Production Sharing Contracts Signed (SNEPCO).
1993: Sixth participation agreement (NNPC=55%, Shell=30%, Elf=10%, Agip=5%).
1993: The coming onstream of Elf's Odudu blend, offshore OML100.
1995: SNEPCO started drilling first exploration well.
1995: NLNG's final investment decision taken.
1999: NNLG's first shipment of gas out of Bonny terminal.
2000: NPDC/NAOC service contract signed.
2001: Production of Okono offshore field.
2002: New PSCs agreement signed.
Liberalisation of the downstream sector.
NNPC commenced retail outlet scheme.
2004: Project PACE (NNPC's transformation initiative) was commenced.

Appendix 5

Hydrocarbon activities

Primary sedimentary basins and associated hydrocarbon activities

Main basin	*Size and prospectivity*	*Main reservoirs Hydrocarbon habitat*	*Status*	*Operator*	*Remarks*
1 Anambra (cretaceous)	Oil and condensate; 1billion SIB; Gas over 1 trillion TCF	Ajali sandstones	Proven oil and 705 bill. cu. ft. of gas from two discovery wells; 18 wells drilled to date	Elf, Shell	High geothermal gradient. 19 Blocks
2 Benin (Dahomey) (cretaceous)	Proven hydrocarbon 1 million barrels of oil and 1 trillion TCF of gas	Mainly stratigraic with minor structural/fault control	Interest/scramble by many operators. One discovery well in 1996	Yinka Folawiyo (indigenous operator) backed by Abacan Resources of Canada	Extends across the Nigeria–Benin border. Seme field 18 Blocks
Benue (cretaceous)	60,000 km^2. Estimated hydrocarbon volume: 500 million barrels	Asu river group	Relatively unexplored. Over 3,000 km of 2D seismic. Shell's well Kalmani-1 spudded mid-March, 1999	Chevron, Elf and Shell commenced exploration in the Gongola sub-basin	Ongoing exploration activity – 29 Blocks
Bida (cretaceous)	20,000 km^2		Relatively unexplored although large acreage holdings have been established		18 Blocks

Main basin	*Size and prospectivity*	*Main reservoirs Hydrocarbon habitat*	*Status*	*Operator*	*Remarks*
Chad (cretaceous)	35,000 km^2	Tight reservoirs	Active exploration over 7,000 km of 2D seismic data, and 24 wells with marginal show in well Wadi-1	NNPCs Frontier Exploration Services (FES)	Improved seismic techniques required for further basic evaluation of 48 Blocks
Niger Delta (including the relatively new concession in the Nigerian Deep Waters) Tertiary	75,000 km^2 of sedimentary basin fill and over 12 km thickness in its central part. Recoverable hydrocarbon over 20 billion barrels of oil and 120 trillion TCF of gas	The paralic Agbada formation which lies between the base continental and the source rock Akata formation	Mature hydrocarbon province with vast potential still	57 indigenous and all major MNCs including Shell, Mobil, Elf, Texaco and Agip	One of the most prolific hydrocarbon provinces in the world. Five major deposits with a few major sub-divisions – 164 Blocks
Sokoto (cretaceous)	30,000 km^2	Shallow basin sedimentary thickness between 500 and 1,500 metres	Exploration activities by Elf and Mobil. Over 1200 km of 2D seismic acquired by Elf		

Source: NNPC Profile Publication – World Petroleum Congress 2005

Appendix 6

Global oil spill

Global level: Number of oil spills over 7 mt

Year	*7–700 MT*	*>700 MT*	*Total quantity spilled (MT)*
1970	6	29	330,000
1971	18	14	138,000
1972	48	27	297,000
1973	27	32	164,000
1974	89	28	175,000
1975	95	22	357,000
1976	67	26	364,000
1977	68	17	291,000
1978	58	23	386,000
1979	60	34	640,000
1980	52	13	206,000
1981	54	7	48,000
1982	45	4	12,000
1983	52	13	384,000
1984	25	8	28,000
1985	31	8	85,000
1986	27	7	19,000
1987	27	10	30,000
1988	11	10	190,000
1989	32	13	174,000
1990	51	14	61,000
1991	29	7	430,000
1992	31	10	172,000
1993	31	11	139,000
1994	26	9	130,000
1995	20	3	12,000
1996	20	3	80,000
1997	28	10	72,000
1998	25	5	13,000
1999	19	6	31,000
2000	19	4	14,000
2001	16	3	8,000
2002	12	3	67,000
2003	15	4	42,000
2004	16	5	15,000
2005	21	3	17,000

Source: ITOPF Statistical Data (www.itopf.com/stats)

Appendix 7

Oil industry conversions*

Crude petroleum, and the refined products made from crude oil, are normally measured either by volume in gallons and US barrels, or by weight in tons or tonnes. The relationship between volume and weight is usually measured by density in the United Kingdom (the alternative measure is relative density or specific gravity). American oilmen usually reckon quantities of oil produced, moved or processed in barrels per day (b/d or bpd). The loose but simple rule of thumb for conversion is that a barrel a day is roughly 50 tonnes a year, but the relationship varies according to density and so according to product.

Conversion factors for oil

	At 86°F (30°C) Approximation			
Product	*A. Barrels per E. Ton*	*Litres per M. Ton*	*Litres per M. Ton*	*A. Gallons per M. Ton*
L.P.G	1,864	1,835	484.6	11.54
JP.4	1,355	1,333	352.4	8.39
Jet A-1	1,274	1,254	331.2	7.89
Premium	1,375	1,353	357.5	8.51
Regular	1,440	1,418	374.5	8.92
Kerosene	1,293	1,273	336.2	8.00
Gas Oil	1,197	1,177	311.2	7.41
Diesel Fuel	1,177	1,159	306.1	7.29
Fuel oil 80 CST	1,082	1,065	281.2	6.70
Fuel oil 180 CST	1,067	1,050	277.4	6.60
Fuel oil 230 CST	1,064	1,047	276.6	6.59
Fuel oil 280 CST	1,061	1,044	275.9	6.57
Bitumen	994	979	258.5	6.15

* These conversions are based on the assumption that all weights are weights in air, which is the correct basis for computing bulk commercial quantities of petroleum.

1 MMSCF of natural gas	= 172.3 barrels of crude oil equivalent
	= 365 × 1,000,000 scf
1 million cu.ft. of natural gas	= 18.91 tons liquid
	= 1,598.69 cu.ft. liquid
1 std.cu.ft. of natural gas	= 1000 BTU = 252 kilocalories
1 m.ton of coal	= 4.879 barrels of crude oil equivalent
1 m.ton of lignite	= 2.053 barrels of crude oil equivalent
1 ltr of fuel oil 1500 sec	= 38.9 cubic feet of natural gas
1 kg of LPG	= 47.0 cubic feet of natural gas
1 normal cu.m. per day (Nm^3/d) [flow rate of gas]	= 37.33 standard cu.ft. per day (SCFD)
1 ton of LNG (LNG conversions)	= 1.14 1.4 × 10^3 normal cu.m. natural gas (Nm^3)
	= 52.3 × 10^3 standard cubic feet natural gas (SCF)
	= 55.0 × 10^9 joules (HHV)
1 ton of LNG (energy equivalents)	= 1.22 tonne crude oil
	= 0.80 tonne heavy fuel oil
	= 0.91 tonne LPG (commercial composition)
	= 1.91 tonne coal
1 barrel per day (b/d)	= 50 tonnes per year (approx.)
1 barrel of oil equivalent	= 1 barrel of crude oil
	= 5,487 cubic feet of gas

Natural gas is converted to barrels of oil equivalent using a ratio of 5,487 cubic feet of natural gas per one barrel of crude oil. This ratio is based on the actual average equivalent energy content of Total's natural gas reserves.

1 barrel of crude oil per day	= appr. 50 tons of crude oil per year
1 ton of crude oil	= 1 metric ton of crude oil
	= appr. 7.3 barrels of crude oil (assuming a spec.gravity of 33 API)
	= 6.6–8.0 bbl. of crude oil with 7.333 bbl taken as average
	= 1.16 kl. of crude oil (average)
1 ton of oil equivalent gas	= appr. 1,125 cubic metres of natural gas

MMSCF	Million standard cubic feet
CST	Centi-Stroke
A.	American
E.	English

VOLUME	1 American barrel	= 158.984 litres
		= 42 American (US) gallons
		= 34.9726 Imperial (UK) gallons
		= 5.6146 cubic feet
		= 0.15899 cubic metre
		= 3.78541 cubic decimetres(dm^3)
		= 0.136 tonne (approx)
	1 American gallon	= 231 cubic inches
		= 0.133681 cubic foot
		= 3.7854 litres
		= 0.0238095 American barrel
		= 0.83268 Imperial gallon
		= 0.0037854 cubic metre

	1 Imperial gallon	= 277.42 cubic inches
		= 1.20094 American gallons
		= 4.54609 litres
		= 0.028594 American barrel
		= 0.160544 cubic foot
		= 0.004561 cubic metre
	1 litre	= 1,000 cubic centimetres
		= 1 cubic decimetre (dm^3)
		= 0.035314 cubic foot
		= 61.024 cubic inches
		= 1.7597 pints
		= 0.219969 Imperial gallon
		= 0.26417 American gallon
	1 kilolitre	= 1,000 cubic decimetres (dm^3)
		= 6.28981 American barrels
	1 gross ton	= 100 cubic feet or 2.83 cubic metres of permanently enclosed space (shipping)
	1 cubic metre	= 35.315 cubic feet
		= 219.97 Imperial gallons
		= 1,000 litres
		= 6.2898 American barrels
		= 264.17 American gallons
WEIGHT	1 oz	= 28.35 grams
	1 lb	= 0.453592 kilograms
		= 0.009 cwt
	1 kg	= 2.20462 lb
		= 0.01 quintal
	1 cwt	= 112 lb
		= 50.802 kg
	1 metric ton	= 0.98421 long ton or English ton
		= 1.10231 short ton
		= 2,204.6 lb
	1 English ton	= 1.01605 metric tons
	or long ton	= 1.12 short tons
	1 short ton	= 0.907186 metric ton
		= 0.892857 long ton
		= 2,000 lb
PRICE	1 US cent/AG	= 0.42 US$/BBL
	1 US$/BBL	= 0.163 Baht/Litres (Based on exchange rate of 25.9 Baht/US$)

PETROLEUM SPECIFIC GRAVITY VOLUME PER TON

Degree API	*Specific gravity*	*Barrels* per metric tonne*	*Long tonne*
25	0.904	6.98	7.09
26	0.898	7.02	7.13
27	0.893	7.06	7.18
28	0.887	7.10	7.22
29	0.882	7.15	7.27
30	0.876	7.19	7.31
31	0.871	7.24	7.36
32	0.865	7.28	7.40
33	0.860	7.33	7.45
34	0.855	7.37	7.49
35	0.850	7.42	7.54
36	0.845	7.46	7.58
37	0.840	7.51	7.63
38	0.835	7.55	7.67
39	0.830	7.60	7.72
40	0.825	7.64	7.76
41	0.820	7.69	7.81
42	0.816	7.73	7.85

* Approx. figures 60°F

$$\text{Specific gravity} = \frac{141.5}{131.5 + {}^\circ\text{API}}$$

Power and heat units

1 HP (UK Horsepower)	= 550 foot pounds per second
	= 0.7457 kilowatt
	= 1.014 PS (or Cheval Vapeur)
1 PS (Pferdestaerke)	= 542 foot pounds per second
or CV (Cheval Vapeur)	= 0.986 hp
	= 0.736 kW
1 kW (Kilowatt)	= 1,000 watts
	= 1,340 hp
	= 1.359 PS or CV
	= 737 foot pounds per second
1 foot pound per second	= 0.00136 kilowatt
	= 0.00182 hp
	= 0.00184 PS or CV
1,000 Btu	= 0.293 kWh
100,000 Btu	= 1 therm
1 calorie (dieticians')	= 4.1855 kilojoules
1 Therm	= 100,000 BTU (British Thermal Units)
	= 25,200 kilocalories
	= 25.2 thermies
	= 29.3 kilowatt hours

1000 kilocalories (Large Calories)	= 3,968 Btu
	= 1.163 kilowatt hours
	= 1 thermie
1 kilowatt hour	= 3,411 Btu
	= 1.340 hp hours
	= 859.6 kilocalories

Liquefied methane

1 Ton of Liquefied Methane	= approximately 16 barrels
	= approximately 50,000 cubic feet (1,400 cubic metres) of natural gas, depending on methane content

Product specific gravity ranges

	Spec. Gravity	*Barrels per metric ton*
Crude oils	0.80–0.97	8.0–6.6
Aviation gasolines	0.70–0.78	9.1–8.2
Motor gasolines	0.71–0.79	9.0–8.1
Kerosines	0.78–0.84	8.2–7.6
Gas oils	0.82–0.90	7.8–7.1
Diesel oils	0.82–0.92	7.8–6.9
Lubricating oils	0.85–0.95	7.5–6.7
Fuel oils	0.92–0.99	6.9–6.5
Asphaltic bitumens	1.00–1.10	6.4–5.8

Calorific value of fuels

Rough gross values in Btu per lb

Crude oils	18, 300–19,500
Gasolines	20,500
Kerosines	19,800
Benzole	18,100
Ethyl alcohol	11,600
Gas oils	19,200
Fuel oils (bunker)	18,300
Coal (bituminous)	10,200–14,600
LNG	22,300
LPG (butane)	21,300

TYPICAL PROPERTIES OF EXPORT STREAMS

	MIRI	*Bintulu*	*Labuan*
API Gravity	33°	35°	32°
Pour Point	45°F	15°F	45°F
Wt. % Sulphur	0.07	0.1	0.07

Pressure

1 Bar	= 0.987 atmospheres
	= 750.1 mm Hg
	= 14.50 pound/sq inch
	= 1.020 kg force/sq cm
1 kg force/sq cm	= 14.22 pounds/sq inch
1 pound/sq inch	= 0.070 kg force/sq inch

Heat and energy

1 megajoule	= 238.8 kilocalories
	= 947.8 British Thermal Units
	= 0.278 kilowatt hours
1 terajoule	= 0.00002388 Mtoe
1 kilocalorie	= 3.968 Btu
	= 4186 joules
	= 1.163×10^{-3} kWh
1 kilowatt hour	= 859.8 kilocalories
	= 3,412 Btu
1 million Btu	= 1,055 megajoule
	= 2,520 megacalories
	= 293.1 kilowatt hours
1 ton of oil equivalent (toe)	= 10,000,000 kilocalories
	= 41.868 gigajoules
	= 40.047×10^6 BTU
	= 42.244 GJ

INTERFUEL CONVERSION FACTOR

Whilst individual crudes, coals and gases vary widely in quality, certain standard qualities are often assumed for statistical purposes

Reference Natural Fuel	*Barrel Oil Equivalent*	*Ton Oil Equivalent*	*Ton Coal Equivalent*	*1,000 Cubic Feet Gas*
Calorific Value	5.8×10^6 Btu gross	43×10^6 Btu gross	7×10^6 Kcal net	1×10^6Btu gross
Conversion Factors	1	0.14	0.21	5.8
	7.41	1	1.57	43.0
	4.71	0.64	1	27.3
	0.17	0.02	0.04	1

Based on these qualities, the following equivalent rates of consumption can be used with reasonable accuracy:

LNG t/y	*NG 10^6 mmcF/d* *10^9 Nm^3/y*	*OIL toe/y*	*OIL boe/y*	*COAL tce/y*
1	1.41	1.22	25	1.9
0.71	1	0.87	18	39
0.82	1.15	1	20	1.55
0.04	0.056	0.049	1	0.076
0.53	0.74	0.64	13	1

Calorific value

1 magajoule/cu m	= 25.84 Btu/cu.ft
1 megajoule per normal cubic meter	=25.368 BTU per standard cubic feet (BTU/SCF) = 2.388×10^2 kilocalorie per normal cubic metre ($kcal/Nm^3$)
1 kilocalorie/cu m	= 0.016 Btu/cu ft
1 megajoule/kg	= 429.9 Btu/pound
1 kilocalorie	= 1.8 Btu/pound (exact)
1 Btu/cu ft	= 37.38 kilojoules/cu m = 9.410 kilocalories/cu m
1 Btu/pound	= 2.326 kilojoules/kg = 0.556 kilocalories/kg

Energy values

Btu per 1b:

wood: green	4,400
dry	7,310
peat, sod	6,200
lignite	9,000
coke	12,100
coal: bituminous	13,000
anthracite	14,500
oil: fuel	18,500
gas	19,500

NATURAL GAS AND OTHER ENERGY EQUIVALENTS

The following are quick-reference equivalents. All figures are *approximate values* only for use where precision is not required. They are based on

i for natural gas:
1,000 Btu/ft^3 = 9,500 kcal/m^3 (Groningen gas 8,400 kcal/m^3)

ii for LPG:
an assumed 50/50 propane/butane mixture with (r) or (p) indicating that the LPG is either refrigerated or pressurised.

iii calorific values, MMBtu (gross):

per tonne	– LNG 51.8; LPG 47.3; oil 42.3; coal 27.3
per barrel	– LNG 3.8; LPG (r) 4.45; LPG (p) 4.1; oil 5.8
per cubic metre	– LNG 23.8; LPG (r) 28; LPG (p) 25.8

Natural gas: Cubic metre equivalents

1 mrd m^3 Natural Gas per year

per year	= 0.04 Tcf gas (38 trillion Btu)
	= 890,000 tonnes oil
	= 800,000 tonnes LPG
	= 725,000 tonnes LNG
	= 1.4 million tonnes coal
per day	= 100 million ft3 gas
	= 17,800 barrels oil
	= 23,200 barrels LPG (r)
	= 25,200 barrels LPG (p)
	= 27,200 barrels LNG

1 million m^3 Natural Gas per day

per year	= 0.014 Tcf gas (14 trillion Btu)
	= 325,000 tonnes oil
	= 290,000 tonnes LPG
	= 265,000 tonnes LNG
	= 500,000 tonnes coal
per day	= 37 million ft3 gas
	= 6,500 barrels oil
	= 8,500 barrels LPG (r)
	= 9,200 barrels LPG (p)
	= 9,900 barrels LNG

1 m^3 Groningen gas = 0.88 m^3 (9,500 kcal)
1 m^3 (9,500 kcal) = 1.13 m^3
Groningen gas

Natural gas: Cubic foot equivalents

1 Tcf Natural Gas per Year	
per year	= 27 mrd m^3 gas (30 mrd Groningen)
	= 24 million tonnes oil
	= 37 million tonnes coal
per day	= 2,700 million ft^3 gas
	= 470,000 barrels oil

100 MMcf Natural Gas per Day	
per year	= 0.04 Tcf (37 trillion Btu)
	= 1 mrd m^3 gas (1.1 mrd Groningen)
	= 860,000 tonnes oil
	= 770,000 tonnes LPG
	= 700,000 tonnes LNG
	= 1.35 million tonnes coal
per day	= 2.7 million m^3 gas (3 million Groningen)
	= 17,250 barrels oil
	= 22,500 barrels LPG (r)
	= 24,400 barrels LPG (p)
	= 26,300 barrels LNG

LNG: Volumetric equivalents

1 million tonnes LNG per Year	
per year	= 77 million ft^3 (liquid)
	= 2.2 million m^3 (liquid)
	= 14 million barrels (liquid)
	= 0.05 Tcf (gas)
	= 1.4 mrd m^3 (gas)
	= 1.1 million tonnes LPG
	= 1.2 million tonnes oil
	= 52 trillion Btu
	= 1.9 million tonnes coal
per day	= 140 million ft^3 (gas)
	= 4 million m^3 (gas)
	= 37,500 barrels LNG
	= 31,900 barrels LPG (r)
	= 34,600 barrels LPG (p)
	= 24,500 barrels oil

1 million m^3 LNG per Year	
per year	= 460,000 tonnes LNG
	= 6.3 million barrels LNG
	=0.2 Tcf (gas)
	= 0.6 mrd m^3 (gas)
	= 500,000 tonnes oil
	= 560,000 tonnes oil
	= 24 trillion Btu
	= 870,000 tonnes coal
per day	= 65 million ft^3 gas
	= 14,700 barrels LPG (r)
	= 15,900 barrels LPG (p)
	= 17,200 barrels LNG
	= 11,200 barrels oil
m^3	= kilo-litre

LPG and Ethane: Weight, volume, heat, conversions

Barrels Per Tonne				
	Ethane	Propane	n-Butane	C3/C4 mix
Pressured	17.6	12.4	10.8	11.6
Refrigerated	11.5	10.8	10.4	10.6

Cubic Metres Per Tonne				
	Ethane	Propane	n-Butane	C3/C4 mix
Pressured	2.80	1.97	1.71	1.84
Refrigerated	1.83	1.72	1.66	1.69

10^6 Btu Per Barrel				
	Ethane	Propane	n-Butane	C3/C4 mix
Pressured	2.79	3.85	4.35	4.10
Refrigerated	4.27	4.41	4.49	4.45

10^6 Btu Per Cubic Metre				
	Ethane	Propane	n-Butane	C3/C4 mix
Pressured	17.6	24.2	27.4	25.8
Refrigerated	26.9	27.7	28.3	28.0

10^6 Btu Per Tonne				
	Ethane	Propane	n-Butane	C3/C4 mix
Pressured } Refrigerated }	49.2	47.7	46.9	47.3

1 barrel/Day = Tonnes Per Annum				
	Ethane	Propane	n-Butane	C3/C4 mix
Pressured	20.7	29.4	33.8	31.6
Refrigerated	31.7	33.8	35.0	34.4

Natural gas: Inter-fuel price equivalents (US currency)

ONE CENT per MMBtu

= 5.8 Cents per Barrel crude oil
= 6.4 Cents per Barrel fuel oil
= 5.2 Cents per Barrel naphtha
= 4.5 Cents per Barrel LPG (r)
= 4.1 Cents per Barrel LPG (P)
= 3.8 Cents per Barrel LNG

ONE CENT per MMBtu

= 0.42 Dollars per Tonne crude oil
= 0.40 Dollars per Tonne fuel oil
= 0.45 Dollars per Tonne naphtha
= 0.47 Dollars per Tonne LPG
= 0.52 Dollars per Tonne LNG

ONE DOLLAR per BARREL

= 17 Cents per MMBtu crude oil
= 16 Cents per MMBtu fuel oil
= 19 Cents per MMBtu naphtha
= 22 Cents per MMBtu LPG (r)
= 24 Cents per MMBtu LPG (p)
= 26 Cents per MMBtu LNG

ONE DOLLAR per TONNE

= 2.4 Cents per MMBtu crude oil
= 2.5 Cents per MMBtu fuel oil
= 2.2 Cents per MMBtu naphtha
= 2.1 Cents per MMBtu LPG
= 1.9 Cents per MMBtu LNG

ONE CENT per US GALLON = 10 cent per MMBtu

Oil and coal equivalent

1 million tonnes oil

per year = 1.1 mrd m^3 gas (1.3 mrd Groningen)
= 1.5 million tonnes coal

Index